GLOBALISING FOOD

Agrarian Questions and Global Restructuring

Edited by David Goodman and Michael Watts

ROUTLEDGE

London and New York

First published 1997
by Routledge
11 New Fetter Lane, London EC4P 4EE

Simultaneously published in the USA and Canada
by Routledge
29 West 35th Street, New York, NY 10001

Typeset in Garamond by Routledge
Printed and bound in Great Britain by Clays Ltd, St. Ives PLC

British Library Cataloguing in Publication Data
A catalogue record for this book is available from the British Library

Library of Congress Cataloguing in Publication Data
Globalising food: agrarian questions and global restructuring/
[edited by] David Goodman and Michael J. Watts
Includes bibliographical references and index.
1. Food industry and trade. 2. Agricultural industries.
I. Goodman, David. II. Watts, Michael.
HD9000.5.G589 1997
97–9254
338.1–dc21
CIP

ISBN 0–415–16252–1 (hbk)
ISBN 0–415–16253–X (pbk)

GLOBALISING FOOD

We h ... ome
global ... of
financ ... sti-
mony ... ire?
The h ... ong
and in ... ncy
of nec ... in
differe ...

Global ... the
Unitec ... to
investi ... ess
from n ... are
reshap ... nd
conseq ... he
respon ... ell
as cha ... ect
contem ... n-
ability. ... ed
and Th ... he
strategi ... a-
nies an ...

Combin ... y,
econom ... e,
Globalising Food is a fascinating investigation of the globalisation and restructuring of localised agricultural sectors and food systems, and provides an innovative contribution to the political economy of agriculture, of food and consumption.

David Goodman is Professor and Chair of Environmental Studies at the University of California, Santa Cruz. **Michael J. Watts** is Professor and Director of the Institute of International Studies at the University of California, Berkeley.

What [agriculture] is spared from overseas competition [it] is threatened by industrial development at home. The transformation of agricultural production into industrial production is still in its infancy. [But] bold prophets, namely those chemists gifted with an imagination, already are dreaming of the day when bread will be made from stones and when all the requirements of the human diet will be assembled in chemical factories. . . . But one thing is certain. Agricultural production has already been transformed into industrial production in a large number of fields. . . . This does not mean that the time has arrived when one can reasonably speak of the imminent demise of agriculture. . . . [But] economic life even in the open countryside, once trapped in such eternally rigid routines, is now caught up in the constant revolution which is the hallmark of the capitalist mode of production. . . . The revolutionizing of agriculture is setting in train a remorseless chase. Its participants are whipped on and on until they collapse exhausted – aside from a small number of aggressive and thrusting types who manage to clamber over the bodies of the fallen and join the ranks of the chief whippers, the big capitalists.

(Karl Kautsky, *The Agrarian Question* [1899] 1988: 297)

CONTENTS

Part II Restructuring, industry and regional dynamics

Part III Globalisation, value and regulation in the commodity system

Part IV Discourse and class, networks and accumulation

Part V Transnational capital and local responses

Part VI Nature, sustainability and the agrarian question

ILLUSTRATIONS

FIGURES

TABLES

CONTRIBUTORS

David Goodman is Chair of the Environmental Studies Board at the University of California, Santa Cruz, and has published widely on biotechnology, food systems, and the agrarian question in Latin America.

Michael Watts is Director of the Institute of International Studies at the University of California, Berkeley, and has worked extensively on agrarian change in Africa and South Asia and the global food system.

William Boyd is a doctoral candidate in the Energy and Resources Group at the University of California, Berkeley, and is currently completing a dissertation on the poultry and timber industries in the US South.

Fred Buttel is Professor of Rural Sociology at the University of Wisconsin, Madison. He is currently concerned with questions of sustainability and agrarian change in the US.

Sharad Chari is a doctoral candidate in the Department of Geography at the University of California, Berkeley, and is currently completing a dissertation on the relations between agriculture and regional industrialisation in southern India.

Margaret FitzSimmons is a Professor of Environmental Studies at the University of California, Santa Cruz, and has a forthcoming book on US agriculture and its restructuring.

William Friedland is Professor Emeritus of Community Studies at the University of California, Santa Cruz, and is a leading figure in the study of Californian agriculture.

Lourdes Gouveia teaches Anthropology and Sociology at the University of Nebraska and currently is working on agrarian restructuring in Venezuela and the American Midwest.

Gillian Hart is a Professor of Geography at the University of California, Berkeley, who has written on state and agrarian labor in Southeast Asia and is currently completing a book on South Africa.

Richard Le Heron teaches in the Geography Department at the University of Auckland and has published extensively on the transformation of New Zealand and Australian agriculture in the 1980s.

Professor Norman Long holds an appointment in Sociology at Wageningen Agricultural University in the Netherlands, writing and researching broadly on questions of rural development, commodification, and human agency in agrarian systems.

Philip Lowe holds a Professorship in the Center for Rural Economy at the University of Newcastle and has published on a wide range of issues pertaining to environment, discourse, state, and agrarian change in the UK.

Professor Terry Marsden holds an appointment in City and Regional Planning at the University of Cardiff and is currently completing a research project on agrarian change in Brazil and the Caribbean.

Brian Page is Professor of Geography at the University of Colorado, Denver and has a book in preparation on recent changes in the US hog industry.

Professor Laura Raynolds teaches Sociology at Colorado State University and has published widely on agrarian restructuring, with a particular emphasis on the growth of non-traditional exports in the Caribbean.

Professor Michael Redclift is associated with the Global Environmental Change Program at Wye College, London University, and is currently working on sustainable development and its meanings.

Michael Roche teaches Geography at Massey University and has written widely on the historical and contemporary geography of agriculture and forestry in New Zealand.

Lorraine Thorne is currently completing her doctoral dissertation in Geography at the University of Bristol on food networks in Britain.

Richard Walker is Chair of the Department of Geography at the University of California, Berkeley, and is currently completing a book on San Francisco.

Dr Neil Ward teaches in the Department of Geography at the University of Newcastle and has research interests encompassing network theory, discourses of the environment, and agrarian change in the UK.

Miriam Wells is Professor of Sociology in the Advanced Behavioral Sciences Program at the University of California, Davis, and has recently completed a major study of the California strawberry industry.

Sarah Whatmore is a Reader in the Department of Geography at Bristol University; her published work addresses gender and the household in European agriculture and agro-industrial restructuring in Britain and other parts of Western Europe.

John Wilkinson is a sociologist associated with the Center for Agricultural Development and Planning at the Federal University of Rio de Janeiro in Brazil and has published widely on various aspects of Brazilian agrarian change, notably agro-industrial development.

Anthony Winson teaches in the Department of Sociology and Anthropology at the University of Guelph in Ontario, Canada and has a long record of working on institutional and political economic restructuring of Canadian agriculture.

ACKNOWLEDGEMENTS

This book emerged from a workshop on 'Agrarian Questions and the Restructuring of the Agro-Food System' held on the Berkeley campus of the University of California, on 28–30 September 1995. We wish to acknowledge the support provided by the Center for the Study of Global Transformations of the University of California, Santa Cruz and the Institute of International Studies of the University of California, Berkeley. In addition critical assistance in the preparation of the book has been provided by Chitra Ayar, Letitia Carper, Cherie Seamans and a number of doctoral students including Susanne Friedberg, Julie Guthman, James McCarthy, and the participants in a graduate seminar entitled 'The Global Agro-Food System' taught in the autumn of 1995. Michael Watts wishes to acknoweledge the financial support of the SSRC/MacArthur Foundation Peace and Security Program. Parts of Chapter 1 appeared in an earlier report of the workshop published in *Progress in Human Geography* (1996) 20, 2: 230–45; we are grateful to the editors for permission to reprint in highly modified form parts of this article.

1

AGRARIAN QUESTIONS

Global appetite, local metabolism: nature, culture, and industry in *fin-de-siècle* agro-food systems

Michael Watts and David Goodman

INTRODUCTION

> [C]apitalist agrarian transition is protean in its manifest diversity.
>
> (Byres 1996)

> The fact is that the food business is entering a period of unprecedented turmoil.
>
> (*The Economist*, 1993)

In 1974, the World Food Conference was held in Rome amidst growing anxiety over food availability and food prices. Poor weather and unprecedented US grain sales to the Soviet Union radically tightened the world wheat market, prompting apocalyptic predictions of mass starvation and famine. Three years later, world prices had fallen to the point where they were lower than at any time since the end of the Second World War. Aggregate crop production in the developing world grew at almost 3 per cent per annum from 1970 to 1990, confounding Malthusian expectations that output could not keep abreast of staple food output. Twenty years after the first food conference, the Rome World Food Summit in late 1996 was held in a quite different political and economic atmosphere. Rather than bristling Third World nationalisms and robust protectionism, the 1990s is a moment of unprecedented deregulation of agriculture (a shift from aid to trade), the hegemony (the so-called 'new realism') of export-oriented neoliberal development strategies, and a recognition that globalisation (a word not even part of the lexicon of the earlier Rome summit) of the world agro-food economy was proceeding apace. Of course the transformation of the world food economy since 1974 has been complex; while output (primarily driven by yield increases) outstripped population growth, a number of countries slid into net food import dependency during the 1970s, while the food security situation in global terms itself proved to be intractable. Despite an increase in food production per capita of 20 per cent in developing countries over the period 1980–90, roughly 800 million people suffer from hunger – of whom 500 million are chronically malnourished – and more than one-third of children are malnourished (UNDP 1996: 20). Indeed, shock therapy in the former socialist bloc, coupled with deregulation and a frontal assault on entitlements in the North Atlantic capitalist economies, has elevated the food question to the front pages of the newspapers in the First and Second Worlds alike. A study published by the California Policy Seminar (Neuhauser and Margen 1995) revealed an astonishing increase in hunger in California since 1980: 8.4 million are currently food insecure and 5 million are hungry. A rather more dyspeptic view of the much vaunted new Pacific century. A forthcoming International Labour Office report reveals that adult unemployment or underemployment worldwide exceeded one billion in 1995, an increase of 180 million over 1994, a condition which the UN describes as a crisis not seen since the Great Depression (*New York Times* 26 November 1996). In the fifteen European Union nations, unemployment averaged 11.3 per cent of the labour force in 1995, compared with 2 per cent in the 1960s; a high proportion of the unemployed has been jobless for over a year. One in five American children live below the poverty line, and this proportion rises to 42 per cent of the country's black children (*New York Times* 30 November 1996: A1).

Curiously, while much has changed since the first World Food Conference – indeed, these transformations represent one of the major concerns of this book – the environment within which the latest UN Food Summit transpired was in some respects resonant of the 'crisis' of twenty years ago.[1] *Business Week* devoted most of a recent issue to 'the new economics of food' in which the central motif was 'as global demand outpaces supply, both haves and have-nots are in for a shock'. At the heart of *Business Week*'s concern was the 'huge Chinese appetite' (currently China imports almost 20 million tons of grain), itself rooted in the rising living standards of the many Chinese in the post-Mao period. Collectively, falling grain production in Russia, massive food imports for Africa and the Middle East, and escalating Chinese demand offer the prospect (according to *Business Week*) of low grain reserves, 'cycles of panic', and price inflation. What is at stake, however, in the 1990s is neither Malthusian dearth nor a replay of the 1970s. As *Business Week* puts it:

> As disposable incomes rise sharply across the developing world, millions are eating better. A fast-growing middle-class from Seoul to Sao Paulo is buying more beef, pork and poultry. In Brazil chicken consumption is up 20% in the last year. McDonald's Corp. has brought hamburgers to 93 countries. . . . Per capita global consumption of beef, pork and poultry has surged 11% in the past decade. . . . Consequently, feed grain use is surging to record highs in developing nations . . .
>
> (*Business Week* 23 September 1996: 78)

The food economy is on the one hand increasingly differentiated in new sorts of ways at the level of *consumption* – some within the LDCs are eating better at a time when others in Africa are descending into a universe of ever-greater food insecurity, millions in California go hungry[2] while others consume 'designer' organic vegetables shuttled around the world in a sophisticated late twentieth-century 'cool chain' – a reflection of deep polarities within the world economy at large. Arresting evidence of rising inequality and polarisation between rich and poor countries over the past fifteen years, exacerbated by the 'lost decade' of the 1980s for Latin America and Sub-Saharan Africa, is inventoried annually by the *Human Development Report 1996* (UNDP 1996). Assets of the world's 358 billionaires exceeded the combined annual incomes of countries with 45 per cent of the world's population. Equally sobering is evidence that developing countries with nearly 80 per cent of the world's population account for 20 per cent of global GDP; the gap in per capita income between the industrial and developing worlds has tripled since 1960 (UNDP 1996: 2). On the other hand, at the level of *production and distribution* the food economy is being restructured in radically new ways; to employ *Business Week*'s language, the food economy is increasingly driven by global demand and internationalisation of the agro-food industry. The giant food companies and the large retailers are aggressively transforming the world agro-food economy, offering a future of what Harriet Friedmann (1993) calls the 'private global regulation': '[transnational corporations] are the major agents attempting to regulate agro-food conditions, that is, to organize stable conditions of production and consumption which allow them to plan investment, sourcing of agricultural materials, and marketing on a global scale' (Friedmann 1993: 52).

A casual stroll through the daily newspapers and weekly magazines over the past fortnight confirms that contemporary food and agriculture is very much a global reality which shapes our lives in profound cultural, ideological, and economic ways. The Sunday magazine of a London 'quality' paper directs its middle-class readership to 'Our Festive Food and Drink Guide, Packed with Recipes from London's Leading Chefs. Plus: The Best Champagne Buys' (*Independent on Sunday* 24 November 1996). An 'Eating Out' review of a new central London restaurant coyly observes that the menu trendily includes 'the words "root vegetable", "rocket", "ricotta", "parfait", and "cabbage" ' (*Independent on Sunday* 24 November 1996: 88). A week earlier, the same newspaper commented on the 'barrage of rhetoric' that had marked the closure of the World Food Summit in Rome, where the US delegation opposed international recognition of the right to food, and reported that the US plans to reduce by two-thirds its contribution to the International Fund for Agricultural Development (IFAD), the only UN agency whose exclusive remit is to help poor Third World farmers (*Independent on Sunday* 17 November 1996: 4). This article also cites a new report by a UN committee of experts who estimate that it will take *200 years* to eradicate malnutrition in the Indian subcontinent. Hunger is increasing in Africa. Closer to home, the lead story in the business section of

The New York Times, 'Biting into Food Industry Profits: Companies Squeezed from All Sides in Battle of Prices', depicts the 'reorganisation' and 'downsizing' efforts that are cutting jobs at such food giants as Kellogg, Unilever, Campbell Soup, Grand Metropolitan, RJR Nabisco Holdings, H. J. Heinz, and Conagra. These corporate 'lean-and-mean' strategies at the expense of the work force intensify the feeding frenzy on Wall Street, helping to raise the Dow Jones industrial average to a new high above 6,500 (*New York Times* 26 November 1996: C1).

This selective trawl of news reports conveys not only the diverse meanings and everyday experiences of food, from lifestyle semiotics to basic metabolic life force, but also the dystopic extremes of world political economy in the late twentieth century. These polarities also are emphasised in more explicitly analytical treatments of globalisation. One scenario presented by the Group of Lisbon (1995) combines the renewed prominence of regionalism – the European Union, NAFTA, APEC – in international political economy with the increasing concentration of international production, foreign direct investment (FDI), and trade in the 'triad' of North America, Western Europe, and Japan. This global regionalism of competing continental trade blocs – 'triadisation' – is premised on centrifugal forces that exacerbate fragmentation and exclusion, leading to the progressive de-linking of the triad from the rest of the world – a form of 'truncated globalisation'.

Rapidly changing conditions of global competition have focused attention on emerging forms of corporate organisation and the concomitant reconfiguration of international production. Thus the textbook world of neoclassical 'inter-national' trade theory, involving open or arm's length exchange conducted in competitive world markets between separate national firms in sovereign national economies, is on the wane. A high (and rising) share of world trade is now internalised by transnational corporations (TNCs) via intra-industry transactions between firms integrated into world-wide production systems and *intra-firm* hierarchies. This shift in corporate organisation toward integrated international production under the governance of TNCs is characterised by the UNCTAD-PTC (1993) as a transition from 'shallow' trade-based linkages to 'deep' international production-based linkages. These sea changes are, as this books reveals, integral to any understanding of the world agro-good system – how, in other words, we are all provisioned – at the *fin de siècle*.

The contributions to this volume investigate the political economies and discursive regimes of late twentieth-century agro-food systems, yet they also engage with larger questions of the many-headed beast called global capitalism and its multiple local trajectories. Namely, the processes of globalisation, economic restructuring, and new space-time dynamics shaping *fin de siècle* capitalism, and how these are theorised. The second question can be recast in terms of the tensions between classically influenced, 'modern' political economy and poststructuralist perspectives, which course through contemporary agro-food studies. Trends and changes within provisioning systems frequently are 'claimed' by rival social theoretic factions in other fields, and paraded as either singular exceptions to, or emblematic exemplars of, some aspect of the new realities of late twentieth-century capitalism. Most notably these arguments have encompassed international food regimes and regulation theory, Fordism/post-Fordism in agrarian restructuring, poststructuralist perspectives, especially actor-network theory, and the 'spaces' in late capitalism for material and discursive struggles around alternative forms of social organisation. Two examples illustrate our point. The first is taken from the efforts of the French regulation theorists – working with researchers from INRA in France – to apply regulation theory to the agricultural sector precisely because it reveals attributes – quality, institutional innovation, flexibility, conventions – which are taken to be constitutive of the new ways in which markets, states and indeed capitalism itself are conceptualised (Allaire and Boyer 1995). Ben Fine's work provides a second illustration. In *The World of Consumption* (Fine and Leopold 1994) and *Consumption in an Age of Affluence* (Fine and Wright 1996) he is concerned to show how provisioning systems under capitalism differ in their internal dynamics, and in the case of the agro-food system the difference that makes a difference is what he calls the *organic* ('nature' or 'biology'), which especially shapes both ends of the food circuit (consumption and production) but indeed ramifies throughout its entire length.

The intellectual currents captured in this collection are testimony to the new-found vitality of agro-food studies, reinvigorated by the infusion of theoretical approaches and controversies from such widely dispersed fields as industrial geography, economic sociology, development studies, neo-institutional economics, cultural studies, and the sociology of scientific knowledge. This engagement is drawing agro-food studies into the mainstream of critical social science, marking a final and definitive rupture with the fettered disciplinary traditions of land-grant university rural sociology and rural geography. This epistemological break had its origins in the revival of academic interest in Marxism in the 1960s as a new generation of scholars reconsidered the 'agrarian question'; that is, the politics and political economy of agrarian transitions to capitalism. Marxisms of various sorts brought renewed attention to the earlier agrarian debates of German Social Democracy in the 1890s, distinguished by Karl Kautsky's classic contribution *Die Agrarfage* (*The Agrarian Question*), and the subsequent Russian Narodnik debates, in which Lenin and his populist critics offered contrasting visions of rural social relations as capitalism took hold of the Russian *mir*. These rich intellectual roots were further nourished by neo-Marxist analyses of Third World peasantries and commoditisation processes, the Brenner debate, the extended controversies that surrounded Althusser's structuralist

Marxism, and the articulation of modes of production literature associated initially with French Marxist anthropologists working *outré mer*.[3]

We have self-consciously returned to the classical discussions of the agrarian question, not because the world is unchanged or because there are simple historical parallels between 1890 and 1990. Rather we do so because of the salience and power of many of the questions posed by Kautsky in particular. In his new book, Terry Byres (1996) has suggested that there are three agrarian questions. The first, posed by Engels, refers to the *politics* of the agrarian transition in which peasants constitute the dominant class. What, in other words, are the politics of the development of agrarian capitalism? The second is about *production* and the ways in which market competition drives the forces of production toward increased yields (surplus creation on the land, in short). And the third speaks to *accumulation* and the flows of surplus and specifically inter-sectoral linkages between agriculture and manufacture. The latter Byres calls 'agrarian transition' and embraces a number of key moments, namely growth, terms of trade, demand for agrarian products, proletarianisation, surplus appropriation, and surplus transfer. Byres is concerned to show that agriculture can contribute to industry without the first two senses of the agrarian question being, as it were, activated, and to assert the multiplicity of agrarian transitions (the diversity of ways in which agriculture contributes to capitalist industrialisation with or without 'full' development of capitalism in the countryside). While Byres' approach has much to offer, it suffers in our view from a peculiar narrowness. On the one hand it is focused on the internal dynamics of change at the expense of what we now refer to as globalisation. And second, the agrarian question for Byres is something that can be 'resolved' (see also Bernstein 1996). Resolved seems to imply that once capitalism in agriculture has 'matured', or if capitalist industrialisation can proceed without agrarian capitalism ('the social formation is dominated by industry and the urban bourgeoisie'), then the agrarian question is somehow dead. This strikes us as curious on a number of counts, not the least of which is that the three senses of the agrarian question are constantly renewed by the contradictory and uneven development of capitalism itself (this is Ben Fine's point). It is for this reason that we return to Kautsky, since his analysis embraced all three dimensions of the agrarian question (something seemingly not acknowledged by Byres), and because he focused so clearly on substantive issues central to the current landscape of agro-food systems: globalisation, vertical integration, the importance of biology in food provisioning, the application of science, the shifts of power off-farm, the intensification of land-based activities, and the new dynamisms associated with agro-processing. Of course, Kautsky could not have predicted the molecular revolution and its implications or the role of intellectual property rights and so on. But it is an engagement with his work that seems to us to remain so central.

In selecting our cases to engage with the legacy of Kautsky we have not attempted to be geographically representative in any sense, or to distinguish between First and Third World case studies. Rather we selected our contributions along two axes. First, they revealed processes which seemed to us to be central to the late twentieth-century agrarian question. And second, the contributions were paired around particular thematic and theoretical issues which address directly the restructuring processes that strike to the heart of the current epoch. Chapters draw upon the experience of the US, Britain, New Zealand, India, Brazil, South Africa, Costa Rica, the Caribbean, and tangentially parts of Western Europe and Africa. None of these case studies and theoretical engagements substantiate Kautsky's predictions or analyses in any simple way – indeed a number diverge radically from them. But all engage with his legacy in palpable ways, and in so doing confirm the prescience and vitality of agrarian questions and agrarian transitions.

Arguably, then, the present juncture represents a significant sea change in agro-food studies. Agriculture and food has been propelled into the heart of debates over capitalist dynamics, discourses and regimes of truth, innovation, and institutional change, and the complex hybridities linking nature and culture. The critical interrogation and reformulation of political economy perspectives certainly are implicated in this transformation. If this new intellectual tide is still on the rise, it is marked by a growing engagement with poststructuralist critiques of totalising modernist epistemologies, and a critical attentiveness to the workings of cultural power and discourse. The chapters in this volume reflect this turbulence, fostering new conversations which advance the project of theoretical revision and renewal in agro-food studies.

KAUTSKY REDUX?

> We can have a form of agrarian transition, a resolution of the agrarian question . . . such that the agrarian question appears to be resolved in neither the Engels [the politics of urban–rural alliances and its relation to democracy] nor the Kautsky-Lenin sense [how capital is taking hold of agriculture and transforming it]. If however the agrarian question is so resolved . . . in such a way that capitalist industrialisation is permitted to proceed then, as the social formation comes to be dominated by industry and by the urban bourgeoisie, there ceases to be an agrarian question with any serious implications. There is no longer an agrarian question in any substantive sense.
>
> (Byres 1996)

As the centennial anniversary of Kautsky's *The Agrarian Question* approaches, the parallels between the 1890s and the late 1990s are compelling. Kautsky's focus on the agrarian question in Western Europe rested on a striking paradox: agriculture (and the rural) came to assume a political gravity precisely at a moment when its weight in the economy was waning. Agriculture's curious political and strategic significance was framed by two key processes: the first was the growth and integration of a world market in agricultural commodities (especially staples) and the international competition which was its handmaiden; and the second was the birth and extension into the countryside of various forms of parliamentary democracy. Both forces originated outside of the agrarian sector but lent to agriculture its particular political and economic visibility. International competition in grains was driven not only by the extension of the agricultural frontier in the US, in Argentina, in Russia, and Eastern Europe (what Kautsky called the 'colonies' and the 'Oriental despotisms') but also by improvements in long-distance shipping, by changes in taste (for example from rye to wheat) and by the inability of domestic grain production to keep up with demand. As a consequence of massive new supplies, grain prices (and rents and profits) fell more or less steadily from the mid 1870s to 1896 (Koning 1994). It was precisely during the last quarter of the nineteenth century when a series of protectionist and tariff policies in France (1885), Germany (1879) and elsewhere were implemented to insulate the farming sector. New World grain exports were but one expression of the headlong integration of world commodity and capital markets on a scale and with an intensity then without precedent and, some would suggest, unrivaled since that period.

Kautsky devoted much time to the Prussian Junkers and their efforts to protect their farm interests. But in reality the structure of protection only biased the composition of production in favour of grains (and rye in particular) grown on the East Elbian estates. Tariffs provided limited insulation in the protectionist countries, while the likes of England, the Netherlands, and Denmark actually adopted free trade (Koning 1994). Protection did not, and could not, save landlordism but was rather a limited buffer for a newly enfranchised peasant agriculture threatened by the world market. The competition from overseas producers ushered in the first wave of agricultural protectionism, and in so doing established the foundations of the European 'farm problem' whose political economic repercussions continue to resonate in the halls of the European Commission, the GATT/WTO, and trade ministries around the world. The past, in short, continues to inform the present.

The agrarian question was, then, a product of a particular political economic conjuncture but was made to speak to a number of key theoretical concerns which arose from Kautsky's careful analysis of the consequences of the European farm crisis: falling prices, rents and profits coupled with global market integration and international competition. In brief he discovered that (1) there was no tendency for the size distribution of farms to change over time (capitalist enterprises were not simply displacing peasant farms, indeed German statistics showed that middle peasants were *increasing* their command of the cultivated area); (2) technical efficiency is not a precondition for survivorship (but self-exploitation might be); and (3) changes driven by competition and market integration did transform agriculture but largely by shaping the production mix of different enterprises, and by deepening debt-burdens and patterns of out-migration rather than by radically reconfiguring the size distribution of farms. The crisis of European peasants and landlords in the late nineteenth century was 'resolved' by *intensification* (cattle and dairying in particular in a new ecological complex) and by the *appropriation of some farming functions by capital* in processing and agro-industry (see also Goodman *et al.* 1987, Hussain and Tribe 1981: 70). These processes of appropriation and intensification are central to the chapters by Page, Boyd and Watts, Wilkinson and a number of the commentaries.

Kautsky concluded that industry was the motor of agricultural development – or more properly, agro-industrial capital was – but that the peculiarities of agriculture, its biological character and rhythms (see Wells 1996 and this volume), coupled with the capacity for family farms to survive through self-exploitation (i.e. working longer and harder to in effect depress 'wage levels'), might hinder some tendencies, namely, the development of classical agrarian capitalism. Indeed agro-industry – which Kautsky saw in the increasing application of science, technology and capital to the food processing, farm input and farm finance systems – might prefer a non-capitalist farm sector. In all of these respects – whether his observations on land and part-time farming, of the folly of land redistribution, his commentary on international competition and its consequences, or on the means by which industry does or does not take hold of land-based production – Kautsky's book was remarkably forward looking and prescient.

Kautsky was of course writing toward the close of an era of protracted crisis for European agriculture, roughly a quarter of a century after the incorporation of New World agricultural frontiers into the world grain market had provoked the great agrarian depressions of the 1870s and 1880s. A century later, during a period in which farming and transportation technologies, diet and agricultural commodity markets are all in flux, the questions of competition, shifting terms of trade for agriculture, and subsidies remain politically central in the debates over the European Union, GATT and the neoliberal reforms currently sweeping through the Third World. Like the 1870s and 1880s, the current phase of agricultural restructuring in the periphery is also marked (sometimes exaggeratedly so) by a phase of 'democratisation' (Kohli 1994, Fox 1995a). Agrarian parallels at the 'centre' can be found in agriculture's reluctant initiation into the GATT/WTO trade liberalisation agreement, albeit with a welter of safeguards, and, relatedly, the dogged rearguard action being fought by Western European farmers against further attempts to renegotiate the postwar agricultural settlement, which reached its protectionist apotheosis in the Common Agricultural Policy (CAP) during the 1980s. It is a picture clouded, however, by the strange bedfellows that the CAP has joined in opposition, including environmentalists, food safety activists, animal liberationists, bird watchers, rural preservationists, and neoconservative free marketeers. All of which is to say that if agrarian restructuring has taken on global dimensions, it is riddled with unevenness and inequalities (and here we find claims that the agrarian question is 'dead' rather curious). The rules of the game may be changing, but the WTO playing field is tilted heavily in favour of the OECD sponsors of this neoliberal spectacle. The different modalities and contingent regional specificities of agrarian restructuring under the banner of neoliberal reforms are examined in several chapters in this volume.

The changing bases and forms of capitalist competition constitute both cause and effect in the dialectics of globalisation, but political economic change at the regional and local level is mediated by inherited structures, creating complex patterns, spatially and temporally differentiated. This territorial endogeneity, its frictions and resistances, imparts highly specific characteristics to local/global relationships that are occluded by totalising analyses of globalisation. That is, analyses that conceptualise globalisation as a set of exogenous structural processes, shaped by a retinue of 'de-territorialised' global actors, and idealised, to borrow Storper's phrase, as 'a sort of delocalised "space of flows" '. Such a theoretical sensitivity to the antinomies and tensions of globalisation and localisation seem especially relevant to agrarian studies which have their own concerns with 'heterogeneity' and the multiplicity of farming styles (van der Ploeg and Long 1994). With these points in mind, we now briefly attempt to situate the dynamics of the contemporary restructuring of agro-food systems in the context of world-scale processes.

HIGH VALUE AGRICULTURES AND NEW AGRICULTURAL COUNTRIES

> For us humans, then, eating is never a 'purely biological' activity. . . . The foods eaten have histories associated with the pasts of those who eat them; the techniques employed to find, process, prepare, serve and consume the foods are all culturally variable, with histories of their own. Nor is the food simply eaten; its consumption is always conditioned by meaning. These meanings are symbolic . . . they also have histories.
>
> (Mintz 1996)

One of the presumptions of the research focused on transnational processes and agrarian-food orders is that the 'old' or classical international division of labour within the agro-food system has been irretrievably altered in the last twenty-five years. Classical export commodities (coffee, tea, sugar, tobacco, cocoa and so on) have been increasingly displaced by so-called 'high value foods' (HVF), such as fruits and vegetables, poultry, dairy products, shellfish. During the 1980s, the aggregate value of world trade in cereals, sugar, and tropical beverages declined, quite dramatically in some cases; conversely HVF grew by 8 per cent per annum. In 1989 HVF represented 5 per cent of world commodity trade, roughly equivalent to crude petroleum (Jaffee

1994). Developing economies currently account for over one-third by value of HVF production, roughly twice the value of Third World exports of coffee, tea, sugar, cotton, cocoa, and tobacco. In 1990 there were twenty-four low- and middle-income countries which annually exported more than $500 million of HVFs, mostly located in Latin America and Asia. But four of these countries actually account for 40 per cent of total HVF exports from developing states. These countries correspond to what Friedmann (1993, 1994a) refers to as 'new agricultural countries' (NACs) – the agro-industrial counterparts of the NICs – who occupy a central location in what she calls the durable foods, fresh fruits and vegetables, and livestock/feed complexes. Archetypical examples of these new agro-food systems are Brazilian citrus, Mexican 'non-traditionals' and 'exotics', Argentinean soy, Kenya off-season vegetables, and Chinese shrimp (see Kimenye 1993, Friedland 1994, Jaffee 1994, Watts 1994, Raynolds, this volume).

Dietary changes, trade reform, and technical changes in the food industry (Friedland 1994a) all contributed to the growth of the HVF sector. At the same time there are issues intrinsic to the sector – perishability, heterogeneity, seasonality, long gestation periods, externalities associated with marketing, and so on – which lead many commentators to focus on the 'major problems related to production and market risk, asymmetric information, logistical bottlenecks, and high transaction costs' (Jaffee 1994: viii). What is striking about the NACs is the extent to which their high value foods strategy rests upon highly favourable international market conditions during the initial boom periods, in some cases precipitated by 'market vacuums' as a result of trade embargoes or problems with traditional suppliers (Raynolds *et al.* 1993). The competitiveness of the HVF sectors clearly rests on the low costs of production – particularly labour costs (Collins 1993) – but it also depends on the extent to which *quality* can be established within heterogeneous commodities as a way of establishing dominance within niche markets. Given the concerns with quality and market niches, contract production is a fundamental way in which the division of labour of these global commodity systems is organised (Jaffee 1994, Little and Watts 1994). These 'post-Fordist' attributes (Raynolds *et al.*, 1993, but see Goodman and Watts 1994) raise important questions about the very notion of 'quality' (or standards or value) in international markets when the organic heterogeneity of commodities is the distinctive feature (see Boyd and Watts this volume, and Marsden this volume), and place considerable weight on the point of consumption insofar as HVFs have to be culturally constituted for particular sorts of taste, diet, and 'vanity' (Cook 1994, Fine and Leopold 1994).

The debate over the rise of the NACs – parallel in some respects to the 1980s work on the Gang of Four – turns on the purported successes of commodities such as Mexican tomatoes, Central American exotics, Brazilian soy, and so on (see McMichael 1995). What is striking in these cases is (1) the extent to which (in some cases) domestic consumption was key in a purportedly export-led strategy, (2) the dominance of private and/or foreign capital, (3) a high degree of concentration in export-oriented production, processing and marketing (Heffernan and Constance 1994), and (4) the prominence of contract production and/or vertical integration in linking farm-level production and downstream processing and trade (Watts 1994). But in the wine and grape or poultry sectors, as much as in the semiconductor or textile industries, state policies have had a key impact in shaping competitive opportunities and constraints within global commodity chains. Comparative advantage and market institutions are always 'socially and politically constructed' (Korzeniewicz *et al.* 1995: 132). This raises the question of whether state-agrarian strategies could usefully be categorised in the way that Evans (1995) classifies degrees of 'embeddedness' in his account of industrial strategies among the NICs.

It needs to be said of course that the emergence of high-value agriculture is highly uneven – like 'Third World' manufacturing itself – and the underbelly of new agricultural countries is agricultural marginality. Much of sub-Saharan Africa has returned to an agro-export model dependent largely on the classical commodities (and to date has been relatively marginal to fresh fruits, poultry, and livestock) whose market future looks extremely grim (World Bank 1995). In other cases, structural adjustment and deregulation has drawn investment out of agriculture altogether (see Marsden 1995 on the Caribbean). Another variant of this global marginalisation is the process described by Wood (1995) in Bangladesh, where agricultural involution under conditions of capital investment has generated an 'agricultural reformation' in which private service networks and associations of various sorts (rather than landholding per se) gain from agricultural productivity increases and compromise the very idea of the family farm as the decision-making unit over a range of decisions on the land formally held by the family.

GLOBAL FILIÈRES, GLOBAL COMMODITIES

> It is at the international level that the extraordinary drama of modernity rises up to its full height. It is at this level . . . that we can glimpse the process of capitalist transformation of humanity as a whole.
>
> (Rosenberg 1995)

> It will be some time before the Chinese consumer drives around in a BMW but not long before he can offer himself a Big Mac.
>
> (*The Economist* 1993)

For over a century world agricultural production has grown faster than demand, leading to a long-run decline in international food prices (Mitchell and Ingco 1995). This tendency has been matched by a declining proportion of the world's population deriving its livelihood from agriculture and the increasing absolute number of rural households living in poverty (Binswanger and Deininger 1995). Debates over HVFs and the so-called 'new agricultures' question the relations between these trends and the globalised and deregulated agro-food system of the 1990s. There are, however, a number of difficult and knotty theoretical questions which surround the relations between globalisation, new trade regimes, and regional political economy. The first concerns the relation between agricultural restructuring and regulation. The food regimes literature (Friedmann 1993, 1994b), which starts from the presumption of a relatively stable, rule-governed food order has seen the period since the oil/wheat crisis in 1972–73 (and the collapse of Bretton Woods) as a transitional period in which the dominance of transnational agro-capital and deregulation are the precursors of a new (if unstable) food regime. This transition has necessitated 'the restructuring of national agricultures and shifts in the regulation of food production and consumption' (Raynolds *et al.*1993: 1106). Implicit in this sort of analysis is the hegemonic role of global capital circuits (transnational agro-capitals), the standardisation of diets, new forms of international division of labour, a distinctive social economy, regional specialisation, global sourcing, the homogenisation of production conditions, and the undermining of state autonomy (Raynolds *et al.* 1993: 1103). The role of GATT, NAFTA and the hegemonic role of multilateral lending agencies signals, in this view, the ascendancy of 'private global regulation' (Friedmann 1993: 52).

While it would be wrong-headed to deny the extent to which agricultures have been deregulated in the last decade (the flight from productivism and agrarian corporatism as Marsden [1992] sees it), the pace and direction of liberalisation remains uneven and underdetermined. The NAFTA reforms are far from an unalloyed championing of tariff reduction and free trade (Goodman and Watts 1994), and the passage of GATT and domestic reforms in Europe and the US are unlikely to produce radical changes. According to the World Bank, GATT reforms do not represent 'significant reductions in border protection or a major increase in access to protected markets' (Hathaway and Ingco 1995: 31). Furthermore, so-called dirty tariffication promises the setting of new bound tariff ceilings at levels far above current applicable rates. According to a new OECD study (*The Economist* 3 June 1995: 97), total state support to agriculture in 1994 was, with one exception, *higher* than the 1979–81 average. By the same token, deregulation in the agrarian sector has typically been accompanied by re-regulation elsewhere within the sector, especially in the area of diet, health, and the environment (Marsden and Wrigley 1994, Marsden this volume, Lowe and Ward this volume, Whatmore and Thorne this volume).[4] In this sense, Raynolds *et al.* (1993) are right to point to the multiple trajectories associated with agrarian internationalisation in which the state continues to play a central role in domestic restructuring and negotiating a competitive global environment.

A second concern speaks to the nature of globalisation and global commodity chains in agro-food. Globalisation of agro-industrial corporations has clearly accelerated; the affiliates of the world's 100 largest firms increased in number from 2,070 in 1974 to 5,173 in 1990 and in value from US$121 billion to US$517 billion. The number of sources and host countries also increased. However, growing competition within this sector has produced increased cross-investments within OECD rather than a search for global sources or new markets in the South (Rama 1995). Indeed, partly due to the 1980s recession there was a *reduction* in direct agro-food investment in Latin America, Africa, and South Asia. While this capital mobility has resulted in the centralisation of power by retailers – the share of the ten largest food retailers in Belgium, the UK, Spain, and the US amounted to 79, 78, 66, and 65 per cent of total sales respectively – it does not suggest the emergence of global commodity chains in agriculture along the lines argued by Gereffi (1994) for automobiles (producer driven) and textiles (retail driven) (see Gouveia this volume).

Indeed the very nature of globalisation with the agro-food system is problematic and often confusing. If globalisation is to refer to the spatial configuration of markets, deterritorialised corporations, new forms of corporate and inter-firm organisation exemplified by strategic alliances and networks – the paradigmatic cases being electronics and automobiles – then the agro-food sector is clearly *not* global in any simple sense. In spite of the claims by Bonanno (1994) and Friedland (1994b) that fresh fruits and vegetables are 'truly transnationalised' and 'global production systems', it is clear that the industry is not characterised by vertically integrated transnational production systems. Neither do key firms centrally coordinate global intra-firm divisions of labour involving global outsourcing (Goodman in press). The likes of Conagra and Cargill are in many cases exemplary of multinational 'multi-domestic' strategies rather than sourcing through centralised, global intra-firm production systems. Work by Gouveia (1994, this volume) suggests that the much vaunted parallel between the world car and the world steer is also misplaced; the key corporate actors have greater similarity with mercantilist trading companies and 'Swift-type ventures minus the direct overseas investment' (Gouveia 1994: 136). Some of the food processors and retailers have been and are aggressively global – Kentucky Fried Chicken, McDonald's, and so on – but they must be located on a much more nuanced and heterogeneous map of commodity *filières* within the agro-food system (see Boyd and Watts this volume).

Examining cross-border integration mechanisms and international production organisation more closely, the UNCTAD-PTC (1993) refers to a spectrum from stand-alone or 'multi-domestic' affiliates to 'simple integration' and, more recently, 'complex integration'. The TNCs provide the formative dynamic element in this movement as they respond to competition, policy developments, and institutional change, and so, in turn, help to shape and deepen international economic integration. Briefly, in the case of 'simple integration', parent firms integrate *specific* production activities performed by their affiliates into their value-added chain, notably through outsourcing. In contrast, the more recent 'complex integration' strategies potentially integrate all parts of the value-added chains of *both* parent firms *and* their affiliates through vertical and horizontal production and functional linkages. Affiliates thus become more highly specialised as their activities are subordinated to the demands of firm-wide strategy. In terms of 'complex integration', UNCTAD-PTC (1993) estimates that roughly 25 per cent of the productive assets in the US and Japan 'are potentially part of integrated international production', while 'the share of world output potentially subject to integrated international production may well be around one-third' (UNCTAD-PTC 1993: 158).

This general typology reveals the varied forms of corporate international production subsumed under the composite label of globalisation: 'multi-domestic' affiliates, simple integration through outsourcing, and complex, vertically and horizontally integrated systems, exemplified by leading TNCs in electronics and automobiles. These differences across sectors are significant in conceptual and empirical terms; nevertheless, the label 'globalisation', with its allusions to outsourced international production and intra-firm integration, has become common currency in agro-food studies. Such debased usage often reflects an uncritical extension of concepts from industrial economics and economic geography. Admittedly, a select group of giant food TNCs – Coca Cola, McDonald's, Kellogg, Nestlé, Unilever – with global brand names has evolved global marketing strategies, albeit with adaptation to local tastes, but *production* typically is locally based. Few food manufacturing companies or retailers conform to the industrial model of transnationalisation; that is, centralised, global intra-firm divisions of labour, with production-based sourcing of intermediate components from specialised sites for final assembly.

Clearly, it is important to examine the notion of globalisation more critically, both its totalising, deterministic conceptualisations and as an empirical referent for international production in agro-food systems. Several authors in this volume respond to this challenge to analyse and clarify the distinctiveness of patterns of internationalisation in different agro-food sectors and their regional hinterlands (Page, Raynolds, and Boyd and Watts, this volume). This is done, explicitly or implicitly, by reviving and extending an older analytical tradition – the agro-food *filière* – whose antecedents are most strongly developed in the Latin American and French literatures. This focuses on commodity-specific or sectoral dynamics, and, consequently, reveals the diversity of agro-industrial and regional trajectories. Now infused with more explicit attention to the techno-ecological base, institutional mediations, including the state, and, most significantly, social agency, the *filières* approach furnishes new theoretical perspectives on the complex articulations between regional agrarian political economies and world-scale processes.

The essays in this volume, by problematising these articulations, effectively re-pose the 'agrarian question' in terms appropriate to the political economic conditions of late twentieth-century capitalism created by neoliberalism and intensified world-scale accumulation. In this process of reformulation, the restructuring of contemporary agro-food systems is examined at different spatial scales and with case studies of distinctive political economic geographies.

Unlike Bernstein (1996) and Byres (1996), who consider globalisation to have sounded the death knell of the agrarian question – the agrarian transition is 'resolved' – we argue that in all respects the agrarian question – as politics, as inter-sectoral flows, as labour process and technology, as social economy – is alive and well. Bernstein cites Hobsbawm's strange views on the death of peasantry – China has after all just reinvented a half a billion peasants – and Friedmann's idea of global private regulation – surely a very large hypothesis at this point in history – as the basis for his claim that the agrarian question of capital has been 'transcended by the generalisation of commodity relations on the global plane' (Bernstein 1996: 46). In our view – and in the view of many of the chapters here – the generalisation of commodity production is always uneven and contradictory, with the result that the agrarian question is never simply 'resolved'.

THE AGRARIAN QUESTION COMES TO TOWN[5]

> The industries producing farm-machinery, tools and supplies, and processing agricultural raw materials (meat packing, leather tanning . . .) were at the centre of the US industrial revolution.
>
> (Post 1995)

Some of the most exciting recent work in agrarian studies draws sustenance from the confluence of two related bodies of research: one focuses on the question of flexibility and networks in agriculture in a way that sheds light on debates within industrial geography (see Goodman and Watts 1994, Whatmore 1995). Another draws upon the growing body of work on rural industrialisation in the Third World and relatedly on rural non-farm work (Hart 1995). Both of these trends point to the continued importance of Kautsky (among others) and the agrarian origins of industrialisation (Byres 1995). Chari (1995, and this volume) and Cawthorne's (1995) work on the industrial districts centred on Tirupur in South India is especially important as a case study of Third World flexible specialisation and of what they call 'amoebic capitalism'. The genesis of this form of industrialisation is inseparable from rural–urban linkages in a regional economy dominated by specialised towns, and specifically how the capture of textiles by an agrarian capitalist caste (the Gounders) brought with it the migration (and consequently refashioning) of a variety of agrarian institutions of labour control and discipline which are central to the contemporary organisation of a dynamic small-firm textile sector. Harriss-White (1995) has explored these rural–urban linkages as part of what she calls the 'rural urbanisation of agrarian economies' (see also Chandrasekhar 1993). Here the economic linkages are regarded as the outcome of social relations shaped by local institutions but embedded in relations of power, trust, and reputation (see also Sanghera and Harriss-White 1995, Wood 1995). This of course speaks to a growing body of work which focuses on networks and the embeddedness of markets (see Wilkinson this volume), the European literature drawing on the work of Latour (1993) and Law (1994), and the American counterpart framed more by work on trust, loyalty, and social capital (Putnam 1993, Sabel 1994, Fukuyama 1995; see Boyd and Watts this volume).

These aspects of the agrarian question are central to an understanding of the new rural industrial districts which are emerging in a variety of social and institutional contexts. Certainly a part of the explanation of the dispersed and decentralised character of Taiwanese flexible 'family capitalism' is rooted in the politics of agriculture and postwar agrarian reform. The Chinese case is also relevant here because it shows how the remarkable rates of rural industrialisation combine collective property rights (at the township level) with market discipline and local institutions emerging from the creation of a post-reform peasantry (Bowles and Dong 1994). Akin to the Chinese and Indian cases is a brilliant account of the relations between agriculture and diffused manufacturing in the much vaunted Terza Italia case by Paloscia (1991). He shows vividly how there are clear elements of *continuity* between the new territorial formation – the urbanised countryside of industrial districts – and the original rural society predicated in large measure on sharecropping. In his words:

> This continuity is evident in various aspects: the organisation of work, and the specific qualities and attitudes of the productive classes of rural origin on the one hand and the spatial organisation of agricultural production and the network of minor centres with an artisan tradition on the other. These are all elements which reappear, adapted to the new economic and territorial system formed in Tuscany in the sixties and seventies.
>
> (Paloscia 1991: 53)

Implicit in this line of work is another key dimension of the agrarian question not mentioned by Byres (1996), namely, the ways in which agrarian institutions, social relations, and social capital are, as it were, carried over into the constitution of industrial districts. The surplus flows intersectorally not just as surplus value via terms of trade (however constituted) but in social, cultural, and institutional terms as well. Indeed, in this respect much of the debate over new industrial organisation – flexibility, networks, and so on – fails to recognise that its origins are prefigured in agrarian settings (and typically pre-date the rise of industry as such) (see Boyd and Watts this volume).

Hart (1995 and this volume) continues this line of thinking by documenting cases of what she calls 'interstitial spaces' – foreign capital investing in the quite specific milieux of rural South Africa – in which local networks and institutions are central to understanding the hybrid and multiple trajectories of capitalist development. These studies go beyond the old rural–urban consumption linkages debate (see Hart 1993 for a review) – small-farm growth produces local demand for services, equipment, or local consumer goods – to an examination of both the role of agrarian investment in industry and the ways in which agriculture, either through the provision of industrial wage labour from 'peasant' households or through local institutions of labour recruitment and discipline, is a key local ingredient in the emergence of globalised rural industrial districts in various parts of the Third World. These developments partly explain why land reform has re-emerged as a central plank of current development policy debates. This is also a function of the recognition of the need to deal with the deepening problems of rural inequality in the wake of structural adjustment which has, in many cases, undercut subsidies to large farms (see Binswanger *et al.* 1993),[6] and in South Africa in particular (Levin and Weiner 1994).

What Hart calls 'interstitial spaces', and what Marsden (1995, and this volume) in describing such locations as the São Francisco valley in Brazil calls 'agricultural districts', are both illustrations of how the hybrid and multiple trajectories of rural and agrarian accumulation take place in globalised sites constituted by complex social networks, and how the agrarian question is a constituent part of the 'flexible' forms of industrialisation that are emerging in newly deregulated and internationalised economies.

FOOD PROVISIONING SYSTEMS AND THE ORGANIC

> In short the organic dependence of agriculture upon landed production entails its integration into particular forms of landed property. These give rise to the appropriation of rents which influence the scale and intensity of accumulation. And the conditions of access to landed property cannot be reduced to ownership and tenancy, but need to be situated in relationship to the functioning of the food [provisioning] system as a whole, since vertical integration . . . may embody a displaced form of the rent relation. Consequently the accumulation of capital in the food system gives rise to particular tendencies in relation to the forms of landed property that it confronts. What distinguishes food . . . from other commodities . . . is the persistence of the organic component along the food system.
>
> (Fine and Wright 1996)

Ben Fine (Fine 1994, Fine and Leopold 1994) has proposed systems of 'provisioning' as a way of rethinking what he takes to be the flawed nature of the political economy of food literature. As a chain of activities linking production to consumption, the food system is rendered distinctive from non-food systems by its dependence upon organic properties at both ends of the chain. For Fine, the production end turns less on the biological properties of agricultural production – in other words the so-called Mann-Dickinson theory (Mann 1990) – than on landed property. At the consumption end, the organic question is especially relevant in terms of diet, health, and the relations between food and identity. Fine's intervention – which is controversial in many respects (see Friedmann 1994a, Goodman and Redclift 1994) – does have the advantage of tracing out the implications of the organic along the length of the food commodity chain, and in so doing problematises anew the question of consumption and how *value* and *quality* is added in particular sites in the food chain. This is a complex field, of course, embracing everything from what Hebdige (1988) calls a cartography of taste, to retailer power, to fast food, to health politics (see Crang in press, Ritzer 1993, Cook 1994, Fine and Leopold 1994, Marsden and Wrigley 1994, also Whatmore and Thorne this volume). The complexity of food resides in its polysemia, that is to say, its constant tendency to 'transform itself into a situation' (Barthes 1975: 568).

Arce and Marsden (1993) pointed out that the organic tends to ramify across the entire food network (his language) in terms of 'quality' and 'value'. These are complex terms, of course, carrying a vast array of meanings – and lend themselves to discursive sorts of analysis – since quality can refer to a wide range of criteria and properties, all of which tend to be contested locally, nationally, and globally (for example, from local regulations of what constitutes organic food to the efforts by the World Trade Organisation to legislate food standards). In this respect, foods are constantly in states of discursive animation, being naturalised and renaturalised in the context of the simultaneous and contradictory tendencies toward health consciousness and industrial production of fast foods (Jarosz in press). In Fine's analysis, the notion of the organic in agriculture should be expanded to include nature, to permit a more complex understanding of the ways in which biology and the organic, to use his language, 'temper' the political economy of the food system. In this sense, the questions of risk, perishability, seasonality, sustainability, and non-identity of production and labour time raised by Mann (1990) are necessary to grasp the commodity-specific dynamics of production systems (see Wells 1996) but should be linked with food consumption as integral elements of global food networks.

In the French literature, quality (*qualité*) refers both to properties of the food itself (safety, red labels, marks of vintage, and so on) and of the institutional relations (the conventions) by which actors in the *filière* are linked in the production of legitimate norms of quality (Sylvander 1995).

Considerable attention has been paid in the EU case to the organisational innovations by which cooperative relations are established in a food chain marked by concerns over perishability, standards, grading, shelf life, and so on (Green, Lanini and Schaller 1994, Montigaud and Ferry 1995). Retailers are key in this set of activities – grading, merchandising and logistics – in their deployment of information technologies, the cool chain, just-in-time management, contracting arrangements, and so on – which not only reshape the entire *filière* in, say, fresh products, but also transform the labour process at the farm level. New forms of product specification, grading, quality, and so on – coupled with the privatisation of information consultancy services (Wolf 1995) – demand of the farmer an entirely new labour process – what Wolf and Buttel (1996) call 'precision farming'[7] (see also Page, and Boyd and Watts in this volume). Much of this new precision, in other words, turns on the ways in which farms are integrated institutionally with retailers, processors and so on, and the demands made by these actors for specific quality (much of which itself reflects the organic properties of the food and agricultural commodities).

Quality is of course a complex term – what it conveys to poultry growers linked into massive integrators is radically different from the high-quality ('craft') specialty meat products described by van der Ploeg in northern Italy. Indeed, recent work by students at the University of California at Berkeley (Buck *et al.* 1997) has shed light on this complexity by an examination of the organics ('natural foods') sector in California agriculture. Typically assumed to be dominated by small enterprises – family farms, boutique farmers, hippies returned to the land, and so on – and catering to a rapidly expanding health-conscious market (from local grocers to top end restaurants in New York), the researchers discovered two important findings. First, the meaning of 'organic' is itself contested in the regulatory sphere – there are four certifying agencies with wildly different standards – and the panoply of actors in the organics sector use such agencies for their own (often marketing) ends. Second, the organics sector is now increasingly dominated by agribusiness which employs the struggles over the meaning of organic to expand their production of 'natural' fruits and vegetables. The California organics industry suggests that cultural questions of meaning and power are relevant to agro-industrial studies but that 'natural agriculture' as a tendency within that system must be situated in the rapid commercialisation of the sector as natural foods and health concerns broaden within the US marketplace.

AGRARIAN POLITICS: NATURE, CULTURE, AND CLASS

Kautsky's agrarian question was a political question framed by the electoral significance of a still rather substantial rural constituency in turn of the century Europe. A century later the politics of agriculture has also been a theme to which analysts have returned in the context of three world systemic processes: trade liberalisation and neoliberal reforms (implying if not the end of 'urban bias' at least its reform, see Varshney 1994), the post-Cold War democratisation movements (Kohli 1994), and the environment-development crisis (Peet and Watts 1996). GATT reforms and the Uruguay Round have proven to be so protracted in part because of the residual power of what admittedly are statistically insignificant farm populations – in the same way that Kautsky talked of the irony of agrarian politics being central at the moment when the economic centre of gravity had shifted to industry. With perhaps the exception of New Zealand (Le Heron 1993, Le Heron and Roche 1995), much of the deregulation of so-called agrarian corporatism has witnessed a re-regulation in the face of resistant farm lobbies and other agrarian interests (see Lowe *et al.* 1994).[8] McMichael (1995) makes a more radical claim, however, namely that the politics of the late twentieth-century agrarian question turns not on the comparative advantage of national economies but on the politics of substitution (the displacement of some agricultural commodities), the limits of replication (i.e. the exhaustion of the Green Revolution) and on what he calls reclamation (post-development visions of which the new social movements are paradigmatic). For McMichael, Chiapas represents the model of the globalised, green-oriented, food-secure, anti-industrial sustainable alternative.

Nonetheless this does raise three other issues which surround farm and food politics in the 1990s and meet up with a number of related development concerns centred on civil society and social capital (Evans 1995, Fukuyama 1995). The first is the so-called 'new farmer movements', particularly well documented in the case of India (see Brass 1994). They are new in the sense that they are farmer not peasant in constitution, in that they focus on prices not land, and insofar as they employ non-party agitational tactics, often encompassing women and environmental concerns. For Varshney (1994) these movements are a product of democracy preceding industrialising (and here rural empowerment represents a counterweight to urban bias) yet they suffer from an internal coherence problem expressed through cross-cutting forms of identification (ethnicity, caste, religion, and so on). Brass (1994) provides a more compelling account of these movements as conservative forms of populism; the constituency is prosperous farmers, largely a product of the Green Revolution, now operating under circumstances in which global trade liberalisation and neoliberal reforms undercut their gains. In this sense, of course, it is perhaps reminiscent in some respects of Kautsky's account of the prosperous German 'peasants' who supported the protectionist German Bund when faced with foreign competition.

The second form of 'farmer politics' concerns the new sorts of regulations stemming from the politics of collective consumption (Murdoch and Marsden 1994, Whatmore 1995, Whatmore and Thorne this volume). Here, agricultural production is shaped by concerns over the environment, health and safety, ethics, and diet (see Ward and Lowe this volume). Of course, these consumer-driven concerns may be imposed upon farmers but the greening of agriculture may also represent possibilities for small farmers who can legitimately present themselves as the purveyors of natural produce or sustainable land use (Murdoch and Marsden 1994). Organic production is increasingly a concern of corporate and transnational agribusinesses, of course, and the very definition of organic and natural commodities within the World Trade Organisation will itself become an object of political contestation (see Moran 1993). Sometimes the weight of these new politics – as a counterweight to productivism or so-called Fordist agriculture – can be exaggerated (Goodman and Watts 1994) but it is incontestable that they represent new spheres of regulation and political struggle within agro-food systems everywhere. These concerns in part turn on the question of value raised by Marsden (1995), and how value is imbued within commodities – and fought over and contested – at various sites along the commodity circuit. As the recent case of US federal legislation over whether chilled chicken at 32° F is frozen or fresh reveals, these questions of value can often be the discursive forms in which fractions of capital – in this case the southern US versus the California integrators – fight for market share (Boyd and Watts this volume).

Finally, there is the question of the ways in which rural and agrarian movements help thicken – the language is Fox's (1995a) – civil society in circumstances in which social capital is built up from below. Fox shows in the Mexican case how this can occur from the base communities – organic grassroots, community initiatives – but also from state–society interactions (say decentralisation of service provision), and from links between civic organisations (say the Church and local community groups). The central point here is that the impact of neoliberal reforms often compels states, in the name of fiscal restraint or market perfection, to work with civic institutions (de Janvry and Sadoulet 1995). These reformulations bring social agency in from the wings, acknowledging developments evident for some time now in other areas of critical social and cultural theory. This reassertion of agency is allied to a renewed emphasis on civil society, and associated notions of place, social embeddedness and trust, to counter the disabling, panoptic vision of structuralist globalisation and the hegemonic, triumphalist discourse of post-1989 neoliberalism. These intellectual currents are most evident in agro-food scholarship emanating from Western Europe, where the influence of actor-network theory and related approaches as a source of new insights on agency and power is more pronounced. This theoretical inspiration can be seen in the chapters by Lowe and Ward, Marsden, Long, and Whatmore and Thorne in this collection.

Several essays collected here examine and elaborate different entry points into the complex configurations of urban food politics, and notably their intersection with the more broadly based environmental movement (see

Buttel, this volume). Such intersections are the vanguard of the gradual but now increasingly perceptible and politically active process to renegotiate the nature–society interactions that have characterised modern agro-food systems for over a century (see Redclift, this volume). The sustainability of these fundamental metabolic relations underlying agro-ecosystems and patterns of food consumption is likely to become the predominant formulation of the 'agrarian question' in the new millennium. Such strategic concerns are explored in the final section of this volume.

These introductory comments identify some key themes explored at length in this volume, and apply some broad brush strokes to sketch the global-political-economic conjuncture that frames the current transformation of agro-food systems. The case studies presented in this volume bring a more intimate, nuanced understanding of this contextual backdrop but also serve a common twofold purpose: to shed light on the diversity and heterogeneity of agro-food systems, and yet address and refine theoretical perspectives, and so provide the framework for more ambitious comparative analysis. With these aims, the chapters in this collection are united in the effort to link wide-ranging debates over globalisation and restructuring in a period of neoliberal reform, with questions of regional dynamics, social agency, institutional change, cultural meanings, and, to employ Polanyi's language, the embeddedness of economic relations. We turn now to give a general overview of the individual chapters in the six thematic categories, which provide the book with its organisational structure.

THE ORGANISATION OF THE BOOK

The book has six sections, each with a common structure of a central 'theme', comprising a set of 'paired chapters', and a 'reflective theoretical essay' by a third contributor. The latter relates the paired chapters broadly to the state of the art and explores directions these suggest for further development of the main theme. The reflective essays therefore are significant contributions in their own right rather than commentaries whose purview is restricted to the theoretical perspectives and case-study materials presented in the paired chapters.

These intentional linkages are the principal *raison d'être* for the paired chapters. Each chapter – while empirically focusing on perhaps quite different case studies and geographical scales – engages with a common set of central theoretical and conceptual issues. This structure emerged from the workshop held at the University of California, Berkeley, in September 1995, which gave rise to the present collection. The purpose of these organisational arrangements is to encourage paired contributors simultaneously to speak to a common set of core concerns and yet raise theoretical questions that take us beyond the current state of the art.

Institutions, embeddedness and agrarian trajectories

This thematic category exceptionally has three chapters yet each is concerned with how best to gain analytical purchase on the forces which, historically and currently, configure patterns of social and institutional change as regional agrarian political economies are exposed to the global imperatives of trade liberalisation (Southern Brazil) and rural industrialisation (South Africa, South India). The theoretical tension illustrated by these chapters turns on the merits of classically inspired agrarian political economy (Gillian Hart, Sharad Chari) vis-à-vis new perspectives from economic sociology, neo-Schumpeterian economics, and actor-network theory (John Wilkinson).

Focusing on family-labour farms in the context of neoliberal deregulation and trade liberalisation following the creation of Mercosul and completion of the Uruguay Round, John Wilkinson explores the political-economic determinants of alternative paths of organisational structure and technological trajectories in milk production in Southern Brazil, where significant numbers of small producers remain competitive and provide a potential base for social mobilisation to redefine regional economic priorities. In criticising neo-institutionalist approaches prominent in the new economic sociology, notably that of Granovetter, Gillian Hart supports her position with recent fieldwork on the new industrial centres established in rural former Bantustans following the massive evictions which accompanied the coercive modernisation of white-owned farms from the 1960s in South Africa. Chari's work on the burgeoning, globally integrated knitwear industry of Tirupur in Coimbatore District in South India offers new formulations of the agrarian question and argues for the continuing prescience of the classical tradition in agrarian studies. Taken together, these contributions fruitfully critique and extend classical agrarian perspectives and debate the merits of neo-institutionalist and heterodox approaches to the transformation of contemporary agro-food systems.

Restructuring, industry, and regional dynamics

The general analytical question addressed by the chapters by Laura Raynolds and Brian Page involves the framing of globalisation in ways which bring out the difference and specificity which are encountered empirically as national and regional food systems negotiate processes of restructuring. Since these processes are iterated at multiple scales, the challenge is to identify patterns and regularities in which difference can meaningfully be situated. In this respect, both authors find that the industrial restructuring literature and Fordism/post-Fordism debates are seriously deficient when extended to agrarian restructuring. Furthermore, case studies of agro-food systems can critically inform and extend these literatures, correcting their neglect of agriculture and claims of universality. These questions are explored in the context of the restructuring of Dominican agriculture by Laura Raynolds, and against the background of the spatial mobility and current transformation (and 'export') of regional hog sectors in the US by Brian Page.

Globalisation, value, and regulation in the commodity system

The chapters in this pairing are centrally concerned with the dynamics of globalisation, and specifically with the social and spatial implications of shifting sites of power along globalised commodity chains. A critical question here is to examine the significance of the sphere of consumption and retail power, particularly in view of claims of the historical 'residualisation' of agriculture as its contribution to value-added continues to diminish. This question, in turn, is related to the determination of 'value' and notions of 'quality' in food systems, and the role of consumption politics as concerns for health, food safety, and environment create tensions with intensive agriculture and the industrial processing of food.

These points are addressed by postulating the genesis of new forms of agrarian space and developing the concept of 'food networks' to capture these important refashionings of global agro-food systems. This work is at the cutting edge of agro-food studies, elaborating the new theoretical approaches needed to comprehend future trajectories of food systems. Terry Marsden's chapter draws on case materials from Europe, the Caribbean, and the agro-industrial export complex for fresh produce in the São Francisco River Valley of northeastern Brazil. The southern California and South US poultry industry provides the spatial foci for the chapter by William Boyd and Michael Watts.

Discourse and class, networks, and accumulation

The tension in this pairing lies in the different theoretical approaches – class analysis versus discursive and actor-network analysis – adopted to investigate uneven change, regulation, and the kinds of conflict generated by restructuring in two local agro-food systems: strawberry production in California's Central Coast region and dairying in southwestern England. The authors examine different forms of struggle – everyday resistance, moral economy, and class conflict – and the ways in which they are locally expressed and differentiated in order to assess their importance in explaining local patterns of conflict, accommodation, and consent. Both case studies engage with the significance of local variability and place as constitutive of the kinds of struggle observed. The chapters also adopt contrasting approaches to regulation, with Ward and Lowe emphasising the key influence of moral economy and discourse on bureaucratic practice to regulate pollution incidents in dairy production, whereas Wells addresses regulation through the legal structures that surround labour and labour relations in California agriculture. These different foci on regulation express the contrasting ways in which nature is invoked and informs practice 'on the ground' in the two localities.

Transnational capital and local responses

The theme engaged in this pairing pivots around the relative weight to give to structure and agency in theorising the globalisation of agro-food systems. Conventional treatments of globalisation have been criticised for their totalising conceptualisations and 'hyper-structuralism', which allegedly involve the erasure of agency and space and the reification of the economic sphere. Alternative approaches seek to give voice to local actors, identify modes and 'spaces' of resistance, illustrate the local embeddedness of institutions and social practices, and emphasise the uneven, spatially differentiated impacts of globalisation. The common purpose of these paired chapters is to explore these competing conceptualisations, their methodological strengths and limitations, and identify elements of a more comprehensive framework.

Whatmore and Thorne draw on Latour's notion of 'hybrid networks' and Law's 'modes of ordering' to develop a conception of agro-food systems that opens up analytical and political spaces for the contestation of corporate configurations of land-food-body assemblages and their environmental and social practices. This analysis is illustrated by a case study of the *Cafédirect* consortium created by UK fair trade NGOs in conjunction with a coffee-exporting farmer cooperative in Chiclayo, northern Peru. In responding to the challenge of actor-network theory, Gouveia analyses the transformation of Venezuela's agro-food system in the macro-context of the Latin American debt crisis and multilateral neoliberal adjustment policies, and argues for the continued relevance and analytical purchase of a structuralist political economy approach to globalisation.

Nature, sustainability, and the agrarian question

Although they take different paths, the papers by Michael Redclift and Fred Buttel share a common problematic: how to transform sustainability into a political project. Redclift approaches this question by exposing the flawed perspectives on sustainability found in conventional theory in the major social sciences, whose epistemological traditions occlude nature from the analysis. Using *Risk Society* and more recent writings by Ulrich Beck as a foil, Redclift also discusses the limitations of critical social theory in providing the foundations for a radical environmental politics. Nevertheless, he takes examples of alternative trading practices, farmer networks, and radical consumer groups, to suggest that the litmus test of sustainability very gradually is gaining wider acceptance, a promising indication that theoretical issues are entering the everyday world of economic and social practice.

Buttel approaches the politics of sustainability by focusing initially on US agro-food systems and the constraints to agro-food activism. This discussion leads into a broader assessment of the mutual advantages of coalition-building between mainstream environmental groups and agro-food movements to promote the goal of sustainable agro-food systems. Buttel examines different currents of farm activism in the US and the formation of the anti-NAFTA coalition to illustrate his argument.

The interrogation of social theory and environmental activism in terms of their limited purchase on sustainability and the conditions for its political realisation is the common thread between these two chapters.

As this overview reveals, 'Agrarian Questions' explores new developments in the political economic and social-theoretic analysis of contemporary agro-food systems. Influenced by debates on a globalising world economy, industrial restructuring, neo-institutionalism, and poststructuralism, agro-food studies increasingly is at the confluence of currents in social and cultural theory, political economy, and heterodox economics and sociology. The present volume seeks to capture and distill this intellectual ferment and contribute to the delineation of future directions for critical agro-food studies.

NOTES

1 *The Economist* (16 November 1996: 18) regards the recent World Food Summit in Rome as the latest manifestation of neo-Malthusianism, and counters with a combination of free markets – 'stop interfering in farming altogether' – and technological optimism: 'the green revolution has not run out of steam, and biotechnology has the capacity to create a new cornucopia'.

2 By contrast, a feature article in *The New York Times* entitled 'Refighting the Battle of the Bulge in America' discusses the relative merits of high-protein, low carbohydrate diets versus low-fat, high carbohydrate regimes to combat the observed rise in obesity in the United States since 1990 (*The New York Times* 27 November 1996: B1).

3 Paradoxically, these influences had much weaker, more filtered effects on the scholarship of agro-food studies at the 'centre'. The rich legacy of Kautsky, particularly his nuanced analysis of the reproductive linkages between family labour forms of production and agro-industrial capitals, was allowed to languish, its contemporary relevance unexploited. The differences between Leninist orthodoxy and Russian populism were awkwardly conflated in the agro-industrial complex thesis of farmers as 'propertied labourers', and later largely forgotten with the emergence in the 1970s of the 'political economy of agriculture' school.

4 Some interesting applications of French regulation theory to the re-regulation of national agricultures have been undertaken by a group at INRA in Toulouse (Allaire and Boyer 1995).

5 I have taken this apposite title from the dissertation project of Sharad Chari,

University of California, Berkeley who is working on the agrarian origins of south Indian industrialisation.

6 It is astonishing to read in a recent World Bank report (Binswanger *et al.* 1993) that this august institution has discovered that the property rights and competitive markets theory does not recognise that rights and ownership 'grow out of power relationships' and that landowning groups have used coercion to extract economic rents!

7 Precision or site-specific farming refers to the application of geo-referencing technologies for purposes of data recording and navigational control of field operations. At the field level it is a production tool and at the sectoral level a coordinating technology linking upstream and downstream. We are employing the term here to encompass the larger process by which science and information technologies are harnessed through private means (processors, retailers, consulting and service companies) to ensure the quality demands we outlined in the text.

8 For example, Germany's central role in instituting a protected agrarian model within CAP and its reluctant adherence to the MacSharry reforms.

BIBLIOGRAPHY

Allaire, G. and Boyer, R. (eds) (1995) *La Grande Transformation de l'agriculture*, Paris: INRA.

Arce, A. and Marsden, T. (1993) 'Social construction of international food: a new research agenda', *Economic Geography* 69, 3: 203–311.

Barthes, R. (1975) 'Toward a psychosociology of contemporary food consumption', in M. Elborg and R. Foster (eds) *European Diet from Preindustrial to Modern Times*, New York: Harper and Row.

Bernstein, (1996) 'Agrarian questions then and now', *Journal of Peasant Studies* 24, 1/2: 22–59.

Binswanger, H. and Deininger, K. (1995) *Towards a Political Economy of Agriculture and Agrarian Relations*, unpublished manuscript, Washington DC: World Bank.

Binswanger, H., Deininger, K. and Feder, G. (1993) *Power Distortions, Revolt and Reform in Agricultural Land Relations*, Policy Research Working Papers, Washington DC: The World Bank.

Bonanno, A., Busch, L. Friedland, B., Gouveia, L. and Mingione, E. (eds) (1994) *From Columbus to Conagra: The Globalization of Agriculture and Food*, Lawrence: University of Kansas Press.

Bowles, P. and Dong, X. (1994) 'Current successes and future challenges in China's economic reform', *New Left Review* 208: 77.

Brass, T. (1994) 'Introduction: the new farmers' movements in India', *Journal of Peasant Studies* 21, 1: 3–26.

Buck, D., Guthman, J. and Getz, C. (1997) 'From farm to table: the organic vegetable commodity chain in California', *Sociologia Ruralis*, 37, 1: in press.

Byres, T. (1995) 'Political economy, the agrarian question and comparative method', *Economic and Political Weekly* XXX, 10: 507–13.

—— (1996) *Capitalism From Above and Capitalism From Below*, London: Macmillan.

Cawthorne, P. (1995) 'Of networks and markets', *World Development*, 23, 1: 34–49.

Chandrasekhar, C. (1993) 'Agrarian change and occupational diversification', *Journal of Peasant Studies* 20, 2: 205–70.

Chari, S. (1995) 'The agrarian question comes to town: rural ties in work and investment in Tirupur, South India', paper delivered to the Workshop on The Political Economy of the Agro-Food System, University of California, Berkeley, 28–30 September.

Collins, J. (1993) 'Gender, contracts and wage work', *Development and Change*, 24, 1: 53–82.

Cook, I. (1994) 'New fruits and vanity: symbolic production in the global food economy', in A. Bonanno *et al.* (eds) *From Columbus to Conagra*, Lawrence: University of Kansas Press.

Crang, P. (in press) *Spaces of Service*, London: Routledge.

de Janvry, A. and Sadoulet, E. (eds) (1995) *State, Market and Civil Society*, Ithaca: Cornell University Press.

Evans, P. (1995) *Embedded Autonomy: States and Industrial Transformation*, Princeton: Princeton University Press.

Fine, B. (1994) 'Toward a political economy of food', *Review of International Political Economy* 1, 3: 519–45.

—— and Leopold, E. (1994) *The World of Consumption*, London: Routledge.

—— and Wright, J. (1996) *Consumption in an Age of Affluence*, London: Routledge.

Fox, J. (1995a) 'Governance and rural development in Mexico', *Journal of Development Studies*: 610–44.

—— (1995b) 'How does civil society thicken?', paper presented to American Academy of Arts and Sciences, Cambridge, Mass.: 5–6 May.

Friedland, W. (1994) 'The new globalization: the case of fresh produce', in A. Bonanno *et al.* (eds) *From Columbus to Conagra*, Lawrence: University of Kansas Press.

Friedmann, H. (1993) 'The political economy of food', *New Left Review* 197: 29–57.

—— (1994a) 'Premature rigor', *Review of International Political Economy* 1, 3: 552–61.

—— (1994b) 'The international relations of food', in B. Harriss-White and R. Hoffenberg (eds) *Food: Multidisciplinary Perspectives*, Oxford: Blackwell.

Fukuyama, F. (1995) *Trust*, New York: Free Press.

Gereffi, G. (1994) 'Capitalism, development and global commodity chains', in L. Sklair (ed.) *Capitalism and Development*, London: Routledge.

Goodman, D. (in press) 'World-scale processes and agro-food systems: critique and agenda', *Review of International Political Economy*.

——, Sorj, B. and Wilkinson, J. (1987) *From Farming to Biotechnology*, Oxford: Blackwell.

—— and Watts, M. (1994) 'Reconfiguring the rural or fording the divide?: capitalist restructuring and the global agro-food system', *Journal of Peasant Studies* 22, 1: 1–49.

—— and Redclift, M. (1994) 'Constructing a political economy of food', *Review of International Political Economy* 1, 3: 547–52.

Gouveia, L. (1994) 'Global strategies and local linkages', in A. Bonanno *et al.* (eds) *From Columbus to Conagra*, Lawrence: University of Kansas Press.

Green R., Lanini, L. and Schaller, B. (1994) 'Technical and organizational innovations in the food system', unpublished paper, Paris: INRA.

—— (1995) *Technological and Organizational Innovation in the Food System*, manuscript, INRA, Paris.

Harriss-White, B. (1995) *A Political Economy of Agricultural Markets in South India*, New Delhi: Sage.

Hart, G. (1993) *Regional Growth Linkages in the Era of Liberalization: A Critique of the New Agrarian Optimism*, WEP Research Working Paper No. 37, Geneva: ILO.

—— (1995) *Beyond the Rural–Urban Dichotomy*, paper delivered to the Congress on the Agrarian Question, Wageningen, 22–24 May.

—— (forthcoming) *Making Spaces: The Agrarian Question and Industrial Dynamics in South Africa*, Berkeley: University of California Press.

Hathaway, D. and Ingco, M. (1995) *Agricultural Liberalization and the Uruguay Round*, paper presented at the World Bank Conference on the Uruguay Round, 26–27 January.

Hebdige, D. (1988) *Hiding in the Light*, London: Comedia.

Heffernan, W. and Constance, D. (1994) 'Transnational corporations and the globalization of food', in A. Bonanno *et al.* (eds) *From Columbus to Conagra*, Lawrence: University of Kansas Press.

Hussain, A. and Tribe, K. (1981) *Marxism and the Agrarian Question* vol. 1, Atlantic Highlands: Humanities Press.

Jaffee, S. (1994) *Exporting High Value Food Commodities*, Washington, DC: The World Bank.

Jarosz, L. (in press) 'Working in the global food system', *Progress in Human Geography*.

Kautsky, K. (1988) *The Agrarian Question*, 2 vols., London: Zwan.

Kimenye, L. (1993) *The Economics of Smallholder Flower and French Bean Production in Kenya*, PhD dissertation, Michigan State University, East Lansing.

Kohli, A. (1994) 'Democracy amid economic orthodoxy', *Third World Quarterly* 14, 4: 671–89.

Konnig, N. (1994) *The Failure of Agricultural Capitalism*, London: Routledge.

Kornai, J. (1992) *The Socialist System*, Princeton: Princeton University Press.

Korzeniewicz, R., Goldfrank, W. and Korzeniewicz, M. (1995) 'Vines and wines in the world economy', in P. McMichael (ed.) *Food and Agrarian Orders in the World Economy*, New York: Praeger.

Latour, B. (1993) *We Have Never Been Modern*, Hemel Hempstead: Wheatsheaf.

Law, J. (1994) *Organizing Modernity*, Oxford: Blackwell.

Le Heron, R. (1993) *Globalized Agriculture*, Oxford, Pergamon.

—— and Roche, M. (1995) *Eco-Commodity Systems*, paper delivered to the Political Economy of the Agro-Food System Workshop, University of California, Berkeley, 28–30 September.

Levin, R. and Weiner, D. (1994) 'Community perspectives on land and agrarian reform in South Africa', MacArthur Foundation Report, Chicago, Illinois.

Little, P. and Watts, M. (eds) (1994) *Living Under Contract: Contract Farming and Agrarian Transformation in Sub Saharan Africa*, Madison: University of Wisconsin Press.

Lowe, P., Marsden, T. and Whatmore, S. (eds) (1994) *Regulating Agriculture*, London: Fulton.

McMichael, P. (ed.) (1995) *The Global Restructuring of Agro-Food Systems*, Ithaca: Cornell University Press.

Mann, S. (1990) *Agrarian Capitalism*, Durham: University of North Carolina Press.

Marsden, T. (1992) 'Exploring a rural sociology of the Fordist transition', *Sociologia Ruralis*, 36: 209–30.

—— (1995) *Globalization, Regionalization and Regulation*, paper presented to the Political Economy of the Agro-Food System Workshop, Berkeley, 28–30 September.

—— and Wrigley, N. (1994) *Regulation, Retailing and Consumption*, paper delivered to the Association of American Geographers Meeting, San Francisco.

Mintz S., (1996) *Tasting Food, Tasting Freedom*, Boston: Beacon Press.

Mitchell, D. and Ingco, M. (1995) *Waiting For Malthus*, Cambridge: Cambridge University Press.

Montigaud J. and Ferry, J. (1995) *La Logistique dans les filières agroalimentaires et ses consequences sur la production agricole*, Serie Etudes et Recherches, No. 102, Montpellier, INRA/ENSA.

Moran, W. (1993) 'Rural space as intellectual property', *Political Geography* 12, 3: 263 – 77.

Murdoch, J. and Marsden, T. (1994) *Reconstituting Rurality*, London, University College: London Press.

Neuhauser, L. and Margen. S. (1995) *Hunger and Food Insecurity in California*, California Policy Seminar Report, University of California, Berkeley.

Paloscia, R. (1991) 'Agriculture and diffused manufacturing in Terza Italia', in S. Whatmore, P. Lowe and T. Marsden (eds) *Rural Enterprise*, London: Fulton.

Peet, R. and Watts, M. (eds) (1996) *Liberation Ecologies*, London: Routledge.

Post, C. (1995) 'The agrarian origins of US capitalism', *Journal of Peasant Studies* 22, 3.

Putnam, R. (1993) *Making Democracy Work: Civic Traditions in Modern Italy*, Princeton: Princeton University Press.

Rama, R. (1995) *Latin America and Geographical Priorities of Multinational Agroindustries*, unpublished manuscript, Paris: OECD.

Raynolds, L., Myhre, D., McMichael, P., Carro-Figueroa, V. and Buttel, F. (1993) 'The new internationalization of agriculture', *World Development* 21, 7: 1101–21.

Ritzer, G. (1993) *The McDonaldization of Society*, London: Pine Forge Press.

Rosenberg, J. (1995) 'Isaac Deutscher and the lost history of international relations', *New Left Review* 215: 3–15.

Sabel, C. (ed.) (1994) *Studied Trust*, New York: Praeger.

Sanghera, B. and Harriss-White, B. (1995) *Themes in Rural Urbanization*, DPP Working Paper No. 34, Milton Keynes: Open University.

Sylvander, B. (1995) 'La qualité', *Etud, Rech, System, Agrarires Devel.*, 28: 27–49.

United Nations Conference on Trade and Development – Program on Transnational Corporations (UNCTAD-PTC) (1993) *The World Investment Report 1993: Transnational Corporations and Integrated International Production*, New York: United Nations.

United Nations Development Program (UNDP) (1996) *Human Development Report 1996*, New York: Oxford University Press.

van der Ploeg, J. and Long, A. (eds) (1994) *Born From Within*, Assen: Van Gorcum.

Varshney, A. (1994) *Democracy, Development and the Countryside*, London: Cambridge University Press.

Watts, M. (1994) 'Life under contract', in P. Little and M. Watts (eds) *Living Under Contract*, Madison: University of Wisconsin Press.

Wells, M. (1996) *Reconfiguring Class: Politics and Work in American Agriculture*, Ithaca: Cornell University Press.

Whatmore, S. (1994) 'Global agro-food complexes and the refashioning of rural Europe', in N. Thrift and A. S. Amin (eds) *Globalization, Institutions and Regional Development in Europe*, London: Oxford University Press.

—— (1995) 'From farming to agribusiness', in R. Johnston, P. Taylor and M. Watts (eds) *Geographies of Global Change*, Oxford: Blackwell.

Wolf, S. (1995) *Privatization of Crop Production Service Markets*, PhD dissertation, University of Wisconsin, Madison.

—— and Buttel, F. (1996) *The Political Economy of Precision Farming*, paper delivered to the American Agricultural Economics Association, San Antonio.

Wood, G. (1995) *From Farms to Services*, Occasional Paper 02/95, Centre for Development Studies, Bath University.
World Bank (1995) *A Continent in Transition: Sub Saharan Africa in the Mid-1990s*, Washington, DC: The World Bank.

PART I

INSTITUTIONS, EMBEDDEDNESS AND AGRARIAN TRAJECTORIES

2

REGIONAL INTEGRATION AND THE FAMILY FARM IN THE MERCOSUL COUNTRIES

New theoretical approaches as supports for alternative strategies

John Wilkinson

INTRODUCTION

The dominant criteria currently determining the character of Southern Cone regional integration in general and of agricultural/agro-industrial integration in particular can be summed up as 'competitiveness' and 'efficiency'. Defence of 'inefficient' sectors is discursively illegitimate and sectoral mobilisations for protection have difficulty generating allies, although temporary concessions may be made, usually when lack of competitiveness affects the trade balance. 'Reconversion'[1] is the remedy for non-competitive sectors, but concerted policies in this direction are not much in evidence and relevant funding even less so. Adopting the spirit of post-GATT regulation, Mercosul support measures are increasingly envisaged in the form of lump-sum payments dissociated from production and not thereby interfering with the 'price-mechanism' (Wilkinson 1995).

In the three states of Southern Brazil – Paraná, Santa Catarina, and Rio Grande do Sul – there were a million family farms based on diversified crop systems and with properties of up to 50 hectares, according to the (latest) 1985 census. Neighbouring Paraguay still has almost half its population in the countryside, with up to 200,000 peasant farms. In Uruguay, small farmers are very much under threat but persist in various sectors and the family farmer of the Southern Brazilian and Paraguayan type is still an important social actor in the northern regions of Argentina, where we are talking perhaps of some 150,000 properties.

What are the chances for this diversified family farm sector in the context of the above dynamic of regional integration? In what follows I will examine some of the current arguments in support of the viability of a family farm alternative as the basis for regional integration. I will develop my own alternative approach in contraposition to two types of arguments:

Economic arguments These have in common the conviction that the family farm represents, or would represent, a competitive productive option if certain institutional biases were removed. The first of these involves little more than a plea for equal treatment with large-scale farming, particularly in terms of access to credit, while the second relates to the bias of existing technology paradigms (FAO/INCRA 1995).

Political arguments Such approaches customarily have two components, the normative and the pragmatic. In relation to the former, the defence of the family farm is posed in terms of justice or equity, while in the latter case the question is reduced to that of power and the 'relation of forces'. In both cases there is a politicisation of the economic and a failure to consider issues relating to productive efficiency. At the same time, the notion of conflict not only assumes analytical priority but tends to obliterate possible spaces for the negotiation of interests and the identification of common ground.

In the first section I deal with the claim that the family farm would prove itself competitive if it had the same access to benefits as the large-holding, and use this as an opportunity to introduce readers to the main features of agricultural modernisation in Brazil. This is followed by a discussion of the technology bias argument, where I draw on insights from the neo-Schumpeterian tradition, together with reflections which I have developed in other texts on the specific dynamic of technology in agriculture and agro-industry (Wilkinson 1993).

To give these arguments more operational purchase, I then develop a profile of the Brazilian family farm structure in the southern states, and conclude the discussion with a review of possible strategies emerging from *the existing range of technological and organisational options*.

I will not provide a specific treatment of the political arguments which we have sketched out schematically above since this would involve a discussion of the 'landless movement' in Brazil, together with the positions of the different rural organisations which would take me well beyond the space allocated for this chapter. Nevertheless, in the elaboration of my own position which involves the simultaneous consideration of economic and social mobilisation criteria, I will be directing my arguments both against over-simplified views of family farm competitiveness and what I would term a political reductionism with regard to the economic preconditions for viable family farm strategies.

In the final section I sketch out the main lines of a more socio-economic argument which simultaneously takes into account a 'full costs' approach to the notion of competitiveness (ecological, social, political, cultural), and also situates the issue of alternative strategies in terms of the different actors involved, thereby 'embedding' the economic options and the notion of market forces within the social processes which underlie the dynamic of economic life.

It is my view that the discourse which sustains the liberal challenge can now be countered most effectively in terms of an emerging synthesis which combines a variety of heterodox economic traditions and a plurality of social science actor-network oriented approaches. I have analysed these convergencies between regulation, convention, and neo-Schumpeterian theories, and between these and different social science perspectives at length in another text (Wilkinson 1996). The principal propositions unifying these different approaches have all the makings of an alternative paradigm. Here I will limit myself to singling out those elements which directly bear on the theme of this chapter:

1 An understanding of the dynamic of economic life requires an interdisciplinary approach since economic exchange presupposes the prior emergence of social institutions as enabling mechanisms.
2 There is a plurality of equally legitimate but heterogeneous forms of economic coordination.
3 The suppression of plurality in favour of the universalisation of a specific mode of coordination – in our case a liberal concept of market organisation – is prejudicial to 'efficiency'.
4 Heterogeneous organisational forms are the norm and not the exception, even in the same fields of economic activity.
5 Economic efficiency can only be analysed over the long term and involves a positive trade-off with perceptions of economic equity on the part of the actors involved.
6 The dominant institutional and organisational patterns at any given time may not represent 'efficient outcomes' but are the result of historical patterns of institutional and technological 'lock-in'.
7 Economic development therefore involves a process of social change which can best be analysed and promoted via an actor-network concept of social mobilisation coupled with a convention theory approach to the emergence and consolidation of different forms of economic coordination.
8 Efficiency and equity, competition and cooperation, conflict and solidarity are indissoluble and irreducible components of economic life.

The above perspectives I believe offer the basis for a more effective critique of liberal discourse and policies currently guiding the regional integration of the Mercosul. At the same time I would argue that they overcome the limitations of the dominant alternative discourses based on either a simplified view of the competitiveness of the family farm or a politicisation of economic life which fails to consider the relevance of notions of productive efficiency.

THE FIRST OF THE 'WOULD BE COMPETITIVE IF . . . ' ARGUMENTS: IF THERE WERE A 'LEVEL PLAYING FIELD'

Of the two 'would be competitive if . . . ' economic arguments the first is primarily applicable to a Latin American context and argues that the pattern of 'conservative modernisation' has differentially favoured the large-holding and undermined the natural competitiveness of the family farm.

In Brazil, specifically, the agrarian reform movement was aborted by the military dictatorship and transformed into marginalised land settlement schemes on the frontier. The dominant image of agricultural modernisation is that of an induced transformation of the *latifundio* as an alternative to agrarian reform. This has two variants – a São Paulo and a Northeastern model. In São Paulo, coffee eradication policies allowed for a subsidised recycling of the large-scale production model into cotton, oranges, sugar, and alcohol. More direct subsidies in the northeastern model stimulated the substitution of extensive cattle raising for crops.

The argument therefore is that the competitiveness of large-scale production is artificial and based on a combination of founders' rent, through privileged access to land, and subsidised credit. This latter facilitated large-scale investments which would otherwise have been unviable and created a bias in favour of mechanisation, undermining the natural competitiveness of labour. Subsidised credit in its turn inflated land prices, making new access to land difficult (in the case of young start-up families) and exacerbated tendencies to expel the small farmer both inside and outside the *latifundio*.

This view requires some measure of qualification. Modernisation in the southern states of Brazil in the late nineteenth and early twentieth century took place on the basis of the massive immigration of European farmers – Italian and German primarily – through both private and official colonisation schemes. They were settled in densely organised colonies with access to equal plots of land (around 20 hectares per settler family) creating first a peasant and then a family farm economy. While it is true that these settlements took place on the margin of an agrarian structure based on large cattle ranching, they were able to recreate, with modifications, a diversified farming system combining livestock and crop production.

A flourishing peasant economy therefore expanded in articulation with merchant capital, which provided access to markets, and was transformed into the modern family farm as this trading capital evolved into agro-industry. The farming system that emerged was to form the backbone of much of Brazil's modern agro-industry – pigs, poultry, wine, tobacco, and milk, together with grain inputs.

Nevertheless it is clear that the full competitive potential of the family farming system even in this context was limited by the prior existence of a large-scale farming structure based on ranching, wheat, and irrigated rice production which dominated both the best lands and the flow of resources to the region.

The above view of 'conservative modernisation' therefore offers only a partial interpretation of Brazilian agrarian dynamics. At the same time, it represents a very optimistic stance in arguing that the consolidation of a 'level playing field' would demonstrate the competitiveness of the diversified family farm vis-à-vis large-scale agriculture. That this applies broadly in the Northeastern case is, however, quite plausible, given the collapse of many large-scale projects as rural credit has become available only at positive interest rates.

Curiously, this view of the institutional preconditions required to demonstrate the competitiveness of the family farm is perfectly consistent with the direction of post-GATT regulation. It also harmonises with the dominant agribusiness view which similarly rejects subsidies, defends social policies for the 'small-scale producer', and assimilates the modernised family farm to the undifferentiated entrepreneurial agricultural sector.

THE SECOND ECONOMIC COMPETITIVENESS ARGUMENT: IF THERE WERE NO TECHNOLOGY BIAS

In moving on to a detailed specification of the basis for family farm competitiveness we are in danger of repeating a centuries-old debate which we, along with many others, have dealt with at length over the years (Goodman, Sorj, Wilkinson 1987). Here I would like to focus only on literature and issues which to my mind throw a new light on this question and which can be broadly referred to as the 'technology paradigm', which constitutes the second of our 'would be competitive if . . . ' arguments.

It should be clear in the first instance that we are referring to one subtype of family farm – the diversified farming system. The family farm is an elastic category ranging from peasant subsistence to highly technified monoculture. To the extent though that it adopts a specialised farming pattern, the advantages of the family farm, or rather the 'single operator', are generically related to scale economies within a common technology paradigm. What we are concerned with here are the advantages accruing from what might be termed 'scope economies', to adopt the language of industrial organisation literature, both in land and labour use.

Before developing this point, though, the general technology bias argument can be posed as follows. A family farming route of agricultural development was blocked in Brazil and many other Latin American countries by the consolidation of relative prices which did not reflect real 'factor endowments', basically as a result of regressive land access and subsidised credit. Transaction cost analysis has shown that family labour has clear advantages in relation to a hired labour force needing external supervision and control (de Janvry and Sadoulet n.d.) The option therefore was more generically that between man and machine. In the first instance, labour competitiveness was displaced by institutional mechanisms favoring mechanisation. The second step towards the consolidation of large-scale production units was a technology paradigm for agricultural machinery inappropriate to the family farm sector.

The potential competitiveness of the family farm in this case is highly hypothetical since the technology source is exogenous (the farming equipment multinationals) and the small farmer market provides little attraction for complementary investments on the part of global dealers, in addition to having insufficient political clout itself to mobilise alternative mechanisation strategies. With time, of course, rural exodus confirms the superiority of the established model.

Irreversibility within a particular technological trajectory therefore is not necessarily a direct reflection of competitive superiority, but can be an outcome of institutional arrangements which predispose towards a particular technological solution. Of course, that technology is not neutral has been amply defended in the Marxist tradition for over a hundred years. Recent literature however, both economic and sociological, has given operational content to the notions of reversibility and irreversibility in relation to technological trajectories, providing an important tool for the dimensioning of alternatives (Callon 1991, Dosi and Metcalfe 1991).

The neo-classical theory of production presupposes a simple reversibility of factor proportions depending on the evolution of relative prices, which is consistent with its view of the ready availability of different options along the technological frontier. As a result the dominant technology paradigm would necessarily represent the optimum solution, and therefore the most competitive alternative.

Over the last decade, however, a broad range of economic literature has built up a very different approach to technological 'progress'. On the one hand, the supply of technology and its subsequent development have been shown to have an important degree of autonomy from market pressures (Rosenberg 1982). The scientific community and the compulsive problem solving internal to a technology trajectory are given pride of place here. An additional important notion is that of sub-optimal 'lock-in' whereby factors *independent* of efficiency criteria lead to the preferential choice of a specific technology trajectory whose subsequent development leads to the elimination of alternatives which, at a given point in time, may have been competitive and perhaps even superior options (Arthur 1989). Convention theory has also adopted the notion of 'path dependency' in its own analysis of the way new social processes at the micro level become transformed into a dynamic of broad social and economic change (Boyer and Orleans 1994).

At this juncture economics joins with history and with sociology. Two sociological traditions deal explicitly with the consolidation of technology paradigms, irreversibility and 'lock-in' – Callon (1986) and Latour (1983), with their socio-technical networks, and Granovetter (1985, 1991), with a more traditional network theory integrated into social construction analysis. Both approaches offer operational theories of dominant economic organisation which transcend notions both of efficiency and simple power relations (capital x labour analysis), designing hypotheses with regard to the creation, maintenance, and consolidation of heterogeneous coalitions which ensure specific styles of economic and technological organisation. These two approaches are able to give full play to the insights developed in the economics tradition which demonstrate the derivative character of markets and price structures. Such an approach represents an important demystification of the competitiveness criteria.

In the light of this economic and sociological literature, the dominant technological paradigm of Brazilian agricultural modernisation (large-scale mechanised monoculture) could usefully be analysed within the parameters of 'sub-optimal lock-in'. On the basis of this insight the achievement of a 'level playing field' may be relatively innocuous (our first scenario of 'competitiveness if . . . ') depending on the degree of (ir)reversibility established.

The mainstream economic tradition has likewise tended to a reductionist 'efficiency' view of organisations and institutions, which is the obverse of an underlying technological determinism. Four such 'schools' can rapidly be identified. In the neoclassical application of induced innovation to agriculture, Hayami and Ruttan interpret institutional innovations, from the English enclosures to the organisation of R&D, as optimal adaptations to shifting relative factor endowments reflected in prices (Hayami and Ruttan 1971). Transaction cost theory popularised by Williamson views changing organisational forms as the efficient response to the varied costs of conducting different types of economic activity (Williamson 1975). Regulation theory on the other hand, in what has been referred to as its 'mark 1' version, has tended to view institutional arrangements as a functional response to accumulation requirements (Favereau 1995). The neo-Schumpeterian current comes closer to a non-reductionist account in its micro analysis of firms in terms of 'routines', although the 'long-wave' macro component generally views institutions as adjustments to the potential of a new technology paradigm (Freeman and Perez 1988).

Probably the most significant development in heterodox economics, however, has been the increasing recognition of institutional and organisational variability. This is true, as we have seen, of neo-Schumpeterian micro analysis, where the notion of 'tacit knowledge' as the basis of a firm's routines gives pride of place to organisational heterogeneity (Dosi 1988). It is equally the case for recent developments in regulation theory which explicitly reject a unilateral relation between dominant organisational forms and efficiency (Boyer 1995). Convention theory however, with its pluralistic view of forms of economic coordination, has perhaps gone furthest in relaxing the notion of correspondence between economic efficiency and any specific organisational form. 'Collective learning', implying a specific perceived trade-off between efficiency and equity, is seen as the key micro level concept underpinning organisational variability (Favereau 1995).

The sociological approaches of Callon/Latour and Granovetter complement this evolution in economic thought with a more autonomous view of actors and organisations. The latter in particular, in his elaboration of McGuire's study of the US electrical supply industry, shows how network theory and his notion of 'the strength of weak ties' can give an account of economic dynamics in which actors at certain moments face a variety of options between equally efficient forms of economic organisation (Granovetter 1991). The differential capacity for mobilisation explains the outcomes and the subsequent degree of closure or 'lock-in', which is eminently social in origin, but which may subsequently determine uniform rules of the game over longer or shorter periods. In these later periods narrower concepts of economic efficiency may prevail.

AGRICULTURAL EXCEPTIONALISM AND FAMILY FARM COMPETITIVENESS

If applied directly to the case of the Brazilian family farm, both the sociological and economic approaches mentioned above would probably emphasise the irreversible nature of the large-scale specialised farming model. To advance a more optimistic view, I think we need to combine the above methods of analysis with insights from within the tradition of agro-debates. Four separate issues need to be distinguished here.

First, the positive association of technology has become so self-evident that it has been consecrated in the very phrase 'technological progress'. Nevertheless, in agriculture, nature is still a competitive alternative to technology. Wheat, soybeans, and corn from the Argentinean pampas are more competitive than Brazilian products not because of higher yields but as a result of lower costs through not using fertiliser. The need to use fertiliser (technical progress), which is only now beginning to make itself felt in the pampas region, is in fact an index of loss of competitiveness. In other publications we have tried to capture this notion with the idea of agriculture as a 'natural factory' (Goodman, Sorj, Wilkinson 1987).

Second, in the same way that nature competes with technology, it can also resist the application of technology. Many harvesting activities for instance are still not mechanised. Technological 'progress' therefore may be partial even within the same production activity and limitations imposed by the need for manual activities in one phase may undermine the potential scale benefits of mechanisation in 'before' or 'after' activities.

This line of inquiry may seem to lead back to the notions of a 'natural' division of labour, or a functional dualism, between family farming and large-scale mechanisation – a permanent theme in the literature but perhaps most identified with Servolin (1972). In fact, however, the crucial challenges for the family farm in the Brazilian case, and perhaps in many other Latin American examples, emerge from situations of direct competition between the two systems, with the increasing threat of 'expropriation' by the large-scale specialised enterprise of activities essential to the survival of the family farm. Here the advantages of 'scope' rather than 'scale' are crucial to the characterisation of the family farm.

Third, where technology is unable to neutralise the benefits of optimising marginal land and labour use, the family farm is able to compete directly on individual product terms with technology-intensive models. This is currently the case for milk production. To date, non-specialised, low input milk production has shown itself to be cost competitive with more intensive specialised production in Brazil, and shows better survival prospects when faced with the downward impact on prices of milk product imports from Argentina and Uruguay (Denardi 1994). Here we have the phenomenon of competing technology routes. Reacting to this competitive threat, the specialised sector resorts to lobbying for exclusionary regulatory and technology trajectory measures in an effort to restore competitiveness. This is perhaps a case where the process of 'sub-optimal lock-in' is presently occurring, demonstrating clearly the need for a complementary economic and sociological analysis. At the level of a single product, we can say that milk production in Brazil (and neighbouring Paraguay, Bolivia, and probably other Latin American countries) reproduces the analytical issues behind the notion of 'bifurcation', the moment preceding technological 'lock-in' and therefore most open to the sort of socio-technical network analysis associated with Callon (1986) and Latour (1983) or Granovetter (1991).

At the same time, the insights of convention theory could be drawn on here, particularly the notion of the genesis of conventions whereby agreement on patterns of economic coordination are reached through the common evaluation of technological parameters which are then negotiated with the wider society. This approach has been developed in France to study the emergence and consolidation of special label products in agro-food (Sylvander 1995). A clear convention-style process based on the principles of industrial economic coordination (productivity, scale economies) is evident in the consolidation of the specialised milk sector. On the other hand, while the low-input, low-yield 'domestic' model of milk production in the context of diversified production systems shows surprising vitality and has the support of commonly valued technological criteria, it lacks an effective movement of legitimation which would parallel the *appellation d'origine* product conventions which have emerged in the French context (Letablier and Delfosse 1995).

Fourth, technological progress can have very specific reversible effects in the relations between agriculture and industry. Miniaturisation of primary processing activities – milk pasteurisation, and a number of processing activities, particularly in fruiticulture – allows for the return of certain agro-industrial phases to an agricultural setting. This can be seen perhaps as part of a more general trend to de-verticalisation and tertiarisation, but in the agricultural sector it has important implications for the local and regional appropriation of 'value-added' in the agro-industrial chain (Wilkinson 1995).

Pulling these considerations together, therefore, we would make the following points. The neo-Schumpeterian economics tradition already provides an analysis of technological development which calls into question any simple equation of this process with notions of efficiency or optimisation. The autonomy of the innovative process, the compulsive nature of scientific inquiry, institutional factors, and contingency, all play their parts. Nevertheless, the overall emphasis in this literature gives pride of place to specifically technological features in the subsequent consolidation of the trajectory or paradigm – their appropriability, cumulativeness, range of applicability, and synergies (Chesnais 1986, Dosi 1988).

The convention approach on the other hand has a much more explicitly organisational theory of economic activity which takes into account, as we have just had occasion to note, the negotiations leading to agreements over forms of economic coordination. In addition, its analysis of collective learning (Favereau 1994) and its insistence on the plurality of justifiable forms of economic coordination lead it to insist on the heterogeneity of equally viable organisational forms (Boltanski and Thévenot 1991). This is particularly important both for the defence of diversified family farming systems and the specificities of organisational forms such as the cooperatives which we will discuss below.

Convention theory is much closer to the sociological traditions of Latour, Callon, and Granovetter in emphasising the weight and diversity of the social actors mobilised in the attempt to reorganise economic activities around determinate technological options. These approaches go beyond the power relations analysis of the Marxist tradition to the extent that they are better able to capture the heterogeneity of the actors involved. More particularly, they can identify the way in which interests emerge and are consolidated in the process of mobilisation (the notion of 'translation' is relevant here) together with the more contestable character of the networks which sustain particular trajectories.

We would add that all these approaches need to be complemented by a sensitivity to the *sui generis* character of agricultural activity (whether viewed from the angle of 'nature' or 'labour', which are simply alternative aspects of the same process) in order to contextualise both the technological dynamic and the range of social options. In the case of strategies to strengthen family farm production, a reintroduction of the notion of constraining factors in the form of the unique features of agricultural production allows the favourable and/or unfavourable conditions for actor mobilisation to be more clearly delimited.

A PROFILE OF THE DIVERSIFIED FAMILY FARM IN THE SOUTHERN CONE AND PATTERNS OF AGRO-INDUSTRIAL INTEGRATION

To understand the relevance of these issues to the analysis of the family farm in the southern states of Brazil and similar regions in the Southern Cone, it may be useful to present a composite 'ideal type' of this farming pattern (Testa *et al.* 1996).

In the first place, we are talking of an almost exclusive dependence on income generated by farming activity. The family farming areas, by and large, are concentrated in hilly terrain quite distant from major urban centres, and regional industrial and service sectors provide few opportunities for alternative employment.

The farming system can be very diversified but is basically characterised by three or four tiers – subsistence production, traditional marketing produce, input crops for livestock, and agro-industrial production, usually on the basis of contract quasi-integration. The structure is hierarchical with pride of place given to the agro-industrial activity, although the monetary equivalent of subsistence production, as also the traditional marketed crops, can often represent higher income sources. As a rough guide, it could be said that each of the three basic components – subsistence, traditional cash crops, and agro-industrial production – have equal weight for the viability of the family farm.

Subsistence activities include small livestock and orchard production (and often additional activities such as honey and wine production), together with basic food crops, typically beans and corn. The food crops are normally an important source of income but are also internalised for on-farm consumption as food or feed. The agro-industrial activities vary by region, size, and type of farm-holding, with wine, tobacco, milk, pigs, and poultry predominating.

Tobacco

Tobacco production is lowest in status, requiring little land (up to 2 hectares) but involving heavy and unhealthy labour commitments and providing only a once-yearly income. This activity is concentrated in only one or two areas but engages a very large number of very small farmers. Grape and wine production is also very regionally specific and has a similar disadvantage in being an annual crop.

Poultry

Family farm contracting has been the basis of Brazil's dynamic poultry sector and has constituted the agricultural arm of its leading agro-industrial firms – Sadia, Perdigão, and Ceval. The enormous productivity of poultry production, however, means that agro-industrial integration is a minority option. (There are large waiting lists for inclusion as contract producers, creating a sort of reserve army limiting producer bargaining power in a situation where small numbers potentially permit easier organisation.) A typical family farmer with a 100-metre-sized aviary can produce more than 100,000 kilos of poultry meat per year, enough for 5,000 Brazilians' yearly consumption. On this basis a little over 30,000 farmers could supply all the needs of the national industry.

In spite of the rigorous productivity requirements, poultry production is a much sought after option partly because the 'droppings' provide excellent and easily applicable fertiliser for the cash crops. It is often claimed that the increased productivity of these cash crops leads to greater returns than the poultry activity itself. The notion of synergies therefore is a double-edged weapon in the case of family farm agro-industrial integration, allowing for downward pressure on poultry prices in return for the free use of an unappropriable by-product from the point of view of the agro-industry. (The unintended consequence may be that all actors benefit – farmer, agro-industry, the ecosystem, and perhaps even the consumer, serving to demonstrate in a distorted manner the systemic competitiveness of this production system.)

Scale economies are an important issue here. In the Brazilian case, there would appear to be no direct relation between the size of aviaries and the various performance indicators. Aviaries of 25 or 50 metres can therefore be equally efficient although agro-industry tends to favour modules of a hundred metres (Testa *et al.* 1996). (Interestingly, though, in its model production system for the family farm in the year 2000, the Sadia Company, Brazil's leading firm in this sector, proposes a 50-metre aviary.) The cooperative system, on the other hand, is now favouring the smaller units because of the dual benefits of poultry production for the family farm – the fertiliser spin-off just referred to, and the important cash-flow factor deriving from the short production cycle, which permits five or even six batches, and therefore payments, per year. Since the farmer has to bear the costs of the poultry installations, smaller modules make this option more widely accessible.

While productivity seems not to be affected, logistics, particularly transport and planning, may well benefit from a greater concentration of production activities. If this were true and insurmountable, private agro-industry would have a more favourable cost structure than the cooperative, even though its model may be sub-optimal when the negative impact of concentrated production activities on the viability of the family farm sector as a whole, and therefore on overall regional efficiency, is taken into account. The lower cost model may therefore camouflage externalities, in this case, out-migration with all its implications, both rural and urban. There may of course be other types of hidden externalities – higher road maintenance through the 'wear and tear' of larger trucks. We will return to this 'full cost' issue later.

However, the logistical barriers to a more spatially dispersed poultry production system may well not be insurmountable. Rather, in adopting a strategy of smaller modules, the cooperative model adopts different priorities for technological and organisational innovation. Specific patterns of producer cooperation and coordination would have to be explored in order to determine an alternative route to efficiency and competitiveness; not, in this case, at the level of individual product systems, but on the terrain of production and distribution logistics.

In the case of poultry production, private agro-industry's strategy of greater concentration is still situated within a diversified family farm perspective. Indeed, agro-industry does not in general permit more than one aviary per farmer. Originally this was justified in terms of disease control, but now it is probably more related to a strategy of maintaining relatively high numbers of producers for more favourable bargaining terms. Competitiveness here, too, is designed on the basis of more long-term interests.

Pigs

Pig-farming, however, which has represented the most important agro-industrial component of the family farm in these regions, both in terms of income and synergies with crop production (in addition to being a crucial subsistence component) finds itself now at a fundamental 'bifurcation' point. Private agro-industry is moving to replace the 'complete cycle' production system with a more specialised division of labour separating out the breeding from the fattening phase. At the same time, feed is now being provided directly by agro-industry (following the poultry model) and no longer from on-farm production. In this process, economies of scale are possible which push pig-farming out of the diversified family farming sector.

When viewed on an individual product system basis, the new technological model, involving specialised pig breeding and fattening activities, would seem to demonstrate scale economies with a high degree of irreversibility in terms of price competitiveness. This model provides both a direct challenge to the 'full cycle' pig production of the family farm, and indirectly further undermines the synergies of the diversified production system through reliance on off-farm supplies of feed, principally corn.

For its part, organisational innovation in pig production within the parameters of the family farm has only been able to anticipate the dynamic of the new dominant model. This new dominant model, on the other hand, would seem to be particularly vulnerable to a 'full ecology costs' type of regulation, and also to a critique along the lines of 'full social and regional costs'. Regulation leading to the internalisation of these latter costs, however, may only accelerate regional relocation.

Milk

The depth of the impasse posed by the evolution of pig farming and attendant crop production patterns can be gauged by the fundamental importance of spontaneous reconversion processes under way in the region, particularly towards milk production. Here, as noted earlier, alternative technology routes are in direct competition, but in this case it is the low input non-specialised model which appears to be favoured. Even more important, a tested technical package exists for improvements within the low input non-specialised trajectory. Organisational innovations have been more successful in this case also, with the development of 'condominiums' for the joint use of pasture land and milk parlours. The Achilles' heel again is logistics, but here producer organisational initiatives can substantially lower milk collection costs. A production level of only 20 litres a day provides the equivalent of a minimum monthly salary, with the advantage of regular fortnightly payments. It is not surprising therefore that milk production is assuming the role of anchor product for the family farm in transition to a new production profile.

STRATEGIES FOR THE FAMILY FARM ON THE BASIS OF CURRENT TECHNOLOGICAL AND ORGANISATIONAL OPTIONS

This brief overview of the situation of the family farm in the southern states of Brazil is sufficient to capture the systemic character of its current crisis. (In Paraguay the situation is similar, with cotton playing the role of principal agro-industrial crop. It is estimated that the increased costs of pesticide use against the *bicudo* will make cotton unviable in the family farm system within the next five years. Here, too, milk presents itself as an alternative anchor product.)

So far we have concentrated on the crisis accompanying the various forms of agro-industrial integration, which in the case of livestock affects in its turn the food-feed-cash crop sector (corn). Other more traditional cash crops are also beginning to be threatened by specialised mechanised production.

The crisis therefore affects all levels of market integration, and points to the crucial role of subsistence crops in strengthening the resilience of this sector, providing minimum levels of protection as new forms of market articulation are tried out.

We have argued that liberalisation and regional integration are accelerating tendencies to specialisation and scale economies on an individual product basis. These tendencies initially led to a concentration of production resources within the family farming sector, but are now threatening a relocation of agro-industrial and cash crop products outside this family farming sector, and even outside this region as a whole. As we will see, the transformation of the family farming crisis into a regional crisis is crucial to strategies for mobilisation.

Our analysis of the technology literature, however, combined with an appreciation of the singularities of technology trajectories in agriculture, has allowed us to identify the heterogeneity of technological modernisation and its specific limitations in different product contexts. Even on an individual product basis, therefore, low-intensive milk production, making use of marginal land and labour inputs, is price competitive at levels which make it an attractive option for the family farm. Two technology routes are in conflict here, providing the basis for alternative strategies for mobilisation (R&D priorities, health and hygiene regulation, credit allocation) and attendant opportunities to impose or resist a 'lock- in' situation or 'convention'.

The stakes in this struggle between alternative technological trajectories are high. Thus the competitive consolidation of milk production within the family farm is crucial to the latter's survival, since it represents a strategic dynamic growth market in the Brazilian context, and a source of regular cash income. Competitiveness at low levels of concentration and yield productivity mean that milk production is an option for large numbers of producers and it can perhaps be seen as performing the anchor role associated with pig production in an earlier period. Although viable strategies will tend to demand greater product diversification than previously, together with insertion into new niche markets, it is important to recognise that competitiveness in certain mass consumption commodities is crucial to the generalised viability of the family farm in Brazil.

'Reconversion' is the alternative currently suggested to overcome family farm competitive inefficiency in the face of the increasingly specialised production of its traditional crops. Although official support measures in this direction have been timid, it is clear that reconversion is a permanent component of long-term viability and adaptability to changing production and consumption structures. In fact, the most important example of reconversion has been the spontaneous shift to milk production over the past ten years. But there are clear limits to reconversion at the individual farm level. Milk is an exception in that it represents the reorientation of a traditional subsistence product towards the market, in circumstances in which regulation and consumer preference impose few or no barriers to such a transition.

Other initiatives, however, usually involve important investments, at all levels – technology adaptation, investment credit, agro-industrial installations, and marketing channels – all of which require institutional underpinning against the risks involved. This is particularly the case in these regions where the most clearly identified alternatives are fruiticulture and forestry products, involving long maturation periods.

Private agro-industry has not shown interest in supporting reconversion, and is increasingly moving to more specialised forms of integration. The cooperative structure therefore assumes a strategic position in any serious initiative towards reconversion. This is already apparent in the cooperative's promotion of diversification to oranges and tea production on the basis of 1 or 2 hectares per farm rather than on a large-scale plantation model. There is currently great pressure on the cooperative form of organisation to adapt to liberalisation through a parallel process of 'entrepreneurialisation'. Clearly, efficient management practices are a prerequisite of market accountability and the demand for professionalisation cannot be countered by an appeal to the 'democratic' ideology of the cooperative movement. The cooperative, however, has a dynamic that distinguishes it from that of private agro-industry to the extent that it is rooted in the resources of a specific region and its *raison d'être* is the valorisation of the agricultural production of its members. For private agro-industry the latter is simply an input for the agro-industrial product, to be obtained on the most favourable conditions, even if that means shifting investments out of the region or importing produce from other regions or other countries.

The cooperative, therefore, must develop competitiveness on the basis of different concepts of efficiency from private agro-industry. The Lash and Urry typology can perhaps be adopted here, with the idea of efficiency being built on the human factor rather than the technological package, a German/Japanese rather than an American model, in their terms (Lash and Urry 1994). Here again, the convention analysis would be appropriate given its sensitivity to differing but equally valid modes of economic organisation not only between organisations but within the same organisation (Boltanski and Thévenot 1991). This specific role of the cooperatives in structural reconversion must, however, receive recognition at the level of state policy and financial institutions, and lead to compensatory mechanisms for the risks and the long maturation periods involved.

A third conclusion emerging from these considerations is that successful reconversion requires a period of transition which could perhaps be negotiated with agro-industry. This might assume the form of a guarantee that during a certain period a given percentage of pig-meat requirements would continue to be supplied by the family farm sector on the basis of 'complete cycle' production, or acceptable levels of the more specialised production system. In the Brazilian case, there are historical precedents for such an agreement in the sugar sector where, in the transition to the modern sugar refinery, 30 per cent of raw material requirements was reserved for family farm suppliers.

Given the historical importance of the phenomenon of 'lock-in' and the heterogeneous dynamic of agricultural technology, with differing impacts on scale economies at the individual product level, a simple elimination of 'institutional and technological bias' is not sufficient to guarantee the competitiveness of the family farm.

On the other hand, the *sui generis* character of agricultural production and the consequent ambiguity of 'technological progress' permit the elaboration of a series of measures to maintain the viability of the family farm combining the three elements identified above: (1) promotion of favourable technological and organisational routes where individual product competitiveness persists combined with resistance to adverse 'lock-in'; (2) mobilisation of support for reconversion and recognition of the privileged role of cooperative organisational structures in this process; and (3) negotiation of voluntary or imposed limitations on individual product scale economies, either indefinitely (in the light of ecology or social 'full cost' considerations) or for a transitional period in which reconversion is consolidated.

ECOLOGICAL AND SOCIAL COSTS AS CRITERIA FOR EFFICIENCY EVALUATION: POTENTIAL AND AMBIGUITIES

The post-GATT environment is one in which market prices are supposed increasingly to reflect production cost efficiency. Redistributive measures are not excluded but are to be 'de-linked' from the productive system. Given the *sui generis* features of agricultural production referred to above, it must be recognised that this approach to the market is not unequivocally favourable to the large-scale mechanised monoculture or intensive livestock model. Natural or ecological comparative advantage remains crucial, and where present logically favours less intensive agricultural production systems which can be either large-scale or on the family farm model. Nevertheless the current determinants of price formation tend to reward specialised production systems and the exploitation of scale economies.

The limits of the market for organising economic activity have long been recognised in the literature on 'market failure' and 'externalities' within the neoclassical tradition, together with the neo-Schumpeterian notion of 'degrees of appropriability'. This literature, moreover, has tended to legitimate, or elucidate the logic of, the public organisation of economic activity where these conditions apply. Current ecological considerations, in a climate favouring the withdrawal of the public sector from direct involvement in productive activity, represent a direct challenge to established notions of price formation. The principle of sustainability involves alternative notions of productive time and space, whose internalisation as costs has radical implications for the organisation of agricultural activity.

The literature on this question is now large and increasingly operational in the sense of defining regulatory measures which would lead both to the internalisation of environmental effects into the cost structure of productive activity and to the definition of new notions of permissible types of economic activity. In principle, the family farm might be thought to be the main beneficiary of developments in this direction in agriculture, to the extent that its inter-generational dynamic implies a preoccupation with sustainability. However, the pressures of agro-industrial integration, or even unfavourable market prices can, and often do, lead the family farm to sacrifice its long-term agro-ecological viability. The large-scale specialised operation on the other hand, in the context of a closed frontier and heavy fixed investments, may well be open to measures which reinforce its long-term productive sustainability.

Here again we would appear to be at a crucial 'bifurcation point' with ecological technology packages being adaptable both to large-scale and family farm production units. In the region under study, Southern Brazil, where a key issue is the pollution effect of pig farming particularly on the water supply, it would appear that the larger farms are the principal beneficiaries as a result of the costs of the technology solution on offer. The penalisation of non-adopters in this case would therefore accelerate the marginalisation of the small-scale farmer. Yet, paradoxically, the 'end of pipe-line' technology solution being offered in fact presupposes the maintenance of non-sustainable production systems at the farm level. A detailed analysis of the relative strengths of the socio-technical networks mobilised around alternative ecological trajectories would therefore seem to be the soundest approach to identifying the likely impact of the internalisation of ecological considerations on the future of the family farm in the region (Guivant 1996).

The internalisation of the full social costs of specific production system options is both more polemical and more difficult to envisage in terms of its operationalisation. Clearly, from a neoclassical point of view, the position is in principle unacceptable since there would always be a tendency towards equilibrium and the reabsorption of people and resources into the economic system. From a more social-democratic viewpoint, however, both the state and the trades unions have developed traditions of compensatory intervention, particularly in the industrial sphere. These have normally taken the form of measures to attenuate the effects of unemployment – work-sharing, temporary pacts against lay-offs, tax incentives to take on or not lay off workers, or payment of retraining schemes. This latter is perhaps the closest that such measures have come to penalising technical progress which leads to lay-offs. In this context, convention theory provides a crucial argument for the long-term causal relation between economic efficiency and equity (Favereau 1994).

In the context of Southern Brazil, the idea of full social costs refers to the impact of greater levels of production concentration and specialisation on the disarticulation of the diversified family farm structure. The negative consequences of this process are already in evidence – rural exodus; under-utilisation of rural investments (schools, local roads, health services, electrification); the emergence of urban slums; a sharp increase in the crime rate as rapid regional urbanisation is based more on the impasses of the family farm than the attraction of industrial or service job opportunities; and regional out-migration.

We are dealing here, therefore, with public, local, and regional costs. A simple ban on production scales above a determinate level may only accelerate a movement of investment out of the region. This may be worth the risk if the consequences of further scale economies were to undermine the family farm profile of the region. It is difficult, however, to see on what basis this could be legitimated if environmental considerations were not at stake. Negative taxes as compensation for possible higher production costs through the non-adoption of greater scale economies (to the extent that this is under the power of the local or regional authority) may be a more effective measure. Probably, however, the most realistic option would be the negotiation of some transitional pact, as suggested in the third point in our conclusion to the previous section.

CONCLUSION

In this critical review of commonly adopted arguments made in defence of the family farm, I have tried to show that its competitiveness is not necessarily guaranteed if 'biases' in the institutional, regulatory, cost-accounting, or technological arrangements were to be eliminated. Nor can competitiveness be ignored in the name of justice or reduced to the question of power relations. Economic outcomes, however, are not the direct result of 'optimising under constraints' guaranteeing superior efficiency, and within the economic literature the 'lock-in' perspective offers a fruitful alternative approach, as does the analysis in terms of conventions which we have also considered.

Although the GATT agreements and processes of regional integration under the aegis of a liberalisation discourse may suggest that the door is rapidly closing, the ecology movement, the proliferation of technology options, the resurgence of regional preoccupations, and new consumer pressures and opportunities, are sufficient to characterise the present conjuncture as one of 'multiple stable equilibrium points', to adopt Granovetter's overture to neoclassical formulae.

In this perspective, the future of the family farm in the southern states depends less on narrowly defined competitiveness within existing technological and organisational limits than on the emergence of a coalition of actors committed to redefining economic priorities for the region based on the productive potential of the resources contained within the family farm system. As a first approximation, such actors would include regional political figures and institutions, cooperatives, particularly the local cooperatives at municipal level, extensionists, NGOs, and environmental interests. Agro-industry, which has been historically identified with the region for forty years, could be formally enlisted and perhaps involved in the type of pact suggested earlier. Local commerce might respond positively to the idea of retaining purchasing power in the region, and urban sentiments could be mobilised around the threat of slums, violence, and health issues. The key for mobilisation would be the identification of the family farm with the future of the region. Two basic programmatic documents along these lines are already beginning to be debated – one produced by the state's small farmer extensionists organisation (Testa *et al.* 1996) and the other by the Association of Municipalities in the Region (AMOSC 1995).

Economic activity therefore becomes integrated into social processes, although the options are far from unlimited, and economics and sociology jointly provide the inter-disciplinary tools for analysis.

NOTE

1 Reconversion is a provision of the Mercosul agreements which allows limited financial support for the development of alternative activities by uncompetitive sectors. In the Brazilian case, wheat would be an obvious example and, in the region more specifically under consideration in this chapter, grapes and wine production.

BIBLIOGRAPHY

AMOSC (1995) *Plano de Desenvolvimento Sustentável para o Oeste Catarinense*, Chapeco.

Arthur, B. (1989) 'Competing technologies, increasing returns and lock-in by historical events', *Economic Journal*, vol. 99, no. 394, pp. 116–131.

Boltanski, L. and Thévenot, L. (1991) *De la Justification*, Paris: Gallimard.

Boyer, R. (1995) 'Vers une théorie originale des institutions économiques?' in R. Boyer and Y. Saillard (eds) *Théorie de la Régulation. L'Etat des Savoirs*, Paris: La Découverte.

—— and Orleans, A. (1994) 'Persistence et changements des conventions', *Analyse Économique des Conventions*, Paris: PUF.

Callon, M. (1986) 'The sociology of an actor network: the case of the electric vehicle', in M. Callon, J. Law and A. Rip (eds) *Mapping the Dynamics of Science and Technology*, London: Macmillan.

—— (1991) 'Reseaux tecnico-économiques et irréversibilités', in R. Boyer (ed.) *Les Figures de L'Irréversibilité en Économie*, Paris: EHESS.

Chesnais, F. (1986) 'Technological Cumulativeness, the Appropriation of Technology and Technological Progressiveness in Concentrated Market Structures', paper presented to the Conference on Technology Diffusion, *Venice: Ca'Dolfin.*

Denardi, R. A. (1994) 'Competitividade do setor leiteiro no Mercosul', unpublished masters thesis, Rio de Janeiro: CPDA/UFRRJ.

Dosi , G. (1988) 'Sources, procedures and microeconomic effects of innovation', *Journal of Economic Literature* 26: 1120–71.

—— and Metcalfe, S. (1991) 'Approches de l'irréversibilité en theorie économique', in R. Boyer (ed.) *Les Figures de L'Irreversibilité en Économie*, Paris: EHESS.

de Janvry, A. and Sadoulet, E. (n/d) *Alternative Approaches to the Political Economy of Agricultural Policies: Convergence of Analytics, Divergence of Implications* (mimeo).

FAO/INCRA (1995) *Reforma Agrária e o Refortalecimento da Produção Familiar*, Brasilia: INCRA.

Favereau, O. (1994) 'Règles, organisation et apprentissage collectif: un paradigme non standard pour trois théories hétérodoxes', in A. Orléans (ed.) *Analyse Économique des Conventions*, Paris: PUF.

– (1995) 'Conventions et Régulation', in R. Boyer and Y. Saillard (eds) *Théorie de la Regulation. L'État des Savoirs*, Paris: La Découverte.

Freeman, C. and Perez, C. (1988) 'Structural crises of adjustment: business cycles and investment behaviour', in G. Dosi *et al.* (eds) *Technical Change and Economic Theory*, London: Pinter.

Goodman, D., Sorj, B., and Wilkinson, J. (1987) *From Farming to Biotechnology*, Oxford: Blackwell.

Granovetter, M. (1985) 'Economic action and social structure: the problem of embeddedness', *American Journal of Sociology*, 91 (3), pp. 481–510.

—— (1991) 'Economic Institutions as Social Constructions: A Framework for Analysis', paper prepared for the CREA Conference, Paris, mimeo.

Guivant, J. S. (1996) *Suinocultura e Poluição: Paradoxos do Controle Ambiental*, Paris, mimeo.

Hayami and Ruttan, V. (1971) *Agricultural Development. An International Perspective*, Baltimore: Johns Hopkins University Press.

Lash, S. and Urry, J. (1994) *Economies of Signs and Space*, London: Sage.

Latour, B. (1983) 'Give me a laboratory and I will raise the world', in Knorr-Cetina and M. Mulkay (eds) *Science Observed*, London: Sage.

Letablier, M. T. and Delfosse, C. (1995) 'Genèse d'une convention de qualité', in G. Allaire and R. Boyer (eds) *La Grande Transformation*, Paris: INRA/Economica.

Rosenberg, N. (1982) *Inside the Black Box: Technology and Economics*, Cambridge: Cambridge University Press.

Servolin, C. (1972) 'Aspects Économiques de l'Absorption de l'Agriculture dans le Mode de Production Capitaliste', in *L'Univers Politique de Paysans*, Paris: A. Colin.

Sylvander, B. (1995) 'Conventions de Qualité, Concurrence et Cooperation', in G. Allaire and R. Boyer (eds) *La Grande Transformation*, Paris: INRA/Economica.

Testa, V. M., Nadal, R., Mior, L. C., Baldissera, I. T., and Cortina, N. (1996) *O Desenvolvimento Sustentável do Oeste Catarinense*, Santa Catarina, Brazil: EPAGRI.

Williamson, O. (1975) *Markets and Hierarchies*, New York: The Free Press.

Wilkinson, J. (1993) 'Adjusting to a demand oriented food system: new directions for biotechnology innovation', *Agriculture and Human Values* Spring: 31–39.

—— (1994) 'Agroindústria e a Produção Familiar', *Política Agrícola* II, II, 1: 101–37, Mexico.

—— (1995) *Setores Agro-industriales Sensibles en el Contexto de la Integración Regional del Mercosur*, Montevideo: FAO (mimeo).

—— (1996) 'A New Paradigm for Economic Analysis? Recent Convergencies in French Social Science: An Exploration of Convention Theory with a Consideration of its Application to the Agrofood System', CEDI/Paris XIII Working Paper (forthcoming in *Economy and Society* August 1997).

3

MULTIPLE TRAJECTORIES OF RURAL INDUSTRIALISATION

An agrarian critique of industrial restructuring and the new institutionalism

Gillian Hart

INTRODUCTION

One of the apparent anachronisms of late twentieth-century capitalism is the dispersal of industry into villages and towns in predominantly rural regions. Rural industrialisation appears almost oxymoronic. Yet some of the most spectacular instances of industrial accumulation in recent decades – including the Third Italy, significant parts of the 'Taiwanese miracle', and much of the stunningly rapid growth in China since the mid-1980s – all exemplify rural industrialisation.[1] So too do a number of less glamorous manifestations of late capitalism, such as those in former bantustan areas of South Africa. The emergence of these globalised sites of accumulation, export production, and 'rural urbanisation' (Sanghera and Harriss-White 1995) underscores the continuing salience of agrarian questions as we move towards the new millennium.

In the article that formed the point of departure for this book, Goodman and Watts (1994) delivered a series of telling blows on efforts to interpret contemporary agro-food systems in terms of concepts imported from the industrial restructuring literature – notably regulation theory and flexible specialisation. These conceptual instruments are also quite blunt in coming to grips with the key issues posed by rural industrialisation: the multiple, nonlinear, and divergent *trajectories* through which industrial capital in various guises encounters and intersects with enormously varied agrarian conditions. These multiple trajectories of rural industrialisation exemplify the multiplicity of capitalisms taking shape in different regions of the world economy, rather than the unfolding of some sort of inexorable post-Fordist logic, locally embedded versions of flexible specialisation or, for that matter, 'bloody Taylorization' (Lipietz 1987).

The inability of the industrial restructuring literature to recognise and address rural industrialisation lies partly in its abstraction from the agrarian origins, contexts, and linkages of industrialisation, and partly in its core-centric presumptions. More fundamentally, the dualistic constructions and 'binary histories' (Williams *et al.* 1987) that underpin both flexible specialisation and regulation theory render them severely flawed in dealing with questions of diversity and dynamics. In recent years, proponents of industrial restructuring have increasingly taken recourse to structuralist conceptions of networks, embeddedness, and trust derived from Granovetter's (1985) remarkably influential article, together with notions of 'path dependency'. Yet this 'new institutionalism' is also severely limited in its capacity to illuminate the questions of socio-spatial change posed by contemporary processes of industrial dispersal. The key limitation is its abstraction from everyday politics and the exercise of power.

Part of my purpose is to suggest how elements of the agrarian literature provide greater analytical leverage than industrial restructuring or the new institutionalism. There are two related ways in which this literature speaks not only to rural industrialisation, but also to multiple, nonlinear, and divergent trajectories of capitalist development. First, it is cast in relation to classical political economy debates which give explicit recognition to multiple paths of agrarian transition, and bear directly on contemporary efforts to grasp the multiple reconfigurations of capitalist development in different regions of the world. Second, scholarship addressed to the enormous diversity, complexity, and fluidity of social institutions in 'Third World' agrarian settings both pre-dates and is more sophisticated than structuralist concepts in the institutionalist literature. Instead of asking *what* are the rules embodied or embedded in particular institutional forms, the processual approach taking shape in the agrarian literature focuses on *how* negotiation and contestation take place within and across multiple social arenas. In the discussion that follows, I develop these arguments and illustrate them with examples from my recent research on industrial dispersal in South Africa in relation to comparable processes and regions in Taiwan, China, and Malaysia.

Debates over industrial restructuring and 'global competition' in South Africa today are simultaneously analytical and political. Under the auspices of the Industrial Strategy Project, important segments of organised labour are invoking the discourse of post-Fordism and flexible specialisation to press for a high road of industrial development marked by technological dynamism, enhanced productivity, inter-firm cooperation, and a multi-skilled, high-wage work force (e.g. Kaplinsky 1991, 1994, Joffe *et al.* 1995). Either implicitly or explicitly, these visions of post-Fordism conjure up images of the Third Italy and Silicon Valley. They also play into narrow yet powerful corporatist alliances between organised labour and certain branches of metropolitan capital.

Proponents of a post-Fordist future for South Africa envisage these new industrial spaces as 'urban-centred growth poles' and explicitly dismiss the industrial decentralisation strategies pursued by the apartheid state in the former bantustans as being 'at variance with modern industrial experience'

(Kaplinsky 1991: 54). In fact, despite overall industrial decline and sharp cuts in state subsidies for industrial decentralisation, industries in some former bantustan areas continued to expand in the 1990s, driven partly by Taiwanese investment (Hart 1996a, Hart and Todes forthcoming). Some of these industrial complexes exhibit many features of ideal-type flexible specialisation – industrial clusters, network production, and close inter-firm cooperation – but combined with low-wage, typically coercive labour relations, and intense conflict between industrialists and the predominantly female work force.

There are important reasons for attending closely to these processes of industrial dispersal, rather than just dismissing them as unpleasant holdovers of apartheid that opened the way for exploitation of cheap, unorganised, female labour. First, they focus attention on how the broader history of agrarian transition in South Africa is directly implicated in contemporary industrial dynamics. The distinctive feature of this history is massive dispossession; this is fundamentally what distinguishes industrial dispersal in South Africa from comparable processes in different regions of East Asia which, although quite distinctive in their own ways, have been predicated on redistributive land reforms, the retention of small property, and various forms of state-sponsored subsidies in the agrarian sphere (Hart 1995, 1996b, forthcoming). Second, the particular local manifestations of industrial dispersal in South Africa are in practice quite varied, even in places that are structurally similar. In other words, interlocking dynamics of dispossession and industrialisation have been constituted and experienced in locally specific ways that bear directly on future possibilities for reconstruction.

In seeking to illuminate these processes and explore their larger significance, I have found the agrarian literature considerably more helpful than industrial restructuring or the new institutionalism.

BEYOND 'BINARY HISTORIES' AND THE NEW INSTITUTIONALISM: INSIGHTS FROM THE AGRARIAN LITERATURE

Regulation theory and flexible specialisation have, of course, been subjected to intensive and wide-ranging critiques.[2] My intention here is not to rehearse these arguments, but rather to indicate key features of this literature that render it incapable of coming to grips with central issues posed by rural industrialisation, and draw attention to those elements of the agrarian literature that provide greater analytical leverage.

The first and most obvious problem with industrial restructuring resides in the argument that forces emanating from the 'core' are determinant.[3] Amsden (1991) and Sayer (1989) offer related and trenchant critiques of these Eurocentric presumptions, pointing to features of Korean and Japanese experience that challenge key assumptions of regulation theory and flexible specialisation. These critiques suggest the importance of understanding global economic restructuring as in part the *product* of distinctively non-Western histories, institutional innovations, and dynamics. Japan is by no means the only non-Western exemplar of the limits of flexible specialisation and post-Fordism. For instance, Taiwan and China exemplify extraordinarily rapid industrial growth in the context of organisational forms that defy the ideal-type categories of the industrial restructuring literature. In China, for example, industries owned and operated by village and township governments have been the chief source of industrial growth since the mid-1980s (Oi 1992, Bowles and Dong 1994).

A second, closely related problem in the industrial restructuring literature is the tendency to derive sweeping systemic claims from quite specific studies or stylisations of firms or industrial sectors.[4] A key part of what is missing from this sort of extrapolation is precisely the agrarian context of industrialisation, and the multi-layered relations between agriculture and industry.[5] It almost goes without saying that any effort to comprehend the forms and dynamics of rural industrialisation in any particular setting has to be situated within this context. Yet in the case of the Third Italy, Paloscia's (1991) brief but highly illuminating exposition of 'the deep-seated, historically determined roots of the links between the agricultural and industrial activities which characterise the region and which have formed the basis of its economic growth' makes clear that precisely this agrarian dimension is missing from the ideal-type constructions and paradigmatic claims of flexible specialisation and post-Fordism. Similar insights emerge from studies of industrial districts in India by Tewari (1991) and Chari (1996). A recent study by MacDonald (1996) reveals how industrial sector studies fail to recognise a key source of 'flexibility' in Japanese industrial restructuring – female farm residents who, through 'an elaborate choreography of farming and factory work' in rural households, were absorbed into industries that moved into rural areas. MacDonald shows how, in effect, farm subsidies served also as labour subsidies for a wide range of industries, and how the state helped industrial management to extend a deliberate locational reach into farming regions for locally rooted labour. This key form of agro-industrial linkage to which MacDonald draws attention is not limited to Japan; it constitutes a far more general feature of agrarian transition in different Asian economies where industrialisation was preceded by land and other redistributive reforms (Hart 1995, 1996b). Yet it is also important to bear in mind that these linkages were not simply conjured up by 'developmental' states and far-seeing capitalists; the ways in which agrarian conditions have come to subsidise industrialisation represent unintended and indirect *outcomes* of particular sorts of agrarian transitions driven by quite particular historically defined forms of contestation at multiple societal levels, from the household to the state.

In short, abstraction from the agrarian context of industrialisation and agro-industrial linkages not only renders any effort to understand rural industrialisation impossible; it is also a far more generally problematic feature of the industrial restructuring literature. Page and Walker (1995) for example show how, by ignoring these elements, regulationist periodisations of Fordism fundamentally misspecify the regionally differentiated history of US capitalism. In South Africa, discussions of the transition from 'racial Fordism' to post-Fordism are similarly uninformed by the broader history of agrarian transformation and dispossession.

In contrast to the crude 'binary histories' of regulation theory and flexible specialisation, the much older literature on agrarian transitions is grounded in recognition of, and debates over, multiple, nonlinear paths of capitalist development (Byres 1995). The debate between Lenin (1899) and Chayanov ([1923] 1966) on the interpretation of Russian rural survey data, along with efforts to reconcile these debates (e.g. Shanin 1972, Harrison 1977), prefigure and in many ways go beyond much of the contemporary 'flexibility debate'. Kautsky's ([1899] 1989) exposition of *The Agrarian Question*, as Watts (1989, 1996) points out, posits both the multilinearity of capitalist development and its inherently political character, and retains considerable contemporary relevance. Brenner's (1976) revisionist analysis of multiple paths of agrarian transition in Europe underscores the limitations of demographic, market, and technological forces in explaining multiple paths of change. The Brenner debate (Aston and Philpin 1985) sounds a strong warning against what Gertler (1992) points to as the technological and market determinism underpinning contemporary analyses of industrial restructuring.

A number of studies of agrarian change in Africa, Asia, and Latin America in the postwar period have built on these classical debates, but have also extended them in important ways. Among the key issues confronting these studies have been the simultaneous persistence, refashioning, and re-emergence of supposedly 'precapitalist' forms – household production, sharecropping, and so forth – in the context of profound agricultural restructuring and reconfigurations of rural space, as well as new and increasingly divergent dynamics of industrialisation and urbanisation. A central theme has been the importance of situating these forms of production explicitly in relation to the state and large-scale circuits of capital (e.g. Bernstein 1979, Friedmann 1980, de Janvry 1981, Goodman and Redclift 1982). There has also been growing attention to the multiple alliances through which these relations are constituted and transformed in different regional and national contexts, as well as the complexity and heterogeneity of state–society relations (e.g. Hart *et al.* 1989, Watts 1989). Particularly in the 1970s, much of the neo-Marxist agrarian literature was cast in fairly narrowly functionalist terms, and various forms of economism and structuralism certainly retain a strong hold in some quarters. Yet within the rubric of agrarian studies, there is a growing group of historians, anthropologists, critical geographers, and others – many of them influenced by feminist theory – whose work extends classical agrarian political economy in new ways by incorporating institutionally grounded understandings from cultural theory.

Since I do not have the space to discuss this work in any detail, I shall focus quite specifically on how it contrasts with the new institutionalism. A key distinction here is between the 'new institutional economics' (NIE), and the 'new economic sociology' (NES) (Granovetter 1985, Granovetter and Swedberg 1992, Swedberg 1993), which has come to figure prominently in the industrial restructuring literature. Proponents of the NES explicitly take aim at the undersocialised economism and functionalism of the NIE – more specifically at the transactions cost version propounded by Williamson (1975, 1985) and his followers. The NIE also figures prominently in the agrarian literature, where it is addressed to the problem of sharecropping and the interlocking of land, labour, and credit relations.[6] Instead of the ad hoc and clumsy notion of transactions costs, NIE models of agrarian institutions deploy the far more elegant device of asymmetrical information.[7] There are instructive differences in the ways Granovetter and others, and scholars in the agrarian literature have engaged with these two versions of the NIE.

Granovetter and Williamson essentially address the same issue – the solution to the problem of order (Granovetter) and the inhibition of opportunism and malfeasance (Williamson). For Williamson, the solution lies in the emergence of organisational forms that minimise malfeasance. Granovetter's 'embeddedness argument stresses [instead] the role of concrete personal relations and structures (or "networks") of such relations in generating trust and discouraging malfeasance' (Granovetter and Swedberg 1992: 60). Granovetter then backs away from charges of replacing one sort of optimistic functionalism with another by conceding that social relations may on occasion provide the means for greater malfeasance: 'The embeddedness approach to the problem of trust and order . . . makes no sweeping predictions of universal order or disorder but rather assumes that the *details of social structure* will determine what is found' (ibid.: 63) (emphasis added). For Granovetter, in other words, social outcomes are determined by, and can be read off, social structures. As Williamson (1993) has recently and correctly pointed out, his approach and Granovetter's are entirely complementary.

A key dimension of this basic similarity is that, to the extent that both brands of institutionalists confront questions of change and dynamics, they do so in much the same way: sources of change are either exogenous, or dynamics proceed by way of path dependency. The only difference is that new economic sociologists invoke networks as part of the initial conditions, whereas economists take these conditions as given. Otherwise, the logic is virtually identical. Granovetter and Swedberg (1992: 19) are quite explicit on this point. Networks play a crucial role at an early stage in the formation of an economic institution; once the development is 'locked in', the strategic importance of networks declines and rule-driven behaviour takes over. This argument is, they concede, closely analogous to Arthur's (1989) point that the peculiarities of path-dependent development derive from increasing returns to scale, and that comparative statics are quite adequate in conditions of constant returns to scale (Granovetter and Swedberg 1992: 22). This concession of similarity is all the more remarkable when one recalls that, in mainstream economic theory, returns to scale are determined by exogenously given technology.

In the agrarian literature, the central debate between NIE modellers and their critics turns around propositions by the former (1) that there is a sharp distinction between purely economic relations and those involving extra-economic coercion or 'non-economic' forces (a polite term for power); and (2) that agrarian institutions can be modelled as strategic behaviour under asymmetric information in abstraction from the exercise of power and the socio-political and legal context in which they occur. My own work has shown how the logic of the asymmetric information models in fact violates these claims – in other words, the NIE agrarian models are inconsistent on their own terms. For example, in order to explain contract enforcement, these models have to invoke a particularly crude form of extra-economic coercion (Hart 1986).[8] In addition, any effort to address institutional change forces confrontation with questions of power for precisely the reasons laid out by Brenner (1976).[9] Similarly, in his review of the NIE literature, Bardhan (1989: 11) points out that the question of *efficiency-improving* institutional change cannot be separated from that of *redistributive* institutional change, and that this 'inevitably confronts us with the question of somehow grappling with the elusive concept of "power" and with political processes that much of neoclassical institutional economics would abhor'.

Efforts to grapple with precisely these questions in agrarian studies have moved in a fundamentally different direction from the social structuralism of the embeddedness approach. Although drawing to some degree on post-structuralism and the discursive turn in social theory, the processual approach taking shape in the agrarian literature grounds the exercise of power in specific institutional and political-economic contexts. A key insight is that struggles over material resources, labour discipline, and surplus appropriation are simultaneously struggles over culturally constructed meanings, definitions, and identities.[10] Social institutions are conceived of not as bounded entities or social structures, but as multiple, intersecting arenas of ongoing debate and negotiation, the boundaries of which are fluid and contested:

> People interact within and among social arenas, in multiple ways, and relations among them are constituted not so much through written or unwritten rules, but rather through multiple processes of negotiation

> or contest which often occur simultaneously, or in close succession, but not necessarily synchronised or even consistent.
>
> (Berry 1996: 6)

Instead of asking *what* are the rules or relational structures that produce particular outcomes, the focus instead is on *how* interaction takes place within and among social arenas, and on how definitions of social institutions and their boundaries are also the object of contestation. Since social processes are the product of interaction at multiple societal levels, they cannot simply be read off or deduced from organisational forms or relational structures. Rather, any effort to illuminate the course of social change and accumulation in any particular setting requires in-depth ethnographic and historically grounded understandings.

This sort of processual understanding complements and extends the classical debates over multiple paths of agrarian transition in several important ways. Instead of the exclusive focus on class struggle as in Brenner (1976), it also compels attention to the constitutive roles of gender, race, ethnicity, caste, and other dimensions of social difference in relation to class. One of the key themes in the agrarian literature, for example, has been the importance of gendered understandings and practices in shaping the course of accumulation and social change in different regions of Africa (Guyer and Peters 1987, Mackintosh 1989, Carney and Watts 1990), Asia (Hart 1991, 1992, Agarwal 1994, Kapadia 1995), and Latin America (Stolcke 1988, Deere 1990). In addition, feminist theory directs attention to the relationships between production and the conditions and everyday practices of social reproduction which, I suggest, are particularly relevant to understanding how rural industrialisation is unfolding in different local and regional contexts.

The processual approach emerging from the agrarian literature is also closely congruent with key themes in critical geography that call for explicitly spatialised understandings of how local and non-local processes continually constitute and reconstitute one another within an increasingly interconnected global system (e.g. Pred and Watts 1992, Massey 1994, 1995). 'A local place' as Massey suggests, is most usefully thought of not as a bounded enclosure, but as 'a particular subset of the interactions which constitute social space, a local articulation within the wider whole' (1995: 115). By the same token, the global processes driving capital mobility are always experienced, constituted, and mediated locally (Pred and Watts 1992).

In the discussion that follows, I illustrate – of necessity quite schematically – the way I have put these ideas to work in illuminating how national and global capital is taking hold in former bantustan areas of South Africa. Following from the arguments laid out above, I develop and illustrate two central propositions. First, at a broad societal level, the distinguishing feature of industrial decentralisation in South Africa is that it was preceded by massive dispossession; this, fundamentally, is what distinguishes it from processes of rural industrialisation in various regions of East Asia which, although quite distinctive in their own ways, have been predicated on the retention of small property as well as various forms of state-sponsored subsidies in the agrarian sphere. Second, these historically constructed property relations set the broad terrain of industrial accumulation in predominantly rural regions, but do not in any unilateral way determine specific local outcomes which are in practice enormously varied. My recent ethnographic research in two localities in northwestern KwaZulu-Natal that are structurally and locationally quite similar reveals clearly how particular local trajectories are the product of ongoing processes of struggle and contestation in multiple, intersecting social arenas, both local and non-local. These diverse local trajectories also display unexpected twists and turns that defy notions of embeddedness and path dependency, but are a central element of conflicting efforts to bring about social change or maintain the status quo.

THE AGRARIAN CONTEXT OF INDUSTRIAL DISPERSAL IN SOUTH AFRICA

In terms of the broad agrarian context of contemporary industrial dispersal, what is particularly salient are the processes that produced huge agglomerations of population in the former bantustans from the 1960s – forced removals of landowners and tenants from African freehold land (the so-called 'black spot' removals) and, even more significant in quantitative terms, the massive evictions of black farm workers and labour tenants from white-owned farms since the 1960s as the latter increasingly mechanised their operations with the help of state-subsidised credit (Platzky and Walker 1985). These processes have assumed regionally specific forms that reflect quite particular agrarian histories and forms of struggle. At the same time, they constitute specific manifestations of the broader history of agrarian transition in South Africa, encompassing racially defined restrictions on property rights, coercive labour practices which assumed an extreme form in agriculture, the political character of the accumulating classes in agriculture, and their relations with the state and with different branches of industrial and mining capital.[11]

The dynamics of dispossession have produced the peculiarly South African phenomenon that Murray (1988) terms 'displaced urbanisation' – huge black townships in rural areas defined as part of the bantustans, often within 15–20 km of towns designated white and Indian under the Group Areas Act. In the 1990s, white farmers' fears of land and tenancy reforms have produced yet another massive spate of farm evictions. Economic decline and urban violence in the 1980s and 1990s generated additional flows of people searching the countryside for a modicum of security, many of whom settled in these rural townships with urban-like densities. These areas, which I term 'interstitial spaces', became the primary locus of decentralised industries in the 1970s and 1980s.

If the speed, brutality, and intensity of dispossession and displacement from the land was a distinctive feature of capitalism under segregation and apartheid, the pattern of industrialisation was a locally specific version of import substitution. The shift away from an exclusively mining and export agriculture-based economy via industrialisation during the 1930s and 1940s, and subsequent intensification of import substitution industrialisation, bore some resemblance to industrial trajectories in other parts of the world (Seidman 1994); so too did the spatially concentrated pattern of industrialisation characteristic of import substitution. State-sponsored efforts to decentralise industry to the borders of the bantustans in the 1960s and 1970s were legitimated in terms of 'growth pole' policies in many other parts of the world (Gore 1984). The main divergence from these broad trends came in the early 1980s. At precisely the point when the debt crisis and the rise of neoliberalism undermined and discredited both industrial protection and regional policies in many parts of the world, the apartheid state offered massive subsidies to industry to locate in or adjacent to densely populated bantustan areas through the Regional Industrial Development Program (RIDP). The RIDP in turn coincided with the crisis of over-accumulation in Taiwan, brought about by the phenomenal growth of huge numbers of tiny, predominantly rural industries. Diplomatic links between Taiwan and South Africa forged through their mutual pariah status facilitated the movement of somewhere in the vicinity of 300 Taiwanese factories into industrial districts within or adjacent to former bantustans (Pickles and Wood 1989).

Over the course of the 1980s, growth of decentralised industries accelerated rapidly in the context of stagnation of the national economy (Platzky 1995).[12] It also provoked a powerful backlash from metropolitan capital in the form of a sustained neoliberal critique of the distortions embodied in industrial decentralisation policies.[13] In response to this critique, the RIDP was modified quite substantially in 1991; subsidies were sharply reduced, were linked to output and productivity instead of inputs and, with the exception of the two main metropolitan areas, were made spatially neutral.

An important dissenting voice that challenged this neoliberal consensus was Bell's (1983) contention that the movement of labour-intensive industries to peripheral regions in the 1970s and 1980s was not simply the result of apartheid policies that distorted incentives, but was driven by industrialists' search for lower production costs in the face of intensified international competition and the growing power of the trade union movement. Ongoing processes of industrial decentralisation in the 1990s lend strong *prima facie* support to this argument.[14] Yet by itself this simple capital logic story is inadequate, both because it abstracts from the historically specific processes that produced these 'cheap labour pools', and because the actual processes of industrial dispersal are experienced, fought over, and reconfigured in distinctively different ways, and generate quite diverse local outcomes.

DIVERGENT LOCAL TRAJECTORIES

Since 1994 I have been engaged in an ongoing effort to trace the histories of dispossession, industrialisation, and local politics in two towns and their adjacent townships in northwestern KwaZulu-Natal – Ladysmith-Ezakheni and Newcastle-Madadeni (see Figure 3.1).[15] On the face of it, the two places appear remarkably similar. Both Ladysmith and Newcastle were established as British military outposts in colonial Natal in the mid-nineteenth century. Both are situated on major transport routes between Johannesburg and Durban. Both are surrounded by what were African freehold farms from which landowners and tenants were removed between the early 1960s and the mid-1980s, as well as by white-owned farms from which huge numbers of workers and labour tenants have been (and continue to be) evicted; the

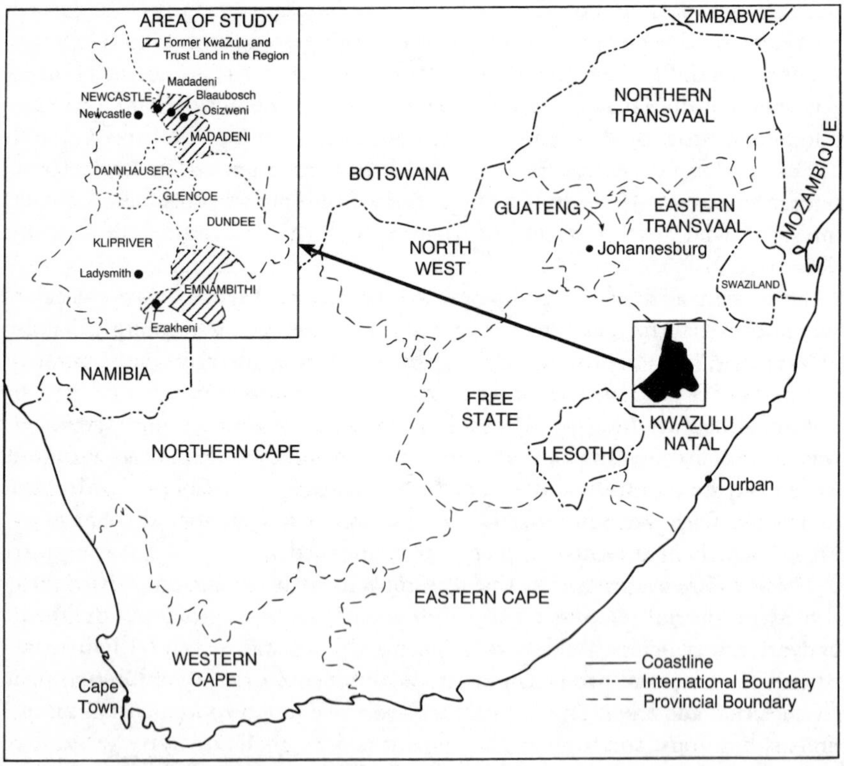

Figure 3.1 Area of study in South Africa

townships are composed mainly of people dispossessed through these processes. Ladysmith is somewhat smaller than Newcastle, but racial demographies and spatial forms are more or less similar. In the late 1960s, both localities were defined as decentralised 'growth poles' to which a few large, heavy import substitution industries were relocated. By the late 1970s, heavy industry in both places was contracting in the face of global economic crisis. In the early 1980s, white local governments in both towns embarked on aggressive and quite successful efforts to attract more labour-intensive industries under the auspices of the Regional Industrial Development Program. Although some industries closed in the 1990s when subsidies fell, new ones have continued to be established.

Yet despite these seeming similarities, both dispossession and industrialisation have been constituted and experienced in entirely different ways in the two places. These divergent dynamics were revealed with great clarity in the June 1996 local government elections, which reconstituted the local state in the two places in strikingly different ways – and in ways that will almost certainly reconfigure the conditions of industrial accumulation. It is possible to provide only a very schematic outline of key sequences of struggles in the townships, industrial work places, and the local state; how they reverberated with one another; and the ways they both reflect and reconfigure forces at work in other, non-local arenas. Yet it will become evident that one cannot even begin to grasp these divergent trajectories with notions of local embeddedness and path dependency.

The origins of dispossession of course stretch back to the nineteenth century. For purposes of this brief exposition I shall start with forced removals, which were far more deeply contested in Ladysmith than in Newcastle. Rural 'black spot' removals in South Africa actually commenced in the vicinity of Newcastle in 1963 in a climate of intense political repression in the country more generally. Forced removals only got going in the Ladysmith district in the early 1970s, by which time there was a general resurgence of organised protest, particularly in Durban where the labour movement was gaining tremendous momentum. Although the pattern of resistance to removals was uneven across different communities and was shaped in part by internal differentiation, these initially non-local struggles intersected with various fights to retain land in the Ladysmith district – some of which were successful (Hart forthcoming).

These differentiated patterns of resistance to removals in turn carried over into township politics, which are significantly more militant in Ezakheni (Ladysmith) than in Madadeni-Osizweni (Newcastle). Until 1994 these townships were administered by the KwaZulu bantustan government, which defined the conditions of access to collective consumption – housing, schools, hospitals, and so forth, as well as the transportation systems linking the townships with the local 'white' towns. The two townships were also connected to the KwaZulu government (and hence the Zulu nationalist

Inkatha Freedom Party) in quite different ways. In Madadeni relatively large amounts of state resources were channelled to a fairly substantial class of local elites who exercised a strong conservative influence in the township more generally, often through reconstituted patronage-type relations between former landowners and tenants (Hart and Tengeni 1997). Ezakheni in contrast became the domain of a particularly brutal Inkatha warlord, whose actions provoked and intensified a highly organised ANC-affiliated youth movement. To cut a long and complicated story short, the 'cheap labour pools' (or, as some would have preferred 'township communities') created through dispossession very quickly became highly differentiated – and differentially contentious – arenas of social interaction.

These diverse patterns of township politics are intimately connected with patterns of labour militancy, which have in turn both affected and been shaped by what are in fact quite distinctive industrial ensembles and trajectories in the two places. To grasp these connections, one has to shift attention to another key arena of social interaction – local governments in the two formerly white towns, and the relations that local politicians and bureaucrats forged with different branches of capital (both national and international) as well as with provincial and national governments. Even during the heyday of so-called 'top-down' regional planning by the apartheid state, industrial decentralisation was a far more locally driven process than is generally recognised. In addition, local bureaucrats and politicians in Ladysmith and Newcastle operated in direct and fierce competition with one another, and their strategies have been forged in mutually conditioning ways. Once again to cut a long story short, the heavy industries that established in the two towns in the 1970s were very different. As a consequence of stronger links into the national government as well as some shady land deals, Newcastle got a branch of Iscor – the state-owned Iron and Steel Corporation – which unleashed an unprecedented and short-lived boom. Local operatives in Ladysmith, who were far more closely connected to capital and the state at the provincial level, persuaded several large private companies headquartered in Durban – by far the most important of which was the Dunlop Tire Company – to open branches in the town. The establishment of these heavy industries, of course, coincided with the resurgence of labour militancy in the country more generally.

What Burawoy (1985) calls production politics were dramatically different in Iscor and Dunlop, and overlapped in interesting ways with township politics. The officials who ran Iscor, operating in effect as part of the apartheid state and in conjunction with the KwaZulu bantustan government, were powerfully determined to prevent the militant metal workers' union from entering Iscor. They brought in men from rural areas of KwaZulu, housed them in closely monitored hostels, and appointed a prince from the Zulu royal house as personnel officer. These efforts to mobilise Zulu nationalism and understandings of masculinity further fragmented the labour movement in Newcastle-Madadeni, which was also being pulled apart by conflicts in the Madadeni youth movement. The intersection of township and production politics in Ladysmith moved in precisely the opposite direction. Both Dunlop and Frame Textiles (another major Durban-based company that had established a branch in Ladysmith in the 1940s) were the major target of the Durban strikes in the 1970s and 1980s. As one of my informants put it, labour militancy quite literally moved up the main road from Durban to Ladysmith. The interactions between labour and township organising were complicated, but they operated in a synergistic fashion. In short, a form of social movement unionism (Seidman 1994) took hold in Ladysmith whereas, from the viewpoint of progressive politics, the larger and more industrialised Newcastle became increasingly disorganised.

Responses to the crisis of 'Fordist' heavy industry in the late 1970s, and the shift to more labour-intensive industrial ensembles, were also forged through locally specific manoeuvring and contestation in multiple, intersecting arenas. The 1980s saw the emergence of particularly aggressive marketing strategies by white local government officials; although these strategies were launched on the basis of resources dispensed by the national state, they operated through a rather different set of non-local linkages. By mobilising their connections in the province, white local officials in Ladysmith forged close relationships with the white male self-styled 'businessmen' who ran the KwaZulu Finance Corporation (KFC), and managed to lay claim to substantial resources for an industrial estate about 5 km from Ezakheni – but still within the borders of the bantustan, where labour organising was prohibited. Newcastle local officials fought bitterly to prevent Ladysmith from monopolising these resources. When it became clear that these ploys had failed, these officials reached directly into the global economy. Starting in the mid-1980s, they established connections in Asia (mainly Taiwan and Hong Kong), and lured approximately sixty-five industrialists and their families to Newcastle with a combination of RIDP incentives, as well as a substantial stock of extremely cheap luxury houses left vacant when a planned expansion of Iscor in the 1970s failed to materialise. The large reservoirs of labour created through forced removals and subsequent in-migration were, of course, a major selling point in the marketing of these places. Partly as a consequence of different modes of recruitment, the mix of factories brought in under the RIDP differed in the two places. Approximately 80 per cent of the factories that came into Newcastle from Taiwan were clothing producers – mainly knitwear – employing an almost exclusively female work force. In Ezakheni, KFC officials were able to recruit a far more varied mix of factories, both South African and Taiwanese.

The locally-specific contexts of struggle and contestation into which these industries were inserted have moulded industrial dynamics in quite distinct ways. Fresh from their victories at Dunlop and Frame, union leaders in

Ladysmith moved quickly to defy the ban on unions and organise the Ezakheni factories. A number of companies (particularly the South African ones) followed Dunlop's and Frame's suit; they capitulated to union organising, but shifted to more capital-intensive technologies and more carefully orchestrated relations with their reduced work forces. Others simply moved to more 'peaceful' bantustans. Taiwanese industrialists in particular tried to sidestep unions by tapping into other spatially and racially defined labour pools – including migrants from Lesotho who, according to union organisers, have often been housed in the factories.[16] Several large Taiwanese industrialists subsequently signed union agreements, raised wages, and adopted the Dunlop strategy of capital intensification and very substantial downsizing of their work forces. In short, Ladysmith industrialists have in general either capitulated to local pressures or departed.

In contrast, production politics in the Taiwanese industrial agglomeration in Newcastle resembles ongoing, low-grade guerrilla warfare which from time to time erupts in open violence, and in the course of which the classic issues of wages, working conditions, and productivity are fought out partly in terms of conflicting understandings of gender, race, and ethnicity. This intense labour conflict is taking place in the context of a Taiwanese production network that is structurally identical to what several observers have hailed as the Asian version of flexible specialisation responsible for rapid and relatively egalitarian growth in Taiwan (e.g. Orru 1991, Hamilton 1994). Elsewhere I discuss in some detail how multiple histories of contestation, both local and non-local, have conditioned the formation of the Newcastle network, the way it operates, and the process through which it is disintegrating – a process accelerated by the influx of identical knitwear from China at prices below costs of production in Newcastle (Hart 1996a and b).[17]

Local arenas of contestation not only shape industrial dynamics, but are continually reconfigured by forces unleashed at the point of production. The struggles and stratagems leading up to the local elections, and their unanticipated outcomes, provide a particularly vivid illustration of these nonlinear dynamics. Political manoeuvring at the local level began in earnest in the latter half of 1994 with the highly contentious formation of Transitional Local Councils (TLCs) which redrew the boundaries to incorporate the formerly white towns and black townships in single administrative entities. Although the white towns were granted disproportionate representation, the TLCs in both Ladysmith and Newcastle appointed black mayors from the former KwaZulu township councils, which had been composed exclusively of Inkatha Freedom Party (IFP) members. Prominent white and Taiwanese business interests in both towns moved quickly to forge links with the IFP, presuming that, as the dominant party in the province, the IFP would sweep the local elections – a presumption that was in fact quite widespread. In both places these efforts to secure the conditions of accumulation backfired in ways that are deliciously revealing of prior histories of struggle both on and beyond the factory floor.

What happened in Ladysmith was that, in 1995, National Party and IFP representatives on the TLC sought to emulate the earlier Newcastle strategy by sending a delegation consisting of the IFP mayor and several prominent white businessmen to Taiwan. They managed to bring in a network of ten small Taiwanese clothing producers, who were set up in a large empty factory vacated several years earlier by an industrialist who had gone off in search of more docile labour. The Taiwanese were promised a non-unionised work force through a strategy designed to kill several birds (including the Clothing and Textile Workers' Union) with one stone: prospective workers were invited to come to the IFP office, register with the party, and obtain a ticket that would guarantee a job in the new factory complex. In fine populist fettle, the ANC promptly leapt on the issue of low wages and tense labour relations to brand the IFP as the lackeys of exploiters of the people. The ANC also swept to a stunning victory in the June 1996 elections, while the IFP gained only one seat on the council (Hart and Tengeni 1997). All the senior white bureaucrats in the Ladysmith municipality promptly resigned, and moved to towns where the IFP won.

In Newcastle, Taiwanese industrialists' efforts to ally with the IFP to secure the conditions of accumulation were even more spectacular, as have been the unanticipated outcomes. The key figure was the leader of the main Taiwanese business faction, the self-proclaimed progenitor of the production network, and widely regarded in the townships as the most exploitative employer. As soon as the TLC was formed, he announced his intention of being elected mayor. One of his visions was to turn Newcastle into a free trade zone, complete with an airport and a ban on union activity. He not only joined the IFP, but donated large amounts of money to the party at both local and provincial levels. His publicly displayed largesse included a large meeting in Madadeni just prior to the elections, at which he personally dispensed huge quantities of lavish gifts. This *noblesse oblige* generated enormous resentment; a key reaction in the townships was why, if he was so wealthy, did he not pay better wages? His performances and ambitions contributed significantly to the defeat of the IFP in the Newcastle elections, which came as a major surprise to all the political parties – not least the victorious ANC, whose leaders now concede that he won the election for them. Part of what happened was that large numbers of IFP supporters in Madadeni and Osizweni simply stayed away from the polls. The IFP did, however, win a larger proportion of seats than in Ladysmith, and this Taiwanese industrialist is now deputy mayor of Newcastle.

If the reconfigurations of social forces prompted by the elections disrupt notions of embeddedness and path dependency, emerging land struggles threaten to explode them. The multiple conflicting efforts to claim and reclaim the land in and around the two towns also serves as a sharp reminder of the ongoing salience of agrarian questions, and of the wise old dictum that people make their own histories in conditions not of their own making. People also use accounts of their own and others' histories to stake claims on resources and authority (Berry 1996), and the future course of social change and accumulation will be shaped at least in part by how these claims are staked and negotiated, as well as by who gets to participate.

CONCLUDING OBSERVATIONS

Whether, how, and by whom the arguments laid out in this chapter are used in practice is, of course, an open question. With this caveat in mind, it seems to me that a processual understanding of multiple trajectories at different societal levels provides a means of navigating between the determinism of 'only one thing is possible' and the voluntarism of 'everything is possible' (Burawoy 1985) that not only illuminates diversity and dynamics, but also challenges neoliberal orthodoxies and opens the way for somewhat more progressive possibilities.

In the present conjuncture in South Africa, these issues assume special intensity. With their narrow focus on the industrial sector and abstraction from agrarian dispossession, post-Fordist visions of the future not only derive from a deracinated understanding of the past. They also serve, in effect, to legitimate key elements of an increasingly influential neoliberal agenda framed in terms of the imperatives of global competition – the consequence of which is the current retreat from redistribution.[18]

The decentralised industries that post-Fordists regard with distaste yield important insights when understood in processual terms as historically constituted expressions of multiple and diverse trajectories of late capitalism. Most obviously, these processes of industrial dispersal compel attention to local and regional histories of dispossession, and the imperatives for broadly based redistribution and expansion of livelihoods – all of which are occluded from post-Fordist accounts and visions. Comparative Asian trajectories are salient *not* because they represent 'models' to be emulated, but rather because the multiple histories of redistribution together with the diversity of institutional forms – here I particularly have in mind the township and village enterprises in parts of China – provide a means for contesting the disabling discourses of globalisation and market triumphalism. On the 'battlefields of knowledge' (Long and Long 1992), these sorts of understandings represent potentially important sources of ammunition.

ACKNOWLEDGEMENTS

The South African research discussed in this chapter was conducted while I held the Nattrass Memorial Fellowship at the University of Natal, Durban, and was also supported by the MacArthur Foundation and the National Science Foundation. The ideas on how the agrarian literature speaks to questions of industrial restructuring were developed in seminars I taught at MIT in the late 1980s, when an extraordinary group of graduate students working on industrial change stimulated me to confront these parallels. These ideas have been further developed in conversation with colleagues and students at Berkeley engaged in similar issues. For careful readings of an earlier draft of this chapter, I am indebted to Donald Moore and David Szanton.

NOTES

1 Of course I am using the term 'rural industrialisation' fairly loosely here to refer to towns that are closely connected with their rural hinterlands.

2 These include Williams *et al.* (1987), Pollert (1988), Sayer (1989), Brenner and Glick (1991), and Walker (1995).

3 Regulation theorists for example trace the crisis of Fordism in part to intensification of international competition since the late 1960s. This in turn is explained in terms of 'peripheral' Fordism which undermined the principles of competition inherent in 'pure' Fordism because it was not accompanied by the expansion of domestic markets (e.g. Lipietz 1987, Schoenberger 1988). A key example of Eurocentricity is Piore and Sabel's (1984: 279) suggestion that mass production be exported to the Third World, opening up spaces for flexible specialisation in the US and Europe.

4 See Sayer and Walker (1992) for an elaboration of this argument.

5 For a critique of the mainstream literature on agriculture–industry linkages, see Hart (1993).

6 For a useful review of this literature, see Bardhan (1989).

7 Among mathematical economists, 'Transactions costs have a well-deserved bad name as a theoretical device, because solutions to problems involving transactions costs are often sensitive to the assumed form of the costs, and because there is a suspicion that almost anything can be rationalised by invoking suitably specified transactions costs' (Fischer, 1977: 322).

8 The distinctive feature of the NIE in relation to orthodox neoclassical economics is that it is non-Walrasian – i.e. it drops the assumption of Pareto optimality. The abandonment of this assumption forces confrontation with questions of power.

9 For example, I show how, although one can discern a similar exclusionary logic of labour control governing agrarian institutions widely separated in space and time – from Chile and Prussia in the nineteenth century to contemporary Asia and California – their dynamics can only be understood in relation to the exercise of power at different societal levels that these forms of production both reflect and alter (Hart 1986).

10 Initial formulations were Peters (1984) and Berry (1989, 1993). Other studies of agrarian change that draw explicitly on these insights include Carney and Watts (1990), Hart (1991, 1992), Moore (1993), Schroeder (1993), Little and Watts (1994), and Agarwal (1994). There are of course many overlaps with other bodies of literature that are moving towards the integration of political economy and cultural studies. In Hart (forthcoming), I discuss the relationship of this type of approach to actor–network theory.

11 In my forthcoming book, I discuss these processes more fully in relation to intense debates around agrarian transition in South Africa; for a useful critical summary of these debates, see Bradford (1990).
12 Platzky's comparative study of industrial patterns in two bantustan estates in 1989 effectively questions major stereotypes, and reveals important local differences.
13 These critiques were sponsored by the Urban Foundation and an overlapping 'Panel of Experts' appointed by the Development Bank of Southern Africa. These basically neoliberal critiques garnered additional legitimacy by underscoring the connections between industrial decentralisation and the apartheid project.
14 For KwaZulu-Natal, these processes are documented in Hart and Todes (1997).
15 The industrial study in Newcastle was conducted jointly with Alison Todes, whose work in Madadeni on migration and the uses of place (Todes 1995) has also informed my own understandings. More recently, Phelele Tengeni and I engaged in collaborative research in both places.
16 One large Taiwanese firm tried employing whites, but dismissed them when he discovered, as he put it, that 'they were even less productive than the blacks'.
17 Comparative labour costs are quite instructive. In mid-1995, comparative money wages – i.e. calculated at market exchange rates – were about 90 per cent higher in Newcastle than in China. If, however, one uses Purchasing Power Parity, real wages in Newcastle were 30–40 per cent below those in China.
18 For a critique of the post-Fordist agenda, see Bell (1995).

BIBLIOGRAPHY

Agarwal, B. (1994) *A Field of One's Own: Gender and Land Rights in South Asia*, New York: Cambridge University Press.

Amsden, A. (1991) 'Third world industrialisation: global Fordism or a new model?', *New Left Review* 182: 5–31.

Arthur, B. (1989) 'Competing technologies, increasing returns, and lock-in by historical events', *Economic Journal* 99, 394: 116–31.

Aston, T. and Philpin, C. (eds) (1985) *The Brenner Debate: Agrarian Class Structure and Economic Development in Preindustrial Europe*, Cambridge: Cambridge University Press.

Bardhan, P. K. (ed.) (1989) *The Economic Theory of Agrarian Institutions*, New York: Oxford University Press.

Bell, T. (1983) 'The Growth and Structure of Manufacturing Employment in Natal', University of Durban-Westville: Institute for Social and Economic Research Occasional Paper No. 7.

—— (1995) 'Improving manufacturing performance in South Africa: a contrary view', *Transformation* 28: 1–34.

Bernstein, H. (1979) 'Concepts for the analysis of contemporary peasantries', *Journal of Peasant Studies* 6, 4: 431–43.

—— (1996) 'South Africa's agrarian question: extreme and exceptional?', *Journal of Peasant Studies* 23, 2/3: 1–52.

Berry, S. (1989) 'Social Institutions and Access to Resources', *Africa* 59, 1: 41–55.

—— (1993) *No Condition is Permanent: The Social Dynamics of Agrarian Change in Sub-Saharan Africa*, Madison: University of Wisconsin Press.

—— (1996) 'Tomatoes, land, and hearsay: property and history in Asante in the time of structural adjustment', unpublished paper.

Bowles, P. and Dong, X. Y. (1994) 'Current successes and future challenges in China's economic reform', *New Left Review* 208: 49–76.

Bradford (1990) 'Highways, byways and culs-de-sacs: the transition to agrarian capitalism in revisionist South African history', *Radical History Review* 46, 7: 59–88.

Brenner, R. (1976) 'Agrarian class structure and economic development in preindustrial Europe', *Past and Present* 70: 30–70.

—— and Glick, M. (1991) 'The regulation approach: theory and history', *New Left Review* 188: 45–120.

Burawoy, M. (1985) *The Politics of Production: Factory Regimes Under Capitalism and Socialism*, London: Verso.

Byres, T. J. (1995) 'Political economy, the agrarian question, and the comparative method', *Journal of Peasant Studies* 22, 4: 561–80.

Carney, J. and Watts, M. (1990) 'Manufacturing dissent: work, gender, and the politics of meaning in a peasant society', *Africa* 60, 2: 207–41.

Chari, S. (1996) 'The agrarian question comes to town: historical geographies and intersectoral linkages in the making of the knitwear industry in Tirupur, India', forthcoming in *World Development*.

Chayanov, A. ([1923] 1966) *The Theory of Peasant Economy*, Illinois: The American Economic Association.

de Janvry, A. (1981) *The Agrarian Question and Reformism in Latin America*, Baltimore: Johns Hopkins University Press.

Deere, C. (1990) *Household and Class Relations: Peasants and Landlords in Northern Peru*, Berkeley: University of California Press.

Fischer (1977) 'Long-term contracting, sticky prices, and monetary policy: a comment', *Journal of Monetary Economics* 3: 317–23.

Friedmann, H. (1980) 'Household production and national economy', *Journal of Peasant Studies* 7, 2: 158–84.

Gertler, M. (1992) 'Flexibility revisited: districts, nation-states, and the forces of production', *Transactions: Institute of British Geographers* 17: 259–78.

Goodman, D. and Redclift, M. (1982) *From Peasant to Proletarian: Capitalist Development and Agrarian Transition*, New York: St Martin's Press.

—— and Watts, M. (1994) 'Reconfiguring the rural or fording the divide? Capitalist restructuring and the global agro-food system', *Journal of Peasant Studies* 22, 1: 1–49.

Gore, C. (1984) *Regions in Question: Space, Development, and Regional Policy*, London: Methuen.

Granovetter, M. (1985) 'Economic action and social structure: the problem of embeddedness', *American Journal of Sociology* 91: 481–510 (reprinted in Granovetter and Swedberg, 1992).

Granovetter, M. and Swedberg, R. (1992) *The Sociology of Economic Life*, Boulder: Westview Press.

Guyer, J. (1981) 'Household and community in African studies', *African Studies Review* 24, 2/3: 87–137.

—— and Peters, P. (eds) (1987) 'Conceptualizing the household: issues of theory and policy in Africa', *Development and Change* 18: 197–214.

Hamilton, G. (1994) 'Organization and Market Processes in Taiwan's Capitalist Economy', Paper prepared for the SSRC Conference on Market Cultures, Boston.

Harrison, M. (1977) 'Resource allocation and agrarian class reform: the problems of social mobility among Russian peasant households, 1880–1930', *Journal of Peasant Studies* 4, 2: 127–61.

Hart, G. (1986) 'Interlocking transactions: obstacles, precursors, or instruments of agrarian capitalism?', *Journal of Development Economics* 23: 177–203.

—— (1991) 'Engendering everyday resistance: gender, patronage, and production politics in rural Malaysia', *Journal of Peasant Studies* 19, 1: 93–121.

—— (1992) 'Household production reconsidered: gender, labor conflict, and technological change in Malaysia's Muda region', *World Development* 20, 6: 809–23.

—— (1993) 'Regional Growth Linkages in the Era of Liberalization: A Critique of the New Agrarian Optimism', Geneva: International Labour Office. World Employment Programme Research, Working Paper 37.

—— (1995) 'Rethinking global competition', *South African Labor Bulletin* 19, 6: 41–7.

—— (1996a) 'Global Connections: The Rise and Fall of a Taiwanese Production Network on the South African Periphery', Working Paper, Institute of International Studies, Berkeley: University of California.

—— (1996b) 'The agrarian question and industrial dispersal in South Africa: agro-industrial linkages through Asian lenses', *Journal of Peasant Studies* 23, 2/3: 245–77.

—— (forthcoming) *Transforming the Landscape of Apartheid: Local Dynamics, Global Connections*.

—— and Tengeni, P. (1997) 'The local elections in KwaZulu-Natal: a tale of two towns', unpublished paper.

—— and Todes, A. (1997) 'Industrial decentralisation revisited', *Transformation* (forthcoming).

——, *Turton, A., White, B. (eds) (1989) Agrarian Transformations: Local Processes and the State in Southeast Asia*, Berkeley: University of California Press.

Joffe, A., Kaplan, D., Kaplinsky, R., and Lewis, D. (1995) *Improving Manufacturing Performance in South Africa: Report of the Industrial Strategy Project*, Cape Town: UCT Press.

Kapadia, K. (1995) *Siva and Her Sisters: Gender, Caste, and Class in Rural South India*, Boulder: Westview Press.

Kaplinsky, R. (1991) 'A growth path for post-apartheid South Africa', *Transformation* 16: 49–55.

—— (1994) 'Economic restructuring in South Africa: the debate continues. A response', *Journal of Southern African Studies* 20: 533–8.

Kautsky, K. ([1899] 1989) *The Agrarian Question*, London: Swan.

Lenin, V. (1899) *The Development of Capitalism in Russia*, Moscow: Progress Publishers.

Lipietz, A. (1987) *Mirages and Miracles: The Crises of Global Fordism*, New York: Verso Press.

Little, P. and Watts, M. (1994) *Living Under Contract*, Madison: University of Wisconsin Press.

Long, N. and Long, A. (eds) (1992) *Battlefields of Knowledge*, London and New York: Routledge.

MacDonald, M. (1996) 'Farmers as workers in Japan's regional economic restructuring, 1965–1985', *Economic Geography* 72, 1: 49–72.

Mackintosh, M. (1989) *Gender, Class, and Rural Transition: Agribusiness and the Food Crisis in Senegal*, New Jersey: Zed Books.

Massey, D. (1994) *Space, Place, and Gender*, Minneapolis: University of Minnesota Press.
—— (1995) 'Double articulation: a place in the world', in A. Bammer (ed.) *Displacements: Cultural Identities in Question*, Bloomington: Indiana University Press.
Moore, D. (1993) 'Contesting terrain in Zimbabwe's eastern highlands: political ecology, ethnography and peasant resource struggles', *Economic Geography* 69, 4: 380–401.
Moore, H. and Vaughan, M. (1994) *Cutting Down Trees: Gender, Nutrition, and Agricultural Change in the Northern Province of Zambia, 1890–1990*, Portsmouth, NH: Heinemann.
Murray, C. (1988) 'Displaced Urbanization', in J. Lonsdale (ed.) *South Africa in Question*, London: James Currey.
Oi, J. (1992) 'Fiscal reform and the economic foundations of local state corporatism in China', *World Politics* 45, 1: 99–126.
Orru, M. (1991) 'The institutional logic of small-firm economics in Italy and Taiwan', *Studies in Comparative International Development* 26, 1: 3–28.
Page, B. and Walker, R. (1995) 'Staple lessons: agriculture, resource industrialization and economic geography', paper presented at the Harold Innis Symposium, Toronto.
Paloscia, R. (1991) 'Agriculture and diffused manufacturing in the *Terza Italia*', in S. Whatmore *et al.* (eds) *Rural Enterprise: Shifting Perspectives on Small-Scale Production*, London: David Fulton Publishers.
Peters, P. (1984) 'Struggles over water, struggles over meaning', *Africa* 54, 3: 29–49.
Pickles, J. and Wood, J. (1989) 'Taiwanese investment in South Africa', *African Affairs* 88: 507–28.
Piore, M. and Sabel, C. (1984) *The Second Industrial Divide*, New York: Basic Books.
Platzky, L. (1995) 'The Development Impact of South Africa's Industrial Location Policies: An Unforeseen Legacy', PhD Thesis, Institute of Social Studies, The Hague.
—— and Walker, C. (1985) *The Surplus People: Forced Removals in South Africa*, Johannesburg: Ravan Press.
Pollert, A. (1988) 'Dismantling flexibility', *Capital and Class* 34: 42–75.
Pred, A. and Watts, M. (1992) *Reworking Modernity: Capitalisms and Symbolic Discontent*, New Brunswick: Rutgers University Press.
Sanghera, B. and Harriss-White, B. (1995) *Themes in Rural Urbanisation*, DPP Working Paper No. 34. Milton Keynes: The Open University.
Sayer, A. (1989) 'Postfordism in question', *International Journal of Urban and Regional Research* 13: 666–95.
—— and Walker, R. (1992) *The New Social Economy: Reworking the Division of Labor*, Cambridge: Basil Blackwell.
Schoenberger, E. (1988) 'From Fordism to flexible accumulation: technology, competitive strategies, and international location', *Environment and Planning D* 6: 245–62.
Schroeder, R. (1993) 'Shady practice: gender and the political ecology of resource stabilization in Gambian garden orchards', *Economic Geography* 69, 4: 349–65.
Seidman, G. (1994) *Manufacturing Militance: Workers' Movements in Brazil and South Africa, 1970–1985*, Berkeley: University of California Press.
Shanin, T. (1972) *The Awkward Class*, Oxford: Clarendon Press.
Stolcke, V. (1988) *Planters, Workers, and Wives: Class Conflict and Gender Relations on São Paolo Plantations, 1850–1980*, New York: St Martin's Press.

Swedberg, R. (ed.) (1993) *Explorations in Economic Sociology*, New York: Russell Sage.

Tewari, M. (1991) 'The state, intersectoral linkages, and the historical conditions of accumulation in Ludhiana's industrial regime', paper prepared for conference on Work in Rural Asia, Honolulu.

Todes, A. (1995) 'Migration, survival strategies and the gendered impact of regional development policies', paper prepared for the GRUPHEL II Workshop, Durban.

Walker, R. (1995) 'Regulation and flexible specialization: challenges to Marx and Schumpeter?', in H. Liggett and S. Perry (eds) *Spatial Practices: Markets, Politics, and Community Life*, Newbury Park, CA: Sage Publications.

Watts, M. (1989) 'The agrarian question in Africa: debating the crisis', *Progress in Human Geography* 13: 1–41.

—— (1996) 'Development III: the global agrofood system and late twentieth-century development (or Kautsky redux)', *Progress in Human Geography* 20, 2: 230–45.

Williams, K., Cutler, T., Williams, J., and Haslam, C. (1987) 'The end of mass production?', *Economy and Society* 16, 3: 405–39.

Williamson, O. (1975) *Markets and Hierarchies*, New York: The Free Press.

—— (1985) *The Economic Institutions of Capitalism: Firms, Markets, Relational Contracting*, London: Macmillan.

—— (1993) 'Transactions cost economics and organization theory', *Industrial and Corporate Change* 2, 2: 107–56.

Wolf, D. (1992) *Factory Daughters: Gender, Household Dynamics, and Rural Industrialization in Java*, Berkeley: University of California Press.

4

AGRARIAN QUESTIONS IN THE MAKING OF THE KNITWEAR INDUSTRY IN TIRUPUR, INDIA

A historical geography of the industrial present

Sharad Chari

INTRODUCTION

> Tirupur, the small, dusty, prosaic, unlovely town in south India has transformed itself into an economic powerhouse, one of the country's most important export centre s and a name to reckon with in the fickle world of the international garment trade. . . . Clearly, Tirupur is a town on the make. It is perhaps a measure of its fixation with the 'here and now' that few people know how it all began. There is no collective memory; no authoritative version of what made the town what it is.
>
> ('Poor Little Rich Town', *Indian Express* 18 December 1994)

A boom town has emerged in south India within the last quarter century. Tirupur,[1] in Coimbatore District of Tamilnadu State, has propelled itself into the centre of the Indian cotton knitwear industry with such a momentum that, in the words of a local reporter, 'this hamlet *breeds* millionaires' (*India Today* 31 March 1994, my emphasis.) Tirupur has captured up to 85 per cent of the Indian market and has expanded its export earnings from $25 million in 1986 to $800 million in 1994 (including indirect exports, about 3.2 per cent of India's foreign exchange earnings).[2] However, the integration of Tirupur into a global knitwear industry has a distinctive character. Specifically, a proliferation of small and mid-sized firms has managed to combine high-tech machinery and cheap, highly adaptable labour in what Cawthorne (1993) calls an 'amoebic capitalism'. As the metaphor indicates, this decentralised organisation of work allows accumulation through expanding firms that break up into smaller units of production connected by subcontracting networks within and between enterprises. Precisely how this industrial form has been forged, and what it means for people in Tirupur and its environs, is unresolved not only in the popular imagination, but also in current work from industrial studies. My contention in this chapter is that while a sectoral focus on these questions does reveal much about Tirupur's genesis, it cannot explain how the agrarian Gounder caste moved from farms to factories in the mid-1960s, subsequently restructuring Tirupur's knitwear industry into networks of small firms. An alternative explanation from historical geography can explain Tirupur as an outcome of regional and *agrarian* processes in interaction with larger political-economic forces. Located on the wider canvas of regional development, with an eye for livelihood and accumulation strategies across the rural–urban divide, Tirupur's industrial present is still resonant with its agrarian past.

Indeed, work in Tirupur is a landscape of sharp contrasts in scale and technique, as new knitting factories work between myriad small dyeing units which extend deep into the countryside. While the recent export boom has heightened conspicuous consumption – witness the lavish houses in the urban periphery, foreign cars, and a three million dollar hotel where local elites negotiate with transnational retailers at the poolside -- affluence lives cheek by jowl with widening public squalor. The urban environment has borne the brunt of industrial pollution, with little reinvestment in maintaining infrastructure or treating polluted water. The River Noyyal flows through the town entirely dye-stained. The streets are clogged and slow, and workers' housing is minimal. These sharp contrasts call for explanation when they are situated within a vibrant political culture in which farmers, industrial workers, and entrepreneurs have at various times formed collective institutions to pursue their different goals. Indeed, the political and economic formation that prevails today is but one act in a larger play that has tied forms of work and accumulation to wider, increasingly global, economic opportunities.

The second section of this chapter reviews explanations of Tirupur's knitwear firms from industrial studies. It demonstrates how quite varied interpretations have emerged from industrial studies explanations, as assumptions about causal mechanisms have been supported by evidence at different phases of Tirupur's growth. I contend that the emphasis on models of industry does not explain the genesis of this particular industrial form.

The third section uses the alternative perspective of historical geography to understand how Tirupur's industry is an outcome of processes linking work and investment across sectors in a form of regional industrialisation based on small towns. This section uses four themes to sketch how production politics and forms of assertion in agrarian Coimbatore District have prefigured the organisation of work in Tirupur, within limits set by state regulation and the regional pattern of industrialisation and unionism. The importance of inter-sectoral linkages in Coimbatore District suggests that the genesis of Tirupur's industrial success owes itself to a refashioning of rural social relations centred on work discipline, within the political circumference of the state and organised labour.

In conclusion, I will suggest how the making of Tirupur's industrial form can be interpreted as one outcome of the 'agrarian question' in India today. This account of the making of Tirupur's knitwear industry reveals the broader relevance of the classical 'agrarian question', and of agrarian classes, to regional industrialisation. My central concern in this chapter is to suggest how and under what conditions agrarian and rural social relations have been refashioned in the making of such an 'economic powerhouse' in which 'few people know how it all began'.

EXPLAINING TIRUPUR'S INDUSTRIAL FORM: POLES OF DEBATE, POINTS OF DEPARTURE

The late 1960s were crucial years both for Tirupur's knitwear industry and for the agrarian Gounders, the dominant caste of Coimbatore District, who moved in large numbers into industrial activities. Krishnaswami's (1989) study centres on changes in the labour process accompanying the Gounder transition from agriculture to industry. During this period, the industrial structure shifted rapidly from large firms run predominantly by members of the Chettiar caste to small-firm networks. Chettiar owners were facing a profitability crisis due to a series of union strikes and shutdowns in the mid-1960s. As the Chettiars started to liquidate their enterprises, rural Gounders responded with an aggressive buying campaign. The Gounders were also facing a combination of social and ecological constraints to profitability in agriculture – in particular, several drought years and problems in securing labour – which facilitated their shift out of farming.[3]

In responding to problems of labour control in Tirupur, Krishnaswami claims that Gounders with some amount of land or savings bought out large firms, replacing them with dispersed small firms, often run by loyal workers. When workers resisted the intensification of work through subcontracting and piece-rated wages, Gounder owners responded by utilising kinship networks to control job access and by distributing loans and fringe benefits to trustworthy workers. Using ethnographic evidence, Krishnaswami demonstrates the different kinds of 'production politics' that prevail across firms in Tirupur – both building 'trust' and 'loyalty' through perquisites to workers and also relying on brute force to keep children working. Finally, Krishnaswami relies on a functionalist argument that caste, gender and age allow owners to cheapen costs of production without investing in technology or deepening workers' stakes in production and growth. On the one hand, this functionalism is quite at odds with Krishnaswami's attempt to get at the active element of struggle between employers and employees in Tirupur. On the other hand, the attempt to draw dynamics from a functionalist model leads to conclusions that stand in contradiction to subsequent developments in Tirupur. Krishnaswami's account is valuable for describing a period of turmoil in Tirupur, but it does not follow on the cue of the Gounder transition from country to city in order to show how the social organisation of work was forged.

By the early 1980s, Tirupur had expanded dramatically into export markets, primarily through indirect exports. Cawthorne's (1990, 1993, 1995) analysis is based on fieldwork in the mid-1980s, by which time technological change and institutions of collective bargaining had been instituted alongside elements of the despotic work relations Krishnaswami describes. In this context, Cawthorne finds that Tirupur's 'amoebic capitalism' allows a combination of rapid expansion and widespread capital accumulation through small firms. Cawthorne views the rise of internal contracting in the mid-1980s as primarily an employers' offensive against the prolonged strikes of 1984. The subsequent growth in number of small-scale industries reflects this 'amoebic' process of differentiated production units within growing large firms. As a corollary, the rise of an intermediary 'semi-autonomous' managerial class and the deepening of subcontracting networks has further decentralised the operation of power and production despite increasing the concentration of capital.

Cawthorne sees Tirupur's growth as primarily based on low wages and on gender and age segmented labour markets. Accumulation turns largely on absolute rather than relative surplus value. Yet Cawthorne goes beyond Krishnaswami's functionalist explanation of labour market segmentation and discrimination as a residual of 'traditional' social relations of kinship and caste. Gender and age segmented divisions of labour, in this account, are facilitated by the particularly rigid social construction of 'skill', determined by a combination of the position in the labour process and the gains of collective mobilisation over what work gets defined as 'skilled'. Jobs classified as skilled include knitting machine attendants, cutters, and machinists. These are entitled to the highest pay and are for the most part monopolised by men. Young boys in knitwear firms are considered to be 'apprentices' in the process of acquiring skills, while young girls are casual, unskilled labour. Skill is mediated in Tirupur by local patriarchal relations which place a clear bias on the transmission of higher skills to men. Moreover, workers have not organised over job classification because the unions have accepted the descriptions offered by the state government, and the generally unregulated nature of production makes it unfeasible for disputes over official classification to actually change wage scales.

On the whole, Cawthorne tends to be pessimistic about Tirupur being a 'Marshallian industrial district' – i.e. horizontal linkages and forms of association between small firms provide 'economies of agglomeration' that allow small firms to be technologically dynamic without relying on sweated labour. Indeed, Cawthorne is critical of those, like Holmstrom (1993), who attempt to graft the First World optimism about small-firm-led industrial alternatives onto Indian realities. At most, argues Cawthorne (1995), Tirupur may be on the 'low road' to industrial flexibility, resting primarily on access to markets rather than on 'internal capacity of individual firms to diversify and improve the quality of their products' (Cawthorne 1995: 51). While the more powerful owners benefit from the intensification of work and the institution of gender-segmented labour markets to the exclusion of working people, Cawthorne (1993) argues that mid-sized firms that can use dependent ties to large-firm networks as 'jobworkers' do hold possibilities for upward mobility.

With Tirupur's 'leapfrogging' into multi-product and direct exports since 1986, Swaminathan and Jeyaranjan (1994) call into question the pessimism of Krishnaswami and Cawthorne. They argue that before 1980 this was a 'cottage industry', with mainly family labour producing inner-wear for domestic markets, but at wage rates comparable with Coimbatore city. During the early 1980s the industry began to export through merchant houses, on 'the dependent subcontractor model' (Swaminathan and Jeyaranjan 1994: 4). Only since the second half of the 1980s has Tirupur launched itself into direct exports, through a diverse range of outerwear garments. During this period, Tirupur has advanced to newer technologies (in few firms, but at every stage of the production process), the top exporters have shifted to higher value products, and an active market for second-hand machines has allowed easier entry into various sections of the production process. Furthermore, the institution of *kashtakoota*, specific to this region, allows workers with skills to enter into partnership with individuals with capital in return for a share of the profit, ranging from 10 to 25 per cent. This has fueled entrepreneurship in the context of high labour mobility. Swaminathan and Jeyaranjan argue that if the first two periods of Tirupur's growth correspond to the first two models in Brusco's (1990) genealogy of small-firm growth in Italy, then Tirupur in the latest phase seems quite similar to the third model: 'the industrial district – Mark I'. Their motivation for this account seems to be in its preference for firm clusters rather than individual firms as the unit of analysis, and its characterisation of firms as those producing the final product, those producing for intermediary stages, and those not in the industry but offering various essential supports such as services (Swaminathan and Jeyaranjan 1994).

Swaminathan and Jeyaranjan see Tirupur as an industrial district in the making because there has been some technological change and upward mobility for the working class. They therefore see a way beyond what they perceive as the confines of a classical Marxist approach; as the 'key to this system of flexible specialisation is labour whose professional skill increases with experience and which allows them to become small independent entrepreneurs' (Swaminathan and Jeyaranjan 1994). Classical Marxism cannot, they argue, capture the ' "owner as worker" and/or "worker as owner" phenomenon' which is pervasive in Tirupur. Finally, Swaminathan and Jeyaranjan argue that Tirupur can only be apprehended as an industrial district if it is understood as a socially and spatially distinct 'local community'. While they have not elaborated at length on this theme, they note the dominance of two communities: agrarian *Gounders* and the traditional weaver caste of *Senguntha Mudaliars* or *Kaikollars*. On the whole, this analysis is an important challenge to the more pessimistic views of Krishnaswami and Cawthorne, but the analogy to the 'Mark I' model based on research in Italy does not explain Tirupur's genesis. Indeed, when it comes to explaining Tirupur's success, these industrial studies approaches tend to fall back on polarised assumptions imported from the debates over choices of backward 'informal sectors' and innovative 'industrial districts', or 'high' and 'low' roads to industrial flexibility (Piore and Sabel 1984, Rasmussen, Schmitz, and van Dijk 1992, Holmstrom 1993). Krishnaswami finds Tirupur mired in the backwardness of the informal sector; Cawthorne (1995) assumes, slightly less pessimistically, that Tirupur is on the 'low road' to industrial flexibility, and Swaminathan and Jeyaranjan conclude with the hope that Tirupur may indeed be a 'Mark I' industrial district. While these approaches are valuable in tracking the growth and transformation of Tirupur's industrial formation and in providing different clues to the dynamics of industrialisation here, they all tend to derive their assumptions about change from models or from industrialisation elsewhere.

A fruitful complement, I contend, would be to retain Tirupur's industrial sector in the context of the historical and regional forces which have given rise to it. Historical geography could respond to some of the unanswered questions in the industrial studies approaches reviewed here. For instance, Krishnaswami's argument suggests that agrarian Gounders brought the means to discipline urban labour militancy in Tirupur in the late 1960s, but does not suggest the means by which they did so. Similarly, Cawthorne suggests that gender-segmented labour markets are accepted as 'natural' by the residents of Tirupur, but does not suggest how they became naturalised. Finally, Swaminathan and Jeyaranjan have referred to the importance of 'local community' and to the fluidity between boundaries of 'worker' and 'owner', but they do not unpack these ideas. In the following section, I use these questions as points of departure in an alternative explanation that seeks the regional and agrarian conditions through which Tirupur's knitwear industry has boomed.

HISTORICAL GEOGRAPHIES IN THE MAKING OF THE INDUSTRIAL PRESENT

Through the lens of historical geography, Tirupur becomes one of a series of towns in the culturally and historically defined region of Kongunad (including Namakkal, Bhavani, Erode, and Tiruchengodu) all of which have grown around particular commodities, primarily under Gounder auspices. Indeed, while Braudel characterises similar 'specialist towns' in early modern Europe as 'comfortable little monopolies protected by local habit and so much a part of normal practice that they cause little protest' (Braudel 1985: 313, in Harriss-White 1996), in Coimbatore District one finds a rich history of peasant and industrial militancy and unionism. This section unpacks the regional and inter-sectoral forces, within the larger political economy, that have given Tirupur its distinctive character.

Accordingly, this section is organised around four themes. The first theme reviews agrarian change in Coimbatore District from sturdy 'yeoman farmers' to rural households diversifying into non-farm activities, and to the rise of capitalist farmers. I will suggest through parallels in production politics and forms of labour market exclusion across agriculture, agro-commerce, and industry that forms of 'unfreedom' have persisted in the context of economic growth and diversification of productive activity. The second theme traces the links between caste and state through rising rural entrepreneurs of middle castes whose forms of assertion through caste and farmers' associations are precursors of today's industrial associations. While these first two themes focus on agrarian resources for the organisation of industry in Tirupur, the final two themes locate the explanation in relation to the actions of state and organised labour. Accordingly, the third theme moves to a wider narrative of state regulation of small-scale industry in relation to pressure from 'textile populism', through which the knitwear industry has managed to maintain its 'smallholder' form. The fourth theme, finally, returns to the history of mills and union politics in Coimbatore District, to suggest how organised labour has persisted despite fragmented work and employers' reworking of rural patronage relations to discipline workers. Together, these narratives sketch the agrarian and historical roots of Tirupur's industrial form to suggest how the knitwear industry has been forged.

Agrarian transition and forms of work discipline

> In matters of cultivation, partnership is still practiced; a number of ryots plough the field of each one in turn; a garden is frequently worked or rented by a partnership of poor ryots, while the owner and cultivator of a garden occasionally shares the expenses of cultivation.
>
> (Nicholson 1887: 265)

Written during his tenure as Collector of Anantapur in colonial Madras Presidency, Nicholson's *Manual of the Coimbatore District* has been a valuable source of information on agrarian change in the district during the late nineteenth century. One is immediately struck by Nicholson's account of the extensive development of well-irrigated garden-farms, called *thottam* farms, particularly by the investment in wells and forms of cooperation between *ryots*, or peasants with long-term rights to land owned by the state. This section traces changes in the countryside to suggest how Tirupur's social form emerges from complex inter-sectoral linkages, not just in savings, investment, and markets, but also in forms of control over labour markets and work.

It is primarily on Nicholson's evidence that Baker's (1984) grand history of rural Tamilnadu from 1880 to 1955 makes claims about the distinctiveness of the region of Kongunad, within which Coimbatore District lies.[4] Baker's account attempts to bridge agro-ecological and social dimensions of capitalist transitions in Madras Presidency through four divergent paths of change discernible in the river valleys, the dry plains, the hills, and semi-arid Kongunad. The rise of Coimbatore as the industrial heartland of Tamilnadu, in this account, is premised on the specificities of its agrarian history: early commercialisation and technical transformation through well irrigation, innovative 'familial' labour relations and financial institutions, and active markets in land. Baker begins with the highly absorbent cultivable soils of Kongunad to show how ecological conditions lent themselves best to well irrigation, with lasting implications for the social organisation of farming. Well-irrigated *thottam* farms required intensive cultivation in order to pay for the sunk cost of reaching through gneissic rock to the deep aquifers below.

These conditions encouraged 'smallish farms . . . worked by family or tied labour', the latter paid mainly in kind and perquisites (Baker 1984: 202). Difficulties in securing labour for the rapid expansion in cotton cash cropping prompted Kongunad farmers to 'a more commercially rational use of labour' than their counterparts in the valleys and plains, who either resurrected old forms of dominance over labour or used debt clientage to secure labour from poor households (Baker 1984: 209). 'Permanent farm servants', whose numbers grew in the second half of the nineteenth century, were unlike the bonded *pannaiyal* labour of the river valleys of Tamilnadu, but 'more like an extension of family labour' (ibid.). Moreover, *thottam* farming in this period and throughout the early part of the twentieth century tended to be 'extremely flexible', through secure access to ground water and an ability to shift crops – cotton, sugar, vegetables, and spices – in relation to market changes. Baker glosses over what it means for Gounder farmers to 'tend more towards economic rationality . . . [and] less [to] the demands of subsistence and tradition', particularly since the ideology of the 'family' is central to control of labour power on the Gounder farm (Baker 1984: 208). Indeed, the evidence of Krishnaswami and Cawthorne suggests continuities in 'patriarchal production politics' between Baker's depiction of Gounder farms and Tirupur's knitwear firms (Burawoy 1985). Nevertheless, Baker's history shows in broad strokes how the rapid expansion of cotton during the early twentieth century made Tirupur and Coimbatore railway hubs in the cotton trade. Both land markets and commercial markets in tenancies had developed by the turn of the century.

Moreover, farmers in Kongunad responded to opportunities from the state to use chemical fertilisers, water pumps and cooperative societies for ginning long-staple Cambodia cotton. The small ginning town of Tirupur became the centre for local trade once Cambodia cotton was encouraged by mills from Madras and Bombay in the 1910s. When the Bombay mills went into long-term decline after 1922, Tirupur cotton shifted more towards Madras and to the expanding mills of Coimbatore District. Ties between countryside and cotton marketing in Tirupur grew throughout this period. Indeed, urban commission agents who met the cartloads of cotton brought in by cultivators were also mainly from farming backgrounds, and growers often maintained tight control over marketing relations through their regular agents in town (Baker 1984: 270). Tirupur was the sole market town for cotton until the late 1930s, and also the fastest growing town in Tamilnadu during the first half of the twentieth century. While other market towns emerged in Kongunad, Tirupur remained at the centre of the cotton trade, dominated by agrarian Kammas and Gounders as producers, traders, and the early industrialists.

Baker suggests that the unique financial institutions emerging out of rural Kongunad, particularly the *nidhis*, or mutual loan societies, grew in numbers alongside the growth of cotton business. Beginning as forms of cooperative finance between small associations of merchants, *nidhis* expanded in the 1920s into lending to non-members, especially in crop dealing, so that they became more like joint-stock banks with precarious relations to formal banks. The turning point in Baker's narrative is the Great Depression. Before the Depression, profits in the rural economy – whether from production, trade, or finance – were primarily cycled back into fixed and working capital requirements for extending or intensifying local cultivation. Following the Great Depression, the swing in the terms of trade against agriculture, the collapse of bazaar banking and uncertainty about produce markets, broke the mechanisms by which capital was ploughed back into agricultural production. Indeed, the rural money market was not to recover during the Second World War and postwar period, and rural credit became more closely tied to local connections and power relations. The decline in lending from town to country meant a large increase in urban finance to feed the rapidly growing towns of Coimbatore District, in sharp contrast to the declining urban centres of the valleys and plains of Tamilnadu. With the subsequent decline in overseas trade, agriculture played second fiddle to a rentier economy based on the towns of Kongunad.

I will return to the mill towns of Coimbatore District later, but want only to note how much of Baker's argument on the peculiarity of Kongunad, culminating in the fast growth of towns, rests on the nature of farming practices in this semi-arid region, on the peculiar kind of patriarchal production politics in *thottam* farms, on the development of unique financial institutions, and on a variety of links between agriculture and industry emerging out of cotton farming, processing and trade.

Nevertheless, agrarian relations in Coimbatore District have changed considerably since 1955, the endpoint of Baker's history. Harriss-White (1996) no longer sees the 'undifferentiated, sturdy entrepreneurial' Gounder farmer in the Kongunad countryside. Instead, Harriss-White's work shows how an agro-commercial class has emerged as master of the countryside 'along the American rather than the Prussian path using agricultural profits rather than agricultural rents as starting capital'. Harriss-White (1985) finds that Coimbatore's merchants tend to be large landholders who do not transfer resources from agriculture directly. Extraction from agriculture via the manipulation of relative prices has been their main mechanism of inter-sectoral resource transfer. Moreover, Coimbatore's merchants have reinvested agro-commercial profits in a diverse portfolio of activities, in response to both the vagaries and opportunities of the agricultural calendar as well as to patchy state intervention.

The home market in Coimbatore District was expanding in the early twentieth century when agrarian rents and profits were reinvested in urban manufacture of goods for local consumption (Harriss-White 1985: 299). Yet Harriss-White notes that the 1940s proved to be a turning point, for state welfare helped provision the cheap labour for a second phase of light engineering and machine manufacture which did not require high entry capital. Rural development was subsequently based on the ruralisation and 'putting out' of industry rather than on commercial agriculture. It seems that Mellor's model had reached its nadir with a 'slender home market of an unreconstructed rural society' (Harriss-White 1985: 299). Yet Harriss-White (1996) notes the expansion of rural industry in Avanashi and Palladam Taluks (sub-districts) of Coimbatore District – in dairy, butter, hand-looms, power-looms, sericulture, cotton ginning, tobacco curing, hosiery, rural engineering works, and foundries (Harriss-White 1996: 70).

Agrarian commercialisation proceeded primarily through public sector irrigation and widespread rural electrification, so as to maintain the conditions for a sizable class of capitalist farmers, with 25 per cent of farmers holding four hectares or more. The number of small holdings of on average half a hectare had also grown, in a process of general pauperisation, and Harriss-White suggests that the figures may conceal growing differentiation (Harriss-White 1996: 66).

In looking closely at the social organisation of agro-commercial firms, Harriss-White (1996) also notes how caste, gender, and locality stratify and segment agricultural markets in terms of entry, ownership, labour participation, and type of crop traded. Gender, for Harriss-White, is not simply a tool for labour market segmentation, but 'is crucial to the social reproduction of a differentiated economic system' (Harriss-White 1996: 232). Indeed, her evidence suggests that agro-commercial firms therefore 'rely on the heads and hips, the fingers and brawn of casual female labour' (Harriss-White 1996: 234). Labour arrangements tend to be highly variegated in Coimbatore's mercantile firms, with reliance on household labour in tobacco and groundnut trade and expansion of wage labour in other kinds of trade. Secure salaried labour, Harriss-White's data reveals, is almost entirely male. Actual wage payments varied considerably based on gender, location, demand for labour from other sectors, experience, 'loyalty', and 'skill and intelligence', and on gifts and bonuses for festivals. Harriss-White also observes that women's control over business operations is rare, and usually limited to married women with older children. Most often, women of trading families provide services to provision the household-firm, especially in making food for the kind-wages paid to attached labour and migrant casual labour; thus, 'unvalorised female labour . . . [reduces] wage costs' (Harriss-White 1996: 247). Elite women, on the other hand, do not sell their labour power, but render labour services for the reproduction of household-firms, and are used in expanding male portfolios through marriage dowries.

Harriss-White has broadened Baker's historical analysis by suggesting how rural entrepreneurs have diversified out of farming into other productive activities, particularly those with local demand. Furthermore, she has furthered the analysis of 'patriarchal production politics' by suggesting that gender relations are not just part of labour market segmentation but are central to the reproduction of the differentiated organisation of agro-commerce as a whole. This evidence is strikingly similar to that of Tirupur's knitwear firms, as both Krishnaswami and Harriss-White suggest that increased competition from petty production fuels 'attempts to cut costs . . . leading to a reversion to unpaid labour of various sorts' (Harriss-White 1996: 265).

This evidence suggests that agrarian processes have *prefigured* the character of production in Tirupur knitwear firms in several ways. First, the progressive Gounder farmer had combined technological innovation and responsiveness to markets with a reworking of 'familial' labour relations. Second, unique sorts of mutual loan societies were early experiments in collective associations for economic growth. Third, the diversification of rural households and the development of rural industry to meet primarily local demand created a pattern of investment linking agriculture, agro-commerce, and industry. Fourth, growing links between cotton farming, ginning, and trade built the towns of Kongunad while towns were in decline in other regions of south India. Fifth, the persistence of fragmented and stratified labour markets and labour processes across sectors suggests the regional solidity of gender relations and ideologies which could mediate and naturalise these forms of discrimination. These are the sorts of social relations, I suggest, that Gounder farmers-turned-industrialists carried with them to the town in the late 1960s. However, I have also implied that something peculiar about the Gounder caste has been useful in forging Tirupur's industry. It is to the social construction of caste and of the forms of assertion vis-à-vis the state developed by rural entrepreneurs in Tamilnadu that I turn next.

Association and assertion among rural entrepreneurs

Bayly (1992) has suggested that an important feature of north Indian development in the nineteenth century was that class formation was tied up with the reworking of caste and community, so that caste communities became 'mobilised associations for the defence of property, trade and status' (Bayly 1992: 481). Rural Tamilnadu reveals similar processes reshaping caste relations alongside the political economic differentiation of middle and lower caste groups (Kapadia 1995: 12). The rise of Gounder elites has in part resulted from their organisation into Bayly's 'mobilised associations' both to control local communities and to access wider opportunities, particularly from the state. This section suggests how Gounders in rural Tamilnadu built certain organisational capacities which prefigure industrial assertion in Tirupur today.

Cultural anthropologists of Coimbatore District have understood caste relations classically within parallel, internally stratified groupings called right hand, or *valangkai*, and left hand, or *idangkai*, castes (Beck 1972). The right hand castes have been thought of as 'interlocked', tied to land-based activities and associated institutions of product sharing, while the left hand castes have been understood as relatively distinct from rural localities and tied to wider social networks. In Kongunad, Gounder farmers have been the dominant right hand caste, while Kaikollars (traditional weavers) and Chettiars (traditional traders and moneylenders) have been the dominant left hand castes (Mines 1984, Beck 1972). I will not dwell on these distinctions, despite the claims made about the different spatial and cultural resources available to each group, because these formulations tend to slip quickly into a kind of idealism. However, it is important to note that in the nineteenth century weaving continued to be largely separate from land-based activities in Coimbatore District. The caste segregation of weavers was maintained, in part, by the early commodification of cloth, even while most goods were bartered within the village economy. This alienation of weavers from the rest of the rural economy presents a markedly different path of rural industrialisation from that in other Asian or European cases in which peasant-weaver households formed the fundamental unit of rural economy (Roy 1993).

One enduring theme in histories of caste association in Tamilnadu is the 'lack of rigidity and of systemisation in caste status . . . [and] in the way that socially mobile groups were able to change caste' (Washbrook 1975: 168). Indeed, nineteenth-century reporters for the Madras census found it difficult to pin down the major castes in the context of highly fluid mega-categories of *Vellala*, *Kamma*, or even *Gounder-Vellala*. Washbrook sees the period of the 1920s and early 1930s in colonial Madras Presidency – the period of 'dyarchy' under which new elected institutions would allow Indians to collaborate in provincial administration – as also a period in which factions were organised in the idiom of caste to access new political resources. In this process of caste-based political organisation, Washbrook argues that 'caste began to secularise itself' (Washbrook 1975: 192). In an alternate rendition, Irschick (1986) demonstrates how 'nativist' constructions of an idealised Tamil past were utilised by Tamilians to access the new elected legislative assemblies, local boards, and municipalities. Irschick notes that caste categories were pliable enough to allow patronage networks extending from the countryside to colonial institutions and categories of political reservation to open up the capabilities not just of rural elites, but also of many of the poor and dispossessed of rural Tamilnadu.

In this context, Gounders were far from holding to the 'essential' features of right handed castes, with primary concerns in maintaining dense local relationships around farming. While it is true that Gounder farmers were slow in taking advantage of the new investment opportunities in the mills of Coimbatore District, certainly in contrast to Kamma Naidu farmers, Gounders did in fact set up cotton gins and cooperatives, as well as a few textile mills, by the early 1930s. Arnold (1974) describes processes of differentiation within agrarian Gounders between those who could invest in electric pump sets and become rich farmers and those who were driven off the land by debt or landlessness to become agricultural or industrial labour. As the towns of Kongunad – Coimbatore, Erode, Pollachi, Dharapuram, Udumalpet, and Tirupur – were granted municipal status, Gounders, whose power was based on wealth rather than on affiliation with Pattagars, or traditional lineage heads, began augmenting their social power through election to the new municipal councils. By the 1930s Gounders were dominant on most local boards in the region, as well as on temple and education committees.

The 1920s and 1930s threw up changes in the society of rural Tamilnadu which have been interpreted in a variety of ways. During this early phase of the Non-Brahmin Movement, the Justice Party was growing in support primarily from forward non-Brahmin castes, such as Vellalas, Naidus, and Chettiars. Both the Justice Party and the Congress Party were encouraging certain Gounders to mobilise their caste, and elite Gounders saw that it was in their own interest to form a caste association, the Kongu Vellala Sangam, in 1921. Arnold describes the Sangam as expressing desires for modernisation as well as for 'sanskritisation', as elite Gounders, it seems, wanted secular advancement through access to new educational and technological advances, as well as advancement of ritual and cultural status through the emulation of upper-caste Hindu values of vegetarianism, abstinence from alcohol, marriage reform, etc., (Arnold 1974: 8). While Arnold is careful in characterising this new Gounder caste association as consented to by all its participants, he claims that they could achieve 'a far greater degree of cohesion' than their rival Kamma Naidus, who chose instead to mobilise their community in industrial associations in Coimbatore city. Arnold's historiography demonstrates some of the complex tensions between the ambitions of

Gounder leaders and of the Congress non-cooperation movement, particularly over superficially common concerns in prohibition and *khadi*, or hand-loom textiles. Tirupur was a hub of activity during the mid-l920s, with its concentration of cotton ginners, traders, and *khadi* merchants, all of whom felt some economic advantages in supporting the drive for a Congress ministry.

The expansion of franchise under the Government of India Act of 1935, which granted the vote to lesser landholders and expanded the electorate by a factor of four or five from the earlier period of dyarchy, made it much more imperative for the Congress Party to gain rural support in Kongunad. The framing of the terms of agrarian reformism was key to both Congress and elite Gounders, who concurred over a mutual silence on issues of land reform. Instead, rural disenchantment over issues of irrigation and land taxation were used against the Justice Party, which had been in power during dyarchy, to the benefit of Congress.

However, Gounder leaders did, as Arnold indicates, keep away from Congress proclivities towards the uplift of *Harijans*, or 'untouchables'. Furthermore, Gounders tended to comply with Congress only to a certain extent, and were more like other rural elites in Tamilnadu in not supporting Congress boycotts of legislatures, colleges, and of government services in the 1920s. Arnold briefly mentions that there was an alternate 'undercurrent of radicalism in Kongunad' which could have been organised by E. V. Ramaswami Naicker's 'Self-Respect Movement', the second phase of the Non-Brahmin Movement. Yet, it seems, Gounder elites became key local power brokers by 1937, and were 'an invaluable asset to the Congress' in the elections of that year (Arnold 1974: 20).

Gounders maintained their Congress connections through the transition to independence, even to the present day. Indeed Gounder connections can be traced to Indira Gandhi's cabinet during the period of bank nationalisation in 1969, after which the landscape of access to institutional credit was dramatically transformed to the benefit of small-scale industrialists in Tirupur. Yet Gounder farmers were also organising politically on non-party lines during the 1960s, as harbingers of the so-called 'new farmers' movements'. Though I will address them more generally at the end of this chapter, it suffices to say that these non-partisan farmers' movements are focused on price issues and are directed at 'non-rural' agents of state and urban society. These movements emerged first in Coimbatore and Tanjavur districts of Tamilnadu, and the new characteristics were more pronounced in the context of capitalist agriculture in Coimbatore. Nadkarni (1987) argues that despite drawing in small peasants and occasionally agricultural labour, they are 'led by the agricultural elite and for the agricultural elite' (Nadkarni 1987: 61). He suggests that part of the inability of farmers' associations to make their mark lay in the divergent interests of these association and of several rich farmers who were also private traders. While Nadkarni's analysis does not reveal whether and how farmers' associations have drawn upon prior caste associations, he does suggest a silence on issues concerning the lowest castes, as well as on caste oppression in general.

Social constructions of caste have therefore been part and parcel of the refashioning of the strategies for advancement by rural elites since at least the 1920s. This is most true for the middle castes of rural Tamilnadu where caste categories have admitted much class diversity. The organisational skills built by the Gounders in rural society have provided, I would suggest, the means for organising as industrialists, and even for agitating in ways that combine elements of class power and mass power. Industrial assertion in Tirupur today is prefigured by forms of assertion around idioms of caste and rurality. While I have suggested in this section and in the previous section that agrarian resources have been refashioned by Gounders in the organisation of the knitwear industry in Tirupur, these narratives must be located within the limits afforded by state regulation and the regional history of industrialisation and unionism. It is to the role of the state in forging Tirupur's smallholder form that I turn to next.

State regulation and textile populism

> [S]mall towns such as Ludhiana, Jalandhar, Moradabad, Tirupur and Panipat . . . have become export successes without, and not because of, government support.
>
> (*India Today* 31 March 1994: 88)

The misconception that the state has not been an active participant in the making of Tirupur's boom conveys a slight understanding of the ways in which state regulation, in conjunction with demands from local elites, has configured the spatial economy of textile production in India. Tirupur's small firms have emerged through this process of regulation and what I have called 'textile populism', through which the smallholder form of today's knitwear industry has been forged. In this section, I sketch the broader forces which have allowed the growth of this local industrial form.

Indeed, state policy was central to the configuration of spatial advantages – in terms of location, size, and spatial linkages – of small-scale industry, and the role of small firms in an independent India was central in nationalist discourse during the struggle against British rule. Tyabji (1989) argues that the portrayal of this discourse as bipolar – as Gandhian rural industry versus Nehruvian central planning – belies a complex history of political bargaining and accommodation.[5] Indeed, 'peasant' capitalist industries as well as small producers were subordinated to merchants' capital in many parts of India *before* the formation of industrial capital, forcing policy to be framed in terms of these groups early on. By the Second Plan of the nation of India in 1956, policy makers sought to expand the production of basic wage goods by supporting 'traditional' sectors, classified as village, cottage, or small-scale industry. Precisely because big industrialists were sufficiently satisfied that Congress reformism would not compromise their interests, they did not organise to oppose this populism.

> The capitalist class . . . came to realize that the administrative structure effectively was their own and they held State power through this structure. There was therefore no need to fear what amounted to populist pronouncements of policy; in fact the shrewder of the capitalists began to welcome these as guaranteeing the cooperation of the petty bourgeoisie for a Government which could, at best, ensure the welfare of only a very small minority of this stratum.
>
> (Tyabji 1989: 110)

This history of populism was in part continuous with the Indian independence movement, in which Congress had successfully represented itself as a multi-class social movement. The early '*khadi* movement', under which foreign cloth was boycotted for hand-spun *khadi* cloth, had already brought in petty bourgeois spinners and weavers under the mantle of Gandhian populism, even if mill owners were the main beneficiaries of the movement. Gandhi's subsequent disillusionment with Congress reformism and his abandonment of formal political involvement paralleled the creation of two important institutions – the All-India Spinners' Association and the All-India Village Industries Association – both of which brought to bear on early Indian planning

> that the small commodity and early capitalist forms of industry would have not only to be protected through State action, but that their contribution to employment and output in the economy had to be increased on an absolute if not relative basis.
>
> (Tyabji 1989: 120)

By the Second World War, Tyabji notes the existence of a range of institutional supports for small-scale production. These supports emerged through a compromise between the need for industrial diversification and a hesitancy to push for growth in basic industry. Moreover, a growing labour movement and an official ideological propensity towards a kind of welfarism that would mitigate the social costs of industrialisation complemented legislative protection of small industry to keep down unemployment. The Small Scale Industries Policy, conceived to modernise production in small industry, was created within this political environment to define and manage the transition to capitalist industry, given the prevailing conditions of the subsumption of producers by merchant and usurer capital which effectively prevented their transformation into industrial capitalists.

Policy with respect to textiles, begun in 1951, was centred on protecting the hand-loom and power-loom sectors from competition by the mill sector. Hand-looms and power-looms were allowed monopoly production of particular items, and were exempt from both excise duties and price and distributional controls. The assumption was that the reservation of items from the small- and large-scale sectors would prevent competition between the two and 'lead to the dissolution of the lower forms of production and the steady growth of small factory production' (Tyabji 1989: 131).

The subsequent passing of the Industrial Development and Regulation Act (IDRA) in the early 1950s was a landmark in defining the parameters – by sector and by employment – over which industries had to be registered with the government, and by exclusion of 'the small-scale sector', which was exempt from these licensing requirements. The Small Scale Industries Board (SSIB) was subsequently (in 1954) created to encourage growth in precisely this newly defined sector, in part as a reaction to a wariness about rising unemployment among urban educated classes. The Ford Foundation at this point undertook a study of the question of small firms, leading to the creation of Regional Institutes of Technology and a Small Industries Corporation, providing owners with institutional mechanisms through which their interests as a group could be communicated to the state.

Yet Tyabji notes that it was with the Second Five Year Plan (1956) that small production was firmly installed in the trajectory of Indian development, at least through the period of the Third Plan (1961–66). Small industry was to supply consumer goods which would be increasingly in demand, driven by the growth of large industry. This support for small business was the most feasible alternative to the creation of a welfare state in the context of a highly unequal economy; welfare would have required steeply progressive taxation on a narrow base.

Support for small business was to accomplish at least three objectives. First, it was to promote small industry in consumer and simple capital goods through reservations on production and the extension of money capital and means of production. Second, the growth of small business was to create employment and encourage a petty bourgeois class. Third, small industry was to be a mechanism for the regional dispersal of industry and the spatial development of the home market to counter the allegedly retarding effects of merchant and usurer capital in rural economies.

However, Tyabji argues that by the end of the Second Plan period, in 1963, industrial dispersal had not taken place to any large degree, nor had there been substantial government support for small firms outside urban areas. State support had become more targeted to capitalists, not to rural artisans. Tyabji's general argument is that the small industries policy failed as a tool for converting merchant capital, or capital in manufacture, into full fledged industrial capital. However, regional patterns of growth in the textile industry could allow rural capitalists to use small-scale industries policy under certain conditions.

Coimbatore and its hinterland were to fulfil such conditions through the specific pattern of textile industrialisation outlined here. While the main mill centres at the time of independence were Bombay, Ahmedabad, Kanpur, and Coimbatore, Coimbatore was unique in not consolidating spinning, weaving, and finishing operations, but concentrating on spinning. This pattern of industry left open opportunities in the region for both the hand-loom sector, to which Coimbatore mills provided yarn, and the decentralised power-loom sector, which would boom in the years to come.

As Goswami (1990) demonstrates, the industry has become more fragmented and disintegrated than ever, largely through protection for weaving in the decentralised (hand-loom and power-loom) sector. Prompted by the state-appointed Kanungo and Karve Commissions on the restructuring of the textile industry, restrictions were imposed on the number of looms in the mill sector during the 1950s to promote weaving in the hand-loom sector. As the latter still depended upon yarn, spinning was not restricted by this new legislation, which meant that Coimbatore's spinning mills were at a relative advantage in relation to other mill centres. Consequently, spindleage increased rapidly between 1971 and 1986, while weaving in mills levelled off after 1966. An unintended outcome of these regulations was the expansion of the power-looms in the decentralised sector. While the state attempted, consequently, to regulate this sector, it was eluded by the technical ingenuity of small workshops. Indeed, by the late 1980s, the decentralised sector was producing more than 78 per cent of all cloth produced in India, about 50 to 60 per cent of which was from the power-loom sector (Khanna 1989). In addition, the protracted Bombay textile mills strike in 1982 provoked a burst of expansion in power-loom capacities (Goswami 1990).

Furthermore, south India witnessed a surprising expansion of mill production in the 1980s, mainly in proximity to regions that had recently come under cotton cultivation. Increased output of long-staple cotton brought down the price of cotton while raising its quality. The 1980s also brought a rise in 'processing houses' in the major power-loom centres, giving a boost to the regional economy of Coimbatore District (Khanna 1989). Indeed, the success of Coimbatore mills is tied to the fact that almost all cotton and blended spun yarn is provided by south Indian spinning mills (Goswami 1990). Furthermore, Coimbatore has moved from being a 'textile town' by diversifying into other light industry so that mills here do not dominate and crowd out small producers.

Indeed, the pattern of industrialisation development in Coimbatore District has kept possibilities open for small-scale industry; the larger regulatory apparatus of the small-scale industries policy has provided resources targeted at capitalists. The small-scale industries policy has allowed the knitwear industry to retain its smallholder form, because of the character of regional industrialisation in Coimbatore District. Tirupur's booming small firms are in part a consequence of specific kinds of state reformism, in sharp contrast to the claims of neoliberal analysts who pose Tirupur's export proficiency in contrast to the government sponsored Export Processing Zones. Instead, the intervention of the state was key to expanding the capabilities of small producers in using their flexibility to access wider networks of power. In the following section I turn to the ways in which the regional trajectory of industrialisation and unionism has contributed to forging Tirupur's industrial form.

Mills, unions, and labour politics

While the politics of state reform provide the broad context within which sections of the Indian textile industry developed its 'smallholder' form, the proliferation of small firms in Tirupur is also a consequence of the particular industrial trajectory in Coimbatore and the highly charged history of trade unionism that emerged within it. This history finds an uncanny culmination in the way in which forms of collective bargaining between capital and labour have been institutionalised in Tirupur's largely unregulated industry. Cawthorne's evidence indicates that most units are unregistered with the Factories Inspectorate, while some are registered with the Small Scale Industries Board in order to procure loans. The role of labour unions in such a context of informalisation is unprecedented. Workers rarely take recourse to the state Labour Office or Labour Court. Beyond the General Agreements which are renegotiated every three years, conflicts at work are usually resolved through 'mutual arrangements'. This outcome – another aspect of what Swaminathan and Jeyaranjan see as 'the balance between cooperation and competition' (Swaminathan and Jeyaranjan 1994: 3) i.e. between workers and employers as opposed to between firms – is a consequence of regional forces which will be explored in this section. If, as I have argued, the reorganisation of the labour process into small firms in the late 1960s was partly a response to labour militancy in towns, it is imperative to understand how this militancy was framed and shaped historically from the mills of Coimbatore city to the rural hinterland.

Coimbatore only had four mills in 1919; this number had increased to twenty-nine by 1930, marking the emergence of Coimbatore as an industrial centre in Madras Presidency. First, rich farmers and those involved in the cotton trade set up ginning plants in Coimbatore. Subsequently, these ginners diversified into milling operations. These classes provided the bulk of investment for the growth of industrial Coimbatore, while labour was provided by a range of castes, including poor farmers, landless labourers, and poor hand-loom workers, and with women comprising about a third of the work force in the 1930s. The deplorable working conditions fuelled militant labour struggles in this period against 'intransigent, ruthless employers' (Krishna 1992).

Part of the successes in unionisation during this period, in Murphy's (1981) narrative, depended on ties forged in the 1920s between organised labour and Indian nationalist politics. The 1910s saw the establishment of labour propagandists within the nationalist movement under the Madras Labour Union, after which a range of unions were formed in Coimbatore District. Yet the links between organised labour and nationalism, which had strengthened the labour movement in other parts of Tamilnadu, were weak because Coimbatore industry in the early 1930s was spatially dispersed. The following years marked not only an increase in unionisation among women as well as men, but the emergence of rival factions of communists and social democrats. Coimbatore labour during the 1930s was broadly divided between older mill hands and younger recruits to the newer country mills on the urban periphery. While workers were given few incentives to remain in the same mill, they seemed to prefer changing jobs until they were close to their homes or villages. This pattern of industry was characterised by high turnover and mobility between mills, within the context of economic growth. However, it seems that this spatially clustered and mobile labour force tended to spread collective consciousness across the dispersed mills so that towns became centres of labour protest.

Country mills, on the other hand, faced a range of obstacles to unionisation, including isolation from other unions and the predominance of patronage ties between workers and owners, many of whom were landlords who had hired the same workers in agriculture. 'Jobbers' or foremen often acted as recruiting agents and received gifts, including sexual favours, from villagers in exchange for highly prized jobs in Coimbatore mills. Indeed, Coimbatore mills were noted for their 'petty tyrant' jobbers, who ran operations in the mills under only token supervision by the owners. Jobbers' relations with managers and unions, on the other hand, were usually ambivalent, and it was only in the large, rationalised mills, where higher level management held most of the disciplinary power, that jobbers took a strong stance alongside workers in their struggles against management. On the whole, Murphy (1981) makes the fascinating claim that Coimbatore was markedly different from other industrial centres in Tamilnadu because 'neither caste nor community was an important variable complicating the history of labour organisation' (Murphy 1981: 53). Indeed, Murphy's evidence suggests that caste and community were elements in the recruitment strategies of mill owners and jobbers in their search for a loyal work force, and workers tended to congregate informally in caste groups. However, these issues hardly seemed to come up in labour disputes, for workers' grievances tended to centre on working conditions and the intransigence of employers. The central problem for labour organisers, argues Murphy, was 'of trying to organise scattered groups of workers throughout the many mills of the district' (Murphy 1981: 55). This could not be more familiar to the problems faced by organised labour in Tirupur today.

It was not until the creation of the Congress Socialist Party in 1934 that the nature of labour mobilisation was transformed, rendering Coimbatore a centre of militant unionism. Worker strikes in 1937 over the retrenchment of workers with the introduction of Japanese technology, or of the nature of specific tasks, crossed sectional and gender lines, with the CSP intervening to wrest concessions from management. Conflict between labour and capital in the late 1930s was marked by occasions of considerable solidarity between the rank-and-file and union leadership. Yet late 1937 brought a divergence of interests between Congress mediators, CSP leaders, and socialists within the labour movement. The next few years saw the establishment of a rival Congress union, the Coimbatore District Textile Workers' Union (TWU), and the rise of violent conflicts between pickets and the police. Since the 1930s, this rift between communist and gradualist tendencies in Coimbatore District has widened. After 1947, the two factions became more antagonistic to each other, to the point of violent conflict. The TWU subsequently broke with the Congress Party, and the years after 1948 witnessed the growth of several other rival unions, forging links with different political parties as with the new regional party, the Dravida Munnetra Kazhagam (DMK). Town workers were most loyal to the communist union (CMWU), which also managed to gain the support of many of the jobbers, who were usually organised in special branches in village unions. Along with the growth in the number of unions in independent India, labour unrest in Coimbatore District took on a distinctive character: labour struggles became largely between rival unions, played off against each other by mill owners (Murphy 1981).

In short, the pattern of industrialisation in Coimbatore District promoted differentiated labour arrangements and complex reworkings of pre-existing ties in controlling access to work, such as existed between workers and jobbers in rural mills. However, mobility between dispersed firms helped in spreading labour ideology in urban mills which became crucibles of labour organisation. Despite spatial dispersal and the differentiation of unions through their complex interminglings with political parties, labour activism had persisted in this early period to combat the fragmentation of work. The persistence of a union presence in the towns of Coimbatore District against such odds provides the context within which the Gounders reorganised work in Tirupur's knitwear industry. If, as I have argued, the reorganisation of work into small-firm networks was an employers' strategy of labour control, the presence of unions in today's Tirupur suggests that a longer history of labour resistance and unionism cannot easily be supplanted.

SHARAD CHARI

CONCLUSION: AGRARIAN TRANSITION AND THE EMERGENCE OF AN INDUSTRIAL DISTRICT

> [T]he nature of agrarian transition not only has implications for the fate of the countryside, but has a decisive influence upon the pace, manner, limits, and very possibility of capitalist industrialization.
>
> (Byres 1995: 509)

Historical geography suggests that what is taken for granted in Tirupur today – an industrial district – has emerged through processes crossing sectoral boundaries. I have suggested that what Cawthorne has called 'amoebic capitalism' was fashioned in the turbulent period of the late 1960s and early 1970s, when the agrarian Gounders entered the industry in large numbers. My central argument is that Gounders brought with them various relationships from the countryside with which they could transform the terms of labour discipline in favour of economic growth, within propitious limits set by state regulation and a pre-existing trajectory of industry and unionism.

Accordingly, my argument has centred on the rural and historical roots of Tirupur's industrial form, seen through four themes. First, the agrarian history of Coimbatore District demonstrates the persistence of fragmented and segmented work and labour markets across agriculture, agro-commerce, and the knitwear industry. I have used this evidence of continuities to suggest that forms of work discipline – especially concerning women's work – have been selectively refashioned in Tirupur from the social organisation of rural production. Second, the different ways in which social constructions of caste have been used to build associations that are at once class- and mass-based across sectors suggests a reworking of forms of assertion from rural society that may explain how Tirupur's industrialists organise so effectively in employers' associations. Third, the socio-spatial pattern of industrialisation in India, and the particular character of textile industrialisation in Coimbatore District, reveal an active history of state intervention in the configuration of 'smallholder' knitwear firms. Fourth, if Gounders brought with them forms of labour discipline from the country to the city, as I have suggested, they did so in a historical context of fragmented yet persistent unionism which has retained a role for organised labour in today's knitwear industry.

If historical geography reveals the agrarian and regional origins of Tirupur's industrial form, agrarian questions are still very much a part of Tirupur's present as, for instance, in what Swaminathan and Jeyaranjan (1994) call 'the "owner as worker" and/or "worker as owner" phenomenon' in today's knitwear firms (Swaminathan and Jeyaranjan 1994: 13). The four narratives on the genesis of regional industrialisation in Coimbatore District are part of my ongoing research on agrarian dynamics in the emergence of Tirupur's industrial form. On the one hand, my argument is that dynamics can be determined through a combination of historical and ethnographic methods (Hart 1996). On the other hand, I suggest that the debates concerning the agrarian question in India can shed light on the problematic of regional industrialisation centred on the knitwear industry in Tirupur. It is to the second point that I turn in the remainder of this chapter.

In broad brush-strokes, then, the agrarian question in India today has three outstanding characteristics: a deepening of economic divergence between prosperous regions in the north, west, and south from the east and centre; the rising political and economic clout of rural capitalists in certain prosperous regions; and the fragmentation of labour militancy and the concomitant proliferation of localised forms of labour control and work organisation (Srivastava 1995). In concluding this chapter, I would like to suggest how Tirupur's knitwear industry can be understood and in turn contribute to the analysis of the agrarian question in India, along these three lines.

First, regional divergence between parts of India calls for a broadening of the notion of agriculture's contribution to industrialisation from the savings contribution of agriculture, to include the expansion of non-farm employment and rural industry (Karshenas 1995). I have sought to suggest through the case of Tirupur that farmers-turned-industrialists may also bring, as Watts (1996) suggests, 'local institutions of labour recruitment and discipline' (Watts 1996: 236). However, following Chandavarkar's (1994) history of the rural ties of Bombay's industrial workers, I would also add that the rich history of peasant and mill-union mobilisation in Coimbatore District might provide resources for employers as well as for workers in ongoing struggles over the nature and product of work.

Second, the rising power of rural capitalists in certain highly commercialised regions such as Coimbatore District in Tamilnadu, and Nasik and Nipani Districts in Maharashtra, has spawned contending interpretations ranging from Omvedt's (1994) grassroots 'new social movements' to Brass's (1994) harbingers of Hindu fascism. These movements centre on price issues concerning commercial, not subsistence crops, and rarely voice the problems of farmers in drought-prone regions, or of the rural poor. Nadkarni (1987) is persuasive in characterising these as class movements as well as mass movements, precisely because numbers are needed for the kind of agitational politics pursued, in rallies, arrests and protests. The tensions and contradictions within the new farmers' movements in India today, and from alternate forms of rural mobilisation – such as Dalit movements contesting the rise of middle-caste power in the countryside – suggest the persistence of classically peasant problems concerning unequal access to land, variations on rural exploitation, the supposed solidity of rural culture, and the many means and forms of peasant assertion. As the Gounders were one of the first players in the new farmers' activism in India, and as they continue to be highly organised as industrialists with recourse to visible mass protests involving large numbers of workers, I have suggested that there are continuities between agrarian politics and the politics of industrial assertion.

Third, many studies of recent changes in rural India note the fragmentation of labour organisation and forms of work discipline (Srivastava 1995) or the reworking of patron–client relations (Nadkarni 1987). Dreze and Mukherjee (1989) emphasise the casualisation of labour transactions, the decline in non-negotiable labour contracts, and the rise of piece-rated work, while Harriss (1992) emphasises the 'stickiness downwards' of rural wage rates and the fragmentation and segmentation of rural labour markets which maintain differential wage rates even for the same tasks across neighbouring villages. Through close ethnographic research in a village in Tamilnadu, Kapadia (1995) argues that capitalist farming receives subsidies from the unwaged or under-waged labour of 'untouchable' Pallar women and girls. Gender and caste mediate both the organisation of work and the forms of mutualism pursued by workers. Similarly, Krishnaswami (1989) and Cawthorne (1990) demonstrate how divisions of labour and access to work are mediated in Tirupur by complex social ties and by gendered social constructions of 'skill'. Indeed, Harriss-White's (1995) suggestion that gender relations are central to the reproduction of the differentiated economic system that prevails in Tirupur and its rural environs brings into question the coherence of what Swaminathan and Jeyaranjan (1994) call 'local culture'. Drawing from their evidence, I have suggested that work in Tirupur's web of firms is caught in a field of power refashioned from the social organisation of agriculture and agro-commerce, within relations of power involving the state and organised labour. The ways in which these historical geographies are brought to bear on present workings of labour markets and labour processes in Tirupur will, however, have to await 'detailed, microlevel sociocultural data' to illuminate actual dynamics (Kapadia 1995).

Finally, the rise of Gounder capitalists from family farms to global knitwear firms supports Byres' (1992) suggestion that transitions to agrarian capitalism in India are primarily 'via a peasant route: a form of "capitalism from below" . . . [which] entails a variety of action by the Indian state' (Byres 1992: 64). The Gounders of Coimbatore District have built an industrial empire for the global sourcing of garments, but they retain elements of their past character as 'progressive peasants' in the relations of work discipline and assertion in Tirupur's industrial present.

ACKNOWLEDGEMENTS

This chapter represents an initial stage of dissertation research, funded by the Fulbright Doctoral Dissertation Research Award during 1996–97. I am also grateful to the Social Science Research Council's International Predissertation Fellowship Program for funding during 1994–95, during which time I was fortunate to have discussions with Barbara Harriss-White, Padmini Swaminathan, S. Neelakantan and S. Janakarajan at the Madras Institute of Development Studies, Madras, India. For suggestions to prior versions of this chapter, I am indebted to Michael J. Watts, Gillian Hart, William Boyd, Navroz Dubash, Niels Fold, and Jehanbux Edulbehram at the University of California, Berkeley.

NOTES

1 Tirupur is located 50 km from the industrial city of Coimbatore, in the semi-arid Coimbatore District of Tamilnadu State, India. The population within Tirupur's municipal limits (27.2 sq. km) was 235,000 in 1991 (census of India), an underestimate, as the town extends to 43.52 sq. km including peripheral villages which are rightfully part of this socio-formation, the population would be around half a million (Tirupur Exporters' Association pamphlet 1995).

2 These figures have been taken from the Textile Exporters' Association in Tirupur, as well as from *India Today* and the *Far Eastern Economic Review* articles cited at the beginning of this chapter. There is quite a bit of variance in numbers claimed, as Cawthorne (1995) notes. I concur with her in taking these numbers as indicative of rapid growth.

3 The Gounder transition to Tirupur is typically explained by the media and by themselves as a response to agrarian decline; indeed the values of prime agrarian land in the region have dropped below dry land values, while marginal and substandard land near Tirupur has skyrocketed through speculation (S. Neelakantan, personal communication, MIDS, Madras, 10 July 1995).

4 I have relied heavily on Baker (1884) in this account of agriculture in Kongunad.

5 I have relied heavily on Tyabji (1989) in the argument that follows.

BIBLIOGRAPHY

Arnold, D. (1974) 'The Gounders and the Congress: political realignment in south India, 1920–1937', *South Asia* 4 (October): 1–20.

Baker, C. J. (1984) *An Indian Rural Economy 1880–1955: The Tamilnadu Countryside*, Oxford: Clarendon Press.

Bayly, C. A. (1992) *Rulers, Townsmen and Bazaars: North Indian Society in the Age of British Expansion, 1770–1870*, Delhi: Oxford University Press.

Beck, B. E. F. (1972) *Peasant Society in Konku: A Study of Right and Left Subcastes in South India*, Vancouver: University of British Columbia Press.

Brass, T. (1994) 'Introduction: the new farmers' movements in India', *Journal of Peasant Studies* 21, 3/4 (April–July): 3–26.

Braudel, F. (1985) *Civilisation and Capitalism in the 15th to 18th Centuries, Vol. 2: The Wheels of Commerce*, London: Fontana.

Brusco, S. (1990) 'The idea of the industrial district: its genesis', in F. Pyke, G. Becattini, and W. Segenberger (eds) *Industrial Districts and Inter-Firm Cooperation in Italy*, Geneva: International Institute for Labor Studies.

Burawoy, M. (1985) *The Politics of Production*, London: Verso.

Byres, T. J. (1992) 'The agrarian question and differing forms of capitalist agrarian transition: an essay with reference to Asia', in J. Breman and S. Mundle (eds) *Rural Transformation in Asia*, Delhi: Oxford University Press.

—— (1995) 'Political economy, agrarian question and comparative method', *Economic and Political Weekly* 30, 10 (11 March): 507–13.

Cawthorne, P. (1990) 'Amoebic capitalism as a form of accumulation: the case of the cotton knitwear industry in a South Indian town', PhD dissertation, Milton Keynes: The Open University.

—— (1993) *The Labour Process Under Amoebic Capitalism: A Case Study of the Garment Industry in a South Indian Town*, DPP Working Paper, Milton Keynes: The Open University.

—— (1995) 'Of networks and markets: the rise and rise of a South Indian town, the example of Tirupur's cotton knitwear industry', *World Development* 23, 1: 43–57.

Chandavarkar, R. (1994) *The Origins of Industrial Capitalism in India: business strategies and the working classes in Bombay, 1900–1940*, Delhi: Cambridge University Press.

Dreze, J. and Mukherjee, A. (1989) 'Labour contracts in rural India: theories and evidence', in S. Chakravarty (ed.) *The Balance Between Industry and Agriculture in Economic Development*, vol. 3, London: Macmillan.

Goswam, O. 'Sickness and growth of India's textile industry – analysis and policy options', *Economic and Political Weekly*, Vol. 5, No. 45 (Nov 10, 1990), pp. 2496–2506.

Harriss, J. (1992) 'Does the depressor still work? Agrarian structure and development in India: a review of evidence and argument', in *Journal of Peasant Studies* 19, 2 (January): 189–227.

Harriss-White, B. (1985) 'Agricultural markets and inter-sectoral. resource transfers: cases from the semi-arid tropics of southeast India', *Proceedings of the International Workshop, 24–28 October 1983*, Patancheru, India: International Crops Research Institute for the Semi-Arid Tropics.

—— (1996) *A Political Economy of Agricultural Markets in South India: Masters of the Countryside*, New Delhi: Sage Publications.

Hart, G. (1996) 'The agrarian question and industrial dispersal in South Africa: agro-industrial linkages through Asian lenses', in H. Bernstein (ed.) *The Agrarian Question in South Africa*, London: Frank Cass.

Holmstrom, M. (1993) 'Flexible specialization in India?', *Economic and Political Weekly* (28 August): M82–M86.

Irschick, E. F. (1986) *Tamil Revivalism in the 1930s*, Madras: Cre-A Publishers.

Kapadia, K. (1995) *Siva and Her Sisters: Gender, Caste and Class in Rural South India*, Boulder: Westview Press.

Karshenas, M. (1995) *Industrialization and Agricultural Surplus*, Oxford: Oxford University Press.

Khanna, S. 'Technical change and competitiveness in the Indian textile industry', *Economic and Political Weekly*, Vol. 24, No. 34 (Aug 1989), pp. m103–m111.

Krishna, C.S. (1992) 'First Congress Ministry and Labour: Struggles of Textile Mill Workers in Coimbatore, 1937–1939,' *Economic and Political Weekly* (July 11): 1497–1506.

Krishnaswami, C. (1989) 'Dynamics of capitalist labor process: knitting industry in Tamilnadu', *Economic and Political Weekly* (17 July): 1353–59.

Mines. M. (1984) *The Warrior Merchants: textiles, trade and territory in South India*, Cambridge: Cambridge University Press.

Murphy, E. (1981) *Unions in Conflict: A Comparative Study of Four South Indian Textile Centres 1918–1939*, Delhi: Manohar Publications.

Nadkarni, M. V. (1987) *Farmers' Movements in India*, Ahmedabad: Allied Publishers.

Nicholson, F. A. (1887) *Manual of the Coimbatore District in the Presidency of Madras*, Madras: Government Press.

Omvedt, G. (1994) 'We want the return of our sweat: the new peasant movement in India and the formation of a national agricultural policy', *Journal of Peasant Studies* 21, 3/4 (April–July): 126–64.

Piore, M. and Sabel, C. (1984) *The Second Industrial Divide: Possibilities for Prosperity*, New York: Basic Books.

Rasmussen, J., Schmitz, H., and van Dijk, M. P. (eds) (1992) 'Flexible specialization: a new view on small industry?', special edition of *IDS Bulletin* 23, 3 (July).

Roy, T. (1993) *Artisans and Industrialization: Indian Weaving in the Twentieth Century*, Delhi: Oxford University Press.

Srivastava, R. (1995) 'India's uneven development and its implications for political processes: an analysis of some recent trends', in T. V. Sathyamurthy (ed.) *Industry and Agriculture in India Since Independence*, Delhi: Oxford University Press.

Swaminathan, P. and Jeyaranjan, J. (1994) *The Knitwear Cluster in Tirupur: An Indian Industrial District in the Making?*, Working Paper No. 126, Madras: Madras Institute of Development Studies, November.

Tyabji, N. (1989) *The Small Industries Policy in India*, Calcutta: Oxford University Press.

Washbrook, D. (1975) 'The development of caste organization in south India, 1880–1925', in C. J. Baker and D. Washbrook (eds) *South India: Political Institutions and Political Change 1880–1940*, Delhi: Macmillan Company of India Limited.

Watts, M. (1996) 'Development III: the global agrofood system and late twentieth-century development (or Kautsky redux)', *Progress in Human Geography* 20, 2: 230–45.

COMMENTARY ON PART I

Theoretical reflections

Norman Long

As the title of this book clearly conveys, we are now confronted with a number of agrarian questions whose scope reaches well beyond the conceptual domains of political economy, neo-modernisation frameworks, or global commodity chains to encompass issues concerning the social, cultural, and institutional embeddedness of agrarian livelihoods, agricultural development, and commodity systems. This is not to say, of course, that these earlier formulations never addressed any important cultural or institutional dimensions or that their contributions are now completely irrelevant to understanding contemporary agrarian processes. Rather, the point is that after postmodern deconstructionism, we are left with the impossible task of picking up the pieces of grand narrative, whatever their genre.

Although the alternative is not at all transparent, several things have become clear. In the first place we need to develop ways of theorising the heterogeneity and complexity of agrarian life. Second, we must deal with how various forms of 'agency' interact in the production of emergent structures that, in turn, set the conditions for future patterns of social action. And third, we must acknowledge that 'no matter how much structure is added, and however minutely the specifications of structure are detailed, there will always be gaps in the pattern' (Harries-Jones 1995: 231, paraphrasing Bateson and Bateson 1987).[1] We are left then with nothing short of exploring the multiplicities and discontinuities of agrarian institutions and revealing their 'ordering' properties.

AN OVERVIEW OF THE CONTRIBUTIONS

Viewed from this theoretical perspective, the contributions of John Wilkinson, Gillian Hart, and Sharad Chari to Part I of this volume offer important insights.

Wilkinson's chapter concentrates on the various policy and theoretical arguments pertaining to the survival and viability of diversified family farms in the Southern Cone countries of Latin America where agricultural and agro-industrial integration policies have been actively pursued. He criticises the types of argument advanced in support of small-scale production which assume that improved access to credit and other services combined with more appropriate technology options would simply remove the obstacles to family farm competitiveness. He also rejects attempts to defend the family farm in terms of principles of justice or equity, or through political mobilisation aimed at shifting the balance of power in favour of the family farmer. These 'populist' approaches, he believes, are naive and likely to 'obliterate possible spaces for the negotiation of interests and the identification of common ground'.

As an alternative analytical approach, Wilkinson suggests that we pay closer attention to the processes by which a plurality of social institutions and modes of coordination, based for example on market, state, and communal forms of regulation, sustain and transform a heterogeneous 'regional' assemblage of modes of economic organisation, in which the diversified family farm is embedded and by which it acquires its livelihood dynamics. Hence economic performance cannot be judged by some abstract yardstick of 'superior efficiency', nor is the so-called competitiveness of the family farm simply secured by removing institutional, technological, and fiscal bias. Instead, analysis must address itself to the understanding of more broadly based livelihood and organisational patterns and options, with a view to identifying important issues and arenas of social coalition. Although points of convergence can be short-lived, certain critical moments can arise in which networks of heterogeneous actors – small farmers, agro-industrial entrepreneurs, local cooperatives, environmental groups, and local politicians – are able to define certain common points of view and develop effective forms of collective action aimed at redirecting regional economic priorities. Wilkinson concludes that what is needed theoretically to address these social and institutional processes is some kind of French-inspired marriage between actor-network analysis and convention theory.

Hart's discussion takes up some similar issues in relation to patterns of rural industrialisation. A general objective of her chapter is to refocus attention on the agrarian context and to extend Goodman and Watts' (1994) trenchant comments on the shortcomings of agro-food system analysis based on concepts drawn from the industrial restructuring literature. She argues that we have to recognise the 'enormous diversity, complexity, and fluidity of social institutions in "Third World" agrarian settings [which] both predates and is more sophisticated than parallel concepts in the industrial restructuring and new institutionalist literature'. She grounds her case in a comparative analysis of agrarian transitions in South Africa where rural industrialisation assumes an important role.

Her analysis combines an account of the processes of land dispossession and displacement of black labour from commercial farms and urban areas to rural ex-bantustans where workers were needed for various new rurally based industrial complexes. Many of these new centres of industry were designed as industrial clusters based on network and inter-firm production, but, unlike the flexible specialisation areas of Southern Europe, they are buttressed by low wage levels and coercive labour relations, often leading to intense struggles between the industrialists (some of them Taiwanese) and a predominantly female labour force. Although similar in structural conditions, the details of labour–management relations and the solidarities between different categories of worker in the different industrial enterprise areas evolved differently. Or, as she puts it, the 'interlocking dynamics of dispossession and industrialisation have been constituted and experienced in locally specific ways that bear directly on future possibilities for reconstruction'. The second part of Hart's chapter is concerned therefore with mapping out these divergent local trajectories in terms of differential levels of resistance to removal, to the new industrial labour regime, and to the emerging centres of local power as represented by township and national political authorities.

The theoretical thrust of Hart's contribution is geared to demonstrating the importance of contextualising issues of rural industrialisation within the wider frameworks of agrarian and regional political economy, and to interpreting socio-political organisation as the product of ongoing struggles and contestations in multiple social arenas, local and extra-local. Highlighting the shortcomings of industrial restructuring arguments and 'new' institutionalism (including recent trends in the new economic sociology) – backed by empirical data from South Africa – she makes the case that these approaches fail because they abstract rural industrialisation from its agrarian roots, variations, and regional historical specificities.

In similar vein, Chari provides evidence of the rural and historical roots of industrial forms – this time in relation to the knitwear industry in South India. Like the previous contributors, he too underlines the need to move away from arguments about post-Fordist industrial restructuring. The development of the knitwear industry and its forms of labour control and work organisation have been closely related, not to the imperatives of new standards of flexibility and efficiency characteristic of industrial enterprise regions, but primarily to the persistence of segmented work and labour markets which for many decades have spanned agriculture, rural commerce, and the knitwear industry. In the South Indian case, this segmentation has been shaped by gender and caste relations and ideologies that cross-cut the different economic sectors, impinging not only upon the lives of workers but also upon those of their employers (who in this case are mostly members of the Gounder agrarian caste).

Hence the evolving narrative of textile industrialisation cannot be isolated from the culture and political-economic histories of particular agrarian groups, nor from the profile of provincial town politics. In addition, the national state has played a critical role in promoting policies in support of small-scale industry: it is this factor rather than simply changing market or economic conditions that has prompted the rise of decentralised forms of enterprise.

Chari's study stresses then the importance of exploring the social organisation of rural industrialisation through a regional analysis of the persisting and changing interconnections between rural and urban fields of activity, and through an understanding of how certain provincially rooted cultural values and social commitments shape this process. His analysis accentuates the differences that arise between similar industries within different urban and agrarian spaces, and thus reveals how social and political struggles have influenced contemporary industrial patterns.

All three contributions delve into the regional contexts within which rural change takes place, thus shifting the analytical focus away from the search for models of 'transition' based on productivist theories of capitalist development, modernisation, or institutional transformation, to context-sensitive studies that aim to reveal the connectivities of different fields of social relations in the making of agrarian and industrial structures. Such a shift resonates with an earlier period when regional analysis was high on the agenda,[2] but it also represents a renewed challenge to enlarge our theoretical understanding of complex structures and processes.

RESPONDING TO THE CHALLENGE[3]

Such a challenge is founded upon the following premises: (1) there are no *given* or *a priori* sets of driving forces (such as technologies, markets, policy imperatives, or cultural values) that generate particular social arrangements or patterns of change, only complex sets of connectivities between material, cognitive, social and non-human elements; (2) the flexibility and changing nature of actors' strategies and interlocking projects and their contingent and sometimes turbulent interactions are central to understanding agency and emergent structures;[4] and (3) representations and future-orientations, based on cultural schemas (whether conscious or not), play an important role in the 'ordering' and constitution of social practice.

Agency, ordering processes, and the future

Central to the notion of agency is the capacity to solve problems, pursue desires, communicate values, process information, and influence others (directly or indirectly) within a network of social relations. While agency is usually attributed to specific actors or acting units, the possibilities, appropriateness, and effectivity of certain courses of actions are significantly ascribed or made meaningful by others; and in this sense agency is socially constructed. This is not to say, however, that actors do not formulate decisions, act upon them, innovate, or experiment. Indeed the co-existence of a multiplicity of discursive means and stocks of knowledge and resources implies a degree of choice which actors (individually or collectively) can exploit strategically in formulating objectives and in deploying specific modes of action.

Dealing with everyday circumstances, dilemmas, and evolving sets of social relations requires some attention not only to present situations but also to future scenarios and options. Hence the future is not to be seen as a void that, in principle, is unknowable and unpredictable, but rather is an idea that is filled in diverse ways by the different images and strategic expectations (however imprecise) that the various actors hold of it. Nor is the future to be conceptualised as a state of affairs determined by 'external imperatives' or 'normative schema' assumed to be embedded in the present context. Rather, it consists of a set of highly differentiated and open-ended representations or schemas[5] that embrace a range of divergent experiences and possibilities experienced and envisaged by the various actors. In this way, agency implies processes of *ordering* the future as well as the present.[6]

We must, therefore, consider the conceptual and discursive means by which the future is constructed. Such representations and schemas, together with the capabilities for mobilising the resources (material and social) necessary for attempting to realise certain goals or states of affairs, are what we mean by *actor projects*.[7] A 'project' is powerful if it is clearly distinguishable from other competing 'projects' and if it is grounded in and/or reflects or translates specific sets of interests, commitments, and prospects. A decisive element here concerns the organisational and information-processing capacities associated with particular projects for realising or creatively transforming specific visions of the future and for enrolling the projects of others. Hence, understanding the interlocking of 'projects' also includes an analysis of the processes of negotiation, confrontation, contestation, and accommodation entailed in *mediation*.[8]

Agency here implies the capacity to translate or mediate, that is, to link or distantiate a particular 'project' to or from other 'projects'. Knowledgeability[9] – expressed in the ability to develop and sustain diverse, interlocking projects and also to construct coalitions or handle contradictions – is crucial to this process. To react to the 'projects' of other actors – whether potential allies or adversaries – and to learn how to link with or distantiate from them is critical. But knowledge should not be reified as a domain in itself, as is often implied in knowledge systems theory. Rather, it should be viewed as integral to the social practices in which it functions and from which it emerges. The construction of relevant knowledge is implicated in both the meta-concept of agency and in the processes by which actors' 'projects' are developed and managed.

In this way, knowledgeability is intrinsic to the ordering or construction of the future. The specific constellation of ideas and practices that evolves is built upon many complex and sometimes turbulent processes of mediation, translation, and transformation in which the 'probable' and the 'unlikely', the 'good' and the 'bad', and the 'winners' and the 'losers' become defined. In turn, these emergent cultural representations become powerful elements in the shaping or 'structuring' of social action. Thus, insofar as social action is strategic action (that is, goal-oriented and identified with specific agents), so notions concerning the future – which way to move, what to do if things fail, how to secure a fall-back position, how to enroll others, etc. – become decisive.[10]

Despite the apparently self-evident nature of all this, we should not forget that much social science has been remarkably inattentive to issues of future orientations and of how these are built into agency and social practice. The bulk of sociological and anthropological work has concentrated on trying to explain social action, in one way or another, as reflecting the relations and conditions of the past. What we often fail to understand is how people's actions derive from their deliberations about the future; that is, from the ways in which they attempt to mould or order the future, and from the encounters, coalitions, and contradictions that emerge from such ordering. In arguing this I do not wish to imply that historical events and 'collective memory', or the present and the relations and contradictions embedded in it, should be treated as irrelevant. 'History' is relevant insofar as it is used in the modes of ordering the present and the future. If the future is seen as an elaboration, as a continuation of the past, if historically-created interests are perceived and represented as rights that are to be recognised or which function as some kind of Archimedean point for the construction of the future, or if collective memory is used in the design of particular projects (that is, if collective memory is intrinsic to knowledgeability), then indeed history matters. But it matters only insofar as it is used in the making of the future. If this is not the case, if history is disconnected from projects or representations that order the future, then historical events become largely irrelevant. Furthermore, ruptures with the past are often deliberately fostered by theories that set out to define future trajectories through the application of models of economic or cultural change based upon some ideal-typical construction of the past: post-Fordism, for example, manifests such a tendency.

Structure as the constitution and transformation of interlocking actors' projects

Let me now turn briefly to consider the significance of 'interlocking projects' for conceptualising the relation of structure to action.

It is through the complex encounter and mediation of actor projects that modes of ordering, including specific visions of the future, are generated. The emergence of such ordering processes is the outcome of the interplay of different self-reflexive strategies (Law 1994: 20), or what we have called 'interlocking actors' projects' (Long and van der Ploeg 1995).

From the point of view of any one of the actors implicated in it, such a structure consists of a network of enabling and constraining entities (both human and 'delegated' non-human such as documents, machines, technology, and stocks of capital and material resources) and is therefore internally heterogeneous (cf. Bijker and Law 1992: 300). That is, it is multiply composed and looks and functions differently for actors situated, as it were, at different locations and adopting different stances within the social landscape.

Such a conception of structure has certain analytical advantages. In the first place, it allows for the co-existence and inter-relatedness of categories of actor that differ markedly in terms of their abilities to draw upon and extract benefits from particular resources, relations, and discourses; second, it acknowledges the emergence – temporarily or more stably – of networks of power and of hegemonic tendencies that imply the dominance of certain values and ordering principles, however fragile; third, it calls for a detailed appreciation of the coalitions, contingent relations, and shifts in meanings concurrently present within the overall network structure; and fourth, it eschews the externality of structural phenomena by insisting that the pattern of interlocking actors' projects constitutes a network of enablement and constraint within which new embodiments of agency and social action take shape.

A further advantage of this approach to actor–structure relations is that it facilitates a systematic exploration of how actors become enrolled in or distantiated from the projects of others; it also offers valuable insights into the kinds of mediating processes delineated above.

CONCLUDING REMARKS

The above remarks are intended to frame in general theoretical terms salient points arising from a reading of the three contributions to Part I. All three chapters address the problem of explaining differential patterns of agrarian and industrial development. None seems convinced that existing post-Fordist and neo-institutionalist approaches can adequately explain such differences. Instead they pursue forms of analysis that give weight, not to general principles of organisation and regulation, but to specific historical, regional and socio-political dimensions. Each is concerned to build an understanding of structure and structural change from the 'bottom-up', through providing detailed accounts of ongoing social struggles over livelihoods, status, and resources, thus avoiding the shortcomings of centralist and generic theories. Running throughout the narratives they present is a keen interest in actor strategies, social organisation, networks, multiple arenas, politics, and the reworking of external circumstances and policy interventions by local groups – elements and concepts that are clearly compatible with an actor-oriented perspective (see Long and Long 1992).

To develop this line of analysis further, however, requires a more systematic elaboration and application of key actor concepts designed to open up the black box of institutions, normative frameworks, policy discourses, markets, political domains, and the like. As I indicated above, a beginning has already been made through the theoretical exploration of agency, ordering processes, interlocking projects, interface dynamics, and emergent structures. What is now urgently needed is a comparative framework within which similar and contrasting circumstances and processes can be examined and interpreted. The three chapters discussed here map out a number of possible routes we might follow.

NOTES

1 This issue of 'gaps', 'interstices', 'discontinuities', or what I have characterised elsewhere as 'social interfaces' (Long 1989) focuses on the central problem of how social structures remain intact when, at the same time, they are shot through with discrepancies in values, interests, knowledge, and sites of authority and power. Such gaps in connectivity also of course play a creative role in generating new patterns of organisation.

2 See Booth's (1994: 18) comments about the lack of contemporary regional studies 'integrating micro-action with the exploration of issues of political economy', which Long and Roberts pioneered in the 1970s (Long and Roberts 1978, 1984).

3 The following text draws freely upon some arguments advanced in Long and van der Ploeg 1995.

4 Social forms are of course essentially self-generating in that they constitute both the medium and the outcome of social action (Giddens 1987: 11). Hence 'both strategies and the *consequences* of those strategies should be treated as emergent phenomena. . . . [Also] when two or more strategies mesh together, the end product is an emergent phenomenon: a game of chess cannot, after all, be reduced to the strategies of either one of the players' (Bijker and Law 1992: 10). It is precisely in this sense that the notion of 'interlocking projects' is used, although, unlike chess, social encounters and inter-relations are less confined by pre-existing rules or rationalities.

5 I am using here the notion of 'schema' to stress the way in which experience is interpreted on the basis of flexible configurations: 'Schemas are not fixed, immutable data structures. Schemas are flexible interpretive states that reflect the mixture of past experience and present circumstances' (Norman 1986: 536).

6 This is consistent with Law's (1994: 138) argument that 'modes of ordering may be seen as *strategies or modes of recursion*' . . . 'self-reflexive ordering *depends* on representation. It depends, that is, on how it is that agents represent both themselves, and their context, *to* themselves' (1994: 25). To this we should add the idea of future-oriented representations that are intrinsic to social action.

7 Actor projects entail both cognitive and practical dimensions, and they implicate others. Hence, from a sociological point of view, the crystallisation of a particular project (whether individually or collectively initiated) depends on the discursive and practical enrolment of a heterogeneous network of others, who adhere to, or simply go along with, the idea and proposed course of action (see Callon 1986, Law 1994). The realisation of a particular project implies

therefore the assembling of a negotiated coalition of interests and values that are potentially contradictory or conflicting.

8 Compare this brief account of mediation processes with Bruno Latour's (1994) explication of the meanings of technical mediation. Latour, whose primary focus is the symmetry between human and non-human actors, distinguishes between four meanings of mediation: *translation*, the creation of a link between goals or agents that did not exist before; *composition*, the association of different entities in the production of a specific action; *black boxing*, the process by which the joint production of actors and artifacts, as well as the contribution of spatially and temporarily distant actors, are obscured; and *delegation*, the passing on to others of the responsibility for carrying out specific actions.

9 Following Giddens (1984: 1–16), I make the assumption that actors process their experiences by evaluating and monitoring their own and other actors' actions; and in this sense are 'knowledgeable' within the limits of information and resources available to them, and given the uncertainties and constraining factors involved.

10 Here one might note certain similarities between our formulation and Bourdieu's (1990: 52–55) notion of *habitus* which he uses to depict how cultural dispositions (or 'embodied history') shape cognition and social action, including actors' conceptualisations and orientations to the future. While we accept the general point he makes, we place greater emphasis on how knowledge and cultural repertoires are created at the 'interface' between different actors' lifeworlds and 'projects', that is on how the 'novel' and 'innovative' involve simultaneously the deconstruction and reconstruction of the *heterogeneity* of cultural life, not the restructuration of a given cultural framework.

BIBLIOGRAPHY

Bateson, G. and Bateson, M. C. (1987) *Angels Fear: Towards an Epistemology of the Sacred*, New York: Macmillan.

Bijker, W. E. and Law, J. (eds) (1992) *Shaping Technology/Building Society: Studies in Sociotechnical Change*, Cambridge Mass.: MIT Press.

Booth, D. (ed.) (1994) *Rethinking Social Development: Theory, Research and Practice*, Harlow: Longman Scientific and Technical.

Bourdieu, P. (1990) *The Logic of Practice*, Cambridge: Polity Press.

Callon, M. (1986) 'Some elements of a sociology of translation: domestication of the scallops and the fishermen of St Brieuc Bay', in J. Law (ed.) *Power, Action and Belief: A New Sociology of Knowledge?*, London: Routledge and Kegan Paul.

Giddens, A. (1984) *The Constitution of Society*, Cambridge: Polity Press.

—— (1987) *Social Theory and Modern Sociology*, Cambridge: Polity Press.

Goodman, D. and Redclift, M. (1982) *From Peasant to Proletarian: Capitalist Development and Agrarian Transition*, New York: St Martin's Press.

—— and Watts, M. (1994) 'Reconfiguring the rural or fording the divide? Capitalist restructuring and the global agro-food system', *Journal of Peasant Studies* 22, 1: 1–49.

Harries-Jones, P. (1995) *A Recursive Vision: Ecological Understanding and Gregory Bateson*, Toronto/Buffalo/London: University of Toronto Press.

Latour, B. (1994) 'On technical mediation – philosophy, sociology, genealogy', *Common Knowledge* 3, 2: 29–64.

Law, J. (1994) *Organising Modernity*, Oxford: Blackwell.

Long, N. (ed.) (1989) *Encounters at the Interface: A Perspective on Social Discontinuities in Rural Development*, Wageningen: Wageningen Agricultural University.

—— and Long, A. (eds) (1992) *Battlefields of Knowledge: The Interlocking of Theory and Practice in Social Research and Development*, London/New York: Routledge.

—— and Roberts, B. (eds) (1978) *Peasant Cooperation and Capitalist Expansion in Central Peru*, Austin, Texas: University of Texas Press.

—— (1984) *Miners, Peasants and Entrepreneurs*, Cambridge: Cambridge University Press.

—— and van der Ploeg, J. D. (1995) 'Reflections on Agency, ordering the Future and Planning', in G. E. Frerks and J. H. B. den Ouden (eds) *In Search of the Middle Ground. Essays on the Sociology of Planned Development*, Wageningen: Wageningen Agricultural University.

Norman, W. T. (1986) 'Reflections on cognition and parallel distributed processing', in J. L. McClelland, D. E. Rumelhart and the PDP Research Group (eds) *Parallel Distributed Processing*, vol. II, Cambridge, Mass.: MIT Press.

PART II

RESTRUCTURING, INDUSTRY AND REGIONAL DYNAMICS

5

RESTRUCTURING NATIONAL AGRICULTURE, AGRO-FOOD TRADE, AND AGRARIAN LIVELIHOODS IN THE CARIBBEAN

Laura Raynolds

INTRODUCTION

Agriculture is being restructured around the world: the types of commodities produced, the way in which production is organised, and the livelihoods which it engenders are all changing. This transformation is often attributed to a process of globalisation. Yet while we may usefully speak of temporally and spatially unified tides which are global in their influence, this does not mean that all sectors, all places, and all peoples are riding the same wave of change.

One of the central challenges in analysing restructuring is to recognise both the unity and the diversity in real world experience. This chapter addresses this challenge by exploring the particular character of recent agrarian restructuring in one Caribbean nation, the Dominican Republic, illuminating the specific, as well as the general, nature of ongoing changes. I consider both the reorganisation of the agro-food system and the transformation of agrarian livelihoods. Restructuring is revealed to be a highly political process contested by a range of social actors where the negotiated outcomes are far from certain. In the major arenas of agrarian transformation, domestic and international forces intersect, together configuring the parameters of change.

The chapter begins by locating my approach to the analysis of agrarian restructuring in two divergent research traditions. I then investigate ongoing changes in the Dominican Republic, examining the complex negotiations which define the central dynamics of agrarian restructuring. The Dominican case is used to help unravel the multiple, and potentially contradictory, strands of agrarian transformation and point to cultural, political, and economic factors which condition concrete restructuring outcomes.

DIVERGENT LITERATURES, COMPLEMENTARY STRENGTHS

My analysis of current agrarian restructuring builds on two divergent, but I argue, complementary, research traditions. The first, what can be called the agrarian political economy approach, emphasises diversity and focuses on the dynamic complexities of rural situations around the world. This approach has a long and illustrative history dating back to the debates between Lenin ([1899] 1960) and Chayanov ([1925] 1966) regarding the nature of Russian peasant culture and production processes, the paths of rural transition, and the character of rural class politics. Over the years, these debates have been reformulated in studies of Europe and North America and, perhaps most fruitfully, of Latin America, Africa, and Asia.[1] This research tradition suggests that an analysis of agrarian restructuring must recognise (1) how historically and regionally variable cultural ideologies and social institutions mediate economic processes; (2) how capitalist economy incorporates different labour processes, often manipulating divisions based on gender, race, and ethnicity; and (3) how individual and collective actors negotiate and resist their conditions of existence. Despite its theoretical and methodological insights and its rich cross-cultural empirical findings, the agrarian political economy approach can be myopic, failing to fully appreciate historical continuities and national and international structural forces.

A second more recent and divergent literature on industrial restructuring and regulation provides complementary insights for our analysis. This research focuses largely on the United States and Europe and investigates how, at the national level, specific social and institutional ensembles may temporarily uphold particular regimes of accumulation and how, at the firm level, common technical and economic organisational changes are institutionalised.[2] Major contributions of this multi-level approach include: the recognition of temporal and spatial regularities, the importance of political institutions, particularly the state, in regulating economic activity, and the way in which firm organisation reflects shifting competitive conditions. When exaggerated these strengths become this literature's central weaknesses. Many studies of industrial restructuring and regulation suffer from a US/Eurocentric bias which glosses over real-world inequalities and variations; a reification of the nation state which down plays local and global forces; and a proclivity for elaborating organisational and national models which are static and all encompassing.

Despite their divergent substantive foci, the industrial restructuring/regulation and the agrarian political economy approaches are, at heart, theoretically and methodologically compatible. At their best, both approaches eschew overly essentialist and deterministic models of development and pursue a dialectical analysis of the complex social forces and institutions shaping concrete economic processes and the multiple, recursive, trajectories of change (Goodman and Redclift 1982, Jessop 1990). Building on this common base, the differences in the two literatures can be seen as complementary strengths which help overcome their respective weaknesses. The agrarian political economy approach emphasises cross-cultural analysis, focuses on local level institutions, highlights diversity across cases, and prioritises cultural influences on economic life; the industrial restructuring/regulation approach emphasises multi-level analysis, focuses on national level institutions, highlights regularities across cases, and prioritises political influences on economic life.

A theoretical synthesis of the agrarian political economy and industrial restructuring/regulation approaches provides a promising starting point for analysing current restructuring dynamics. Such a synthesis can help elucidate the striking unity and diversity of ongoing transformations in either agriculture or industry, since it can be sensitive to sectoral differences as well as commonalities. As elaborated below, this type of analysis is perhaps best pursued via grounded studies which explore restructuring in particular institutional settings as the outcome of negotiations among numerous actors at multiple levels, from local to global.

POLITICAL RESTRUCTURING AND NATIONAL AGRICULTURE

As suggested by the agrarian political economy tradition, the investigation of restructuring must be historically and regionally situated, since current dynamics are overlaid on specific histories shaping local cultures and institutions (Roseberry 1989). Analysis of ongoing transformations in the Dominican Republic must appreciate its unique social make-up while remaining attuned to the influence of forces which extend far beyond the Caribbean. For as demonstrated by studies of industrial restructuring and regulation, economic activity is configured by local and global pressures, mediated in large measure by the nation state (Tickell and Peck 1995). This study illuminates how domestic and international political, as well as economic, forces intersect in transforming national agriculture, and the critical role played by the nation state in regulating ongoing agrarian restructuring.

In the Dominican Republic, European colonialism simultaneously configured the nation state and national agriculture (Mintz 1986). Colonial sugar production integrated the region into the international division of labour, establishing its export-oriented economy, externally linked state, and transnational labour force. At the same time, colonialism curtailed domestic agriculture and initiated a reliance on food imports.

In the post-Second World War era, this prioritisation of agro-exports over domestic food production was reinforced by converging foreign and domestic forces. Internationally, world markets were flooded with surplus US wheat advanced as inexpensive 'food aid' to allies throughout Latin

America (Garst and Barry 1990). The Dominican state, like many of its neighbours, took advantage of cheap imported grains to underwrite national industrial development (de Janvry 1981). Local rice production in the agrarian reform sector was simultaneously encouraged via foreign sponsored high yield technologies and domestically popular consumer price subsidies (Sang Ben 1988). These cheap food policies increased grain consumption in place of traditional tubers – a dietary shift heralded as a sign of modernisation despite its devastating impact on peasant producers.

Over the past twenty years, the Dominican state has abandoned its cheap food strategy, reallocating resources between agricultural and industrial sectors and between rural and urban populations. Declining US surpluses and the cessation of food aid have greatly increased the cost of wheat imports, while rising agricultural input prices have increased the cost of domestic rice production. Faced with a mounting fiscal crisis, the Dominican state first cut producer subsidies, marginalising rice farmers and agrarian reform recipients along with peasants (Sanchez Roa 1989). Championing urban industrial interests, the state maintained debt financed consumer subsidies until the mid-1980s. But these subsidies were cut too when the Dominican Republic was forced to reschedule its debt and enact International Monetary Fund (IMF) austerity and adjustment measures (Raynolds 1994a). Rising food prices sparked some of the most violent 'IMF riots' in the world, reflecting the redefinition of Dominican popular politics as urban, consumer based, and fundamentally global (Lozano 1992, Walton and Seddon 1994). With its legitimacy in tatters, the Dominican state has over the past decade periodically defied the IMF by distributing cheap food to garner votes and ease resistance to socially devastating structural adjustments.

Long privileged over domestic production, traditional Dominican agro-exports have recently also come under heavy attack, with sugar, the country's major export earner, being particularly hard hit. On the national front, the recent crisis has ravaged the state dominated sugar sector. Austerity measures and international privatisation pressures have forced the Dominican state to close sugar mills and rent out sugarcane land (Raynolds 1994a). On the international front, world sugar prices have fallen, while increased US protectionism has severely restricted access to the Dominican Republic's major market. Between 1980 and 1993, earnings from traditional Dominican agro-exports – including sugar, coffee, cocoa, and tobacco – plummeted from US$468 million to US$253 million (CEDOPEX 1981–94).

Despite prevailing free market rhetoric, it is the Dominican state which has been charged with re-establishing the country's competitiveness in world markets. Seeking to reconcile international pressures to increase export earnings for debt repayment and national pressures to bolster local accumulation, the Dominican state defined a development strategy centred on non-traditional agro-exports, free trade zone manufacturing, and tourism (Investment Promotion Council 1987). With bilateral and multilateral financing, non-traditional agricultural producers in the Dominican Republic were granted large tax breaks, cheap credit, and low cost access to state land being retired from sugarcane (UEPA 1990). These subsidies fuelled a boom in new agro-exports, with foreign exchange earnings doubling between 1979 and 1986 (CEDOPEX 1981–94).

Over the past decade, Dominican non-traditional agro-export growth has stagnated due to marketing problems and diminishing subsidies. New agro-exports have encountered volatile world prices, stiff competition from neighbouring countries, and strict tariff and non-tariff barriers to entry into major US markets (Conroy *et al.* 1996). Meanwhile, previously generous US economic assistance for Dominican non-traditional agriculture has evaporated and the IMF has, in its most recent debt negotiations, stipulated an end to export subsidies (Ceara Hatton 1993, USDA 1994). Responding to international pressure and national budget shortfalls, the Dominican state has curtailed most subsidies to non-traditional agriculture. These cutbacks are being challenged by a domestically powerful agribusiness group – created and initially funded by the US Agency for International Development – which has recently forced the Dominican state to reintroduce some export subsidies, despite IMF opposition (JAD 1994).

Dominican agro-export opportunities are being shaped by the tension between global market dynamics and regional trading agreements as well as by domestic politics. Internationally sponsored neoliberal reforms presume the rise of a global free market envisioned by the General Agreement on Tariff and Trade (GATT) and the World Trade Organisation (WTO). But globalisation has proved highly asymmetrical: inequalities in the world market have been reinforced on a global scale and on a triadic regional scale as the United States, the European Union, and Japan each develop strategic peripheral trading spheres (Tussie and Glover 1993, Berkerman and Sirlin 1995).

The first wave of new Dominican agro-exports maintained the country's traditional focus on US markets. Yet due to the declining geopolitical importance of the Caribbean in the post-cold war world, US openness to Dominican trade has eroded (Deere and Melendez 1992). Threatened by further negative trade and investment impacts from tightened US/Mexican links under the North American Free Trade Agreement (NAFTA), the Dominican state and private business associations have collaborated in pursuing entrance into Caribbean and Central American trading blocs aimed, in large part, at presenting a united front in confronting NAFTA (Lewis 1995).

The current wave of agro-export growth depends largely on newly acquired preferential access to European markets, negotiated by the Dominican state with its 1991 entrance into the Lome Convention (Exportador

Dominicano 1993). This trade is threatened though by US charges that European trade rules violate GATT free market provisions. Lome's preferential trading system is likely to be curtailed when the agreement is redrafted in 2002 and some commodity provisions may be modified earlier (Chambron 1995). The foundation of current Dominican agro-export growth may thus be undermined by competing US and European trading interests.

While the outcome of the conflict between global free market initiatives and regional trade alliances is unclear, it is clear that the emerging trade regime upholds existing world political and economic inequalities. The Dominican state has struggled to mediate domestic and foreign interests and promote the competitiveness of national agriculture within conflicting global and regional trade arenas. Yet countries like the Dominican Republic remain largely peripheral to the regulatory negotiations which shape the profitability of their export portfolios, the health of their agrarian economies, and their ability to feed their populations.

PRODUCTION RESTRUCTURING AND AGRO-FOOD TRADE

Moving beyond the political regulation of agro-food trade to further specify the dynamics of agrarian restructuring requires shifting our attention to the concrete activities of firms. As demonstrated by studies of industrial restructuring and regulation, firms are the key sites of accumulation and thus disclose in their organisational structure the changing requirements of capitalist competition (Sayer and Walker 1992). Yet as suggested by the agrarian political economy tradition, firms are also social actors which shape and are shaped by cultural, political, and economic forces (Trouillot 1988). This study finds that as the institutional loci of agro-food production, firms both embody changing economic conditions and seek to navigate profitably, and potentially alter, shifting agro-food relations.[3]

In the Dominican Republic, the critical and variable role of firms in agrarian restructuring is demonstrated in the fresh fruit and vegetable sector which is responsible for the largest increase in non-traditional agro-exports. Fruit and vegetable firms accommodate the changing national, regional, and global market conditions outlined above, but in different ways. Rather than following a single post-Fordist model, as some of the industrial restructuring literature might suggest,[4] firms pursue divergent corporate strategies and create different forms of production and marketing flexibility.

Turning first to the commodity area which led the initial boom in Dominican non-traditional agro-exports, pineapples, we find that growth has been spearheaded by two of the world's largest agro-food corporations, Dole Food Corporation and Chiquita Brands International. Both Dole and Chiquita established subsidiaries in the Dominican Republic to supply expanding markets and offset production shortfalls in their Central American plantations. With these new investments, Dominican fresh pineapple export revenues soared from US$3,000 in 1980 to US$7,000,000 in 1990 (CEDOPEX 1981–94).

Dole and Chiquita profited from the previously noted investment and tax incentives accorded to non-traditional producers during the 1980s. Most significantly, they have benefited from their inexpensive access to public land. Both corporations established plantations on state land being retired from sugar production: Chiquita leased 1,000 hectares at a concessionary price; while Dole acquired 5,000 hectares via a lopsided joint venture agreement (Raynolds 1994a). These corporations took advantage of the Dominican state's fiscal crisis and their own strong ties to the US Agency for International Development – which funded and directed the sugar diversification effort – to gain favourable access to public land, despite strong local protests over what was widely viewed as the 'foreign appropriation of national patrimony'.

While the enterprises established by Dole and Chiquita mirrored some elements of traditional plantations – in their large-scale production and sizable labour forces – these new plantations were significantly more flexible since they had no investments in land (Raynolds 1994b). By avoiding land purchases, corporations greatly reduce their fixed costs and increase their geographic mobility. Managers of new plantations in the Dominican Republic point to local antagonism towards foreign land ownership and the possibility of land expropriations to explain their preference for leasing land in a situation in which, as one manager suggested, we 'must rethink our local investments with each surge in nationalist sentiment, with each presidential election, and with each peasant protest'. Dole and Chiquita successfully utilised the threat of a corporate pullout to extract government support and increasingly scarce public subsidies in the 1990s. This proved no idle threat, for Chiquita recently abandoned its Dominican pineapple operation due to limited growth in this product market and the corporation's renewed commitment to its major commodity area, bananas. Dole has been much more successful at expanding pineapple sales, primarily by taking advantage of the Dominican Republic's preferential European market access under the Lome Agreement. For Dole, the world's largest pineapple distributor, access to the expanding EU market is critical (Dole 1994). But if Lome unravels, Dominican production will become unprofitable since, according to local managers, costs are higher here than on Dole's major Central American plantations.

If we turn to the Dominican Republic's newest and most rapidly expanding agro-export, bananas, we find that investment growth is almost entirely due to the opening of preferential European markets. With its recent acceptance into Lome, the Dominican Republic gained access to the European Union's highly protected market for bananas from African, Caribbean, and Pacific (ACP) ex-colonies. Two corporations control the new Dominican banana trade: Fyffes Ltd, a large British corporation which has long been one of the major distributors of ACP bananas, and Chiquita Brands, the world's largest distributor, which trades mostly in Latin American 'dollar bananas' for the US market. Destined almost entirely for Europe, Dominican annual banana export earnings have leaped from US$200,000 to US$7,000,000 since 1989 (CEDOPEX 1981–94).

Neither Fyffes nor Chiquita have invested in major banana plantations in the Dominican Republic, but instead operate largely through contracting systems which embody many post-Fordist flexible production features. These corporations limit their fixed investments and exposure to agricultural risks by contracting with numerous growers who provide the land, engage in cultivation, and deliver the produce to the exporting firm. In the words of one manager, contracting gives their company a way to 'assure quality supplies of a product not readily available on the open market, without the economic *or* political liability of getting into direct production'. In the Dominican context, most production contracting takes place with agrarian reform recipients and other small-scale producers who lack secure access to agricultural inputs or markets (Raynolds 1994b).

The use of contracts has facilitated the rapid growth of banana exports, greatly reducing the time and capital corporations have had to invest to take advantage of newly opened EU markets. Fyffes and Chiquita probably would not have moved into the Dominican Republic if it had meant establishing production facilities, given the uncertainties in Europe's banana market (Chambron 1995). Chiquita perhaps best demonstrates the short-term opportunistic nature of this move. Even though it might appear to have the most to gain from its newly increased access to the profitable ACP banana trade, Chiquita has virulently attacked the EU market system in an effort to expand its competing Latin American dollar banana trade. Claiming that the preferential ACP banana trading system violates free market agreements, Chiquita has sought action directly against the EU in US courts and has pushed the US trade representative to file a WTO grievance against the EU. If, as is likely, the EU banana trade framework falls apart, exporting companies will probably retreat from the Dominican Republic, leaving local banana contract growers to bear the costs of lost markets and declining prices.

The corporations responsible for the recent expansion of non-traditional agro-exports in the Dominican Republic have responded to changing competitive forces by developing flexible production systems and marketing strategies. Major corporations have established new plantation and contract production systems which retain profits, while shedding costs and risks, and market systems based on the attainment of dominant positions in competing global and regional markets. In so doing, seemingly transnational corporations use their ties to national governments to shape both local production conditions and international trade regulations advantageously. While these strategies bolster the profitability of select corporations, they do little to enhance the long-term competitiveness of the Dominican Republic as a producing region.

LABOUR FORCE AND AGRARIAN RESTRUCTURING

Agrarian restructuring involves more than shifts in agro-food trade and production patterns: it entails fundamental transformations in agrarian livelihoods. The industrial restructuring and regulation literature highlights ongoing changes in the labour process, where one of the central ways in which competitive flexibility is achieved is through the casualisation of work and a growing reliance on the most vulnerable workers.[5] My research in the Dominican Republic suggests that while such a pattern of agricultural labour force restructuring is taking place, it has multiple, often contradictory, strands.[6] As revealed by agrarian political economy studies, labour forces are socially constructed in a recursive process involving the manipulation of differences based on gender and race, as well as economic circumstances (Mackintosh 1989, Purcell 1993).

In the Dominican Republic, the production of agro-exports, particularly sugar, has long relied on migrant labour from neighbouring Haiti. These migrants face severe discrimination due to their race, language, and lack of citizenship, and are the most disadvantaged members of the local labour force. In the past, the Dominican state has sponsored seasonal Haitian migration to provide cheap harvest labour for state and private sugar plantations. Yet recent state sugar diversification efforts are designed to control and even reverse Haitian migration, thus 'Dominicanising' the rural labour force (Murphy 1989).

Since it is illegal for the new pineapple plantations to hire Haitians, they have fostered other vulnerable segments of the Dominican labour force. When the largest new pineapple venture was established in the mid-1980s, 55 per cent of workers were women. Dominican women have not traditionally been well represented in agricultural wage labour, but during the economic crisis of the 1980s women broadened their employment search, taking almost any job in an effort to stem falling living standards (Duarte *et al.* 1989). New enterprises have taken advantage of the expanded availability of female workers, since Dominican women are typically willing to work for less than men, given their limited access to high-wage jobs. Even though Dominican minimum wages were already the lowest in the region (Bobbin Consulting Group 1988), workers on the largest new pineapple plantation were paid US$2.87 per day in 1990, only two-thirds of the legal minimum. Plantation management circumvented local wage legislation by registering workers as 'casual labourers' rather than 'full-time employees', even though they worked year round. In an effort to woo further investments, Dominican state officials overlooked this flagrant violation of the law and even provided unsolicited assistance to corporate managers in identifying and blacklisting union activists seeking to increase plantation wages.

Pineapple plantation managers are currently pursuing a new strategy for increasing their flexibility and curtailing wage costs. Hundreds of existing workers have been fired and a labour contracting system is being established. Plantation managers now hire local labour contractors to perform a particular task and they in turn hire the necessary workers. This system allows the corporation to alter the size of the work force to suit seasonal and business demands and to escape legal responsibility for work conditions. No longer covered by labour legislation, most plantation workers toil twelve-hour days and are paid low piece rates upon the completion of a specific job. These jobs may last anywhere from a couple of days to a couple of weeks. Then, according to workers, there is no telling whether they will be out of work or rehired for another short-term job.

The move to labour contracting has entailed a masculinisation of the Dominican pineapple plantation work force. Labour contractors are male and rely on male dominated community networks to procure semi-proletarian workers and ensure their loyalty despite poor working conditions. Local women have individually and collectively resisted their loss of plantation jobs – complaining to labour contractors, community leaders, and company managers and engaging in walkouts and demonstrations. Despite their protests, women have been largely unable to counter converging patriarchal community and corporate interests. Labour contractors and male workers express a clear preference for working with all-male crews and actively seek to exclude women. Corporate managers blame 'Dominican machismo' rather than company hiring practices for expelling women from plantation work, but this interpretation ignores managers' complicity in the recomposition of the work force via their encouragement of men to become labour contractors and their begrudging attitude in working with the only female labour contractor.

Turning to the rapidly expanding banana export sector, we find similar dynamics of work casualisation and worker insecurity embodied in a different labour process. As previously noted, the major corporate exporters of Dominican bananas organise production via contracts with small-scale growers. Exporting firms control production via legal contracts specifying required input usage, production procedures, and product characteristics. The short-term nature of these contracts represents a point of flexibility for exporters, but a locus of insecurity for banana growers who have no guarantees regarding the conditions, or even availability, of future contracts. The extent of control maintained by these production contracts should not be underestimated. As one company manager put it, 'our growers are certainly not independent farmers . . . we organise production on their land, they do a little work, and in the end we get the produce'. Though they often resent it, growers typically acknowledge the tight control inherent in production contracts and the company's proprietary rights over the harvest. As one grower noted, pointing to his field, 'that is not my crop, it belongs to the company'.

Production contracting permits the banana companies to avoid the costs and responsibilities of hiring and managing workers. Wage labourers are used in banana production, but these workers are hired by the contract growers. Most banana workers are day labourers with no job security and wages well below the national minimum. Female day labourers, who are typically paid less than male workers, are hired to work in the packing sheds. Since banana packing tends to take place only a few days a week, women are employed only temporarily and are not covered under Dominican labour legislation. To cut costs in the cultivation of bananas, growers appear to be hiring an increasing number of Haitian men displaced from the sugar sector. Since these workers are employed illegally and could be deported at any time, they work as long a day as required and accept whatever wages are offered, even if they are a fraction of the going wage for Dominicans. As in the labour contracting system in pineapples, production contracting in bananas distances corporations from illegal or unscrupulous labour practices. Company managers, for example, note the use of migrant labour in banana production, but are quick to point out that it is the contract grower, not the company, who is hiring illegal workers and paying 'slave wages'.

A distinctive feature of production contracting is its ability to tap unpaid household labour. Banana contracts in the Dominican Republic are signed almost exclusively with men, yet their fulfilment often requires the work of wives and children. To limit production costs, poorer growers typically engage wives and daughters in the packing sheds and sons in the banana fields. Contracting firms acknowledge the importance of household labour in successful small-scale banana production and may refuse to contract with single men. While the use of family labour significantly cuts production costs, it often raises tensions within the contract grower household. Older children and spouses often echo the views of one wife who reports, 'I would rather sell my labour and earn my own wage rather than work on the parcel for the profit of the company, under conditions set by company bosses'.

In sum, firms in the most dynamic agro-export sectors in the Dominican Republic are increasing their competitive flexibility by establishing multiple informal labour processes which tap various pools of politically and economically vulnerable workers. These new labour processes are largely unregulated by the state both because of government collusion in disenfranchising workers, and contractual arrangements which confound employment relations. Non-traditional agro-exporters profit from the increasingly marginal position of peasant producers unable to make a living without supplemental wages or corporate secured inputs and markets. These enterprises take advantage of female subordination in the economy and the household to employ women in low-wage jobs and access their unpaid family labour. Some expanding agricultural areas are also exploiting illegal Haitian migrants displaced from the crumbling sugar sector. New plantations and contract production systems are thus fuelling competing labour force dynamics, pitting segments of the rural population against each other and further limiting workers' ability to protect their agrarian livelihoods from state and corporate driven processes of restructuring.

CONCLUSIONS

This study has attempted to illuminate the multifaceted and dynamic nature of recent agrarian restructuring using an analytical framework derived from a theoretical synthesis of the agrarian political economy and industrial restructuring/regulation approaches. I suggest that such a framework can orient investigations of restructuring which recognise the complex social forces shaping economic processes and the multiple trajectories of change. My analysis has focused on the factors conditioning concrete restructuring processes in the Dominican Republic in order to highlight the specific, as well as the general, nature of ongoing changes.

Restructuring is found to be a highly political process which involves a range of social actors negotiating changes in interlocking arenas which span the local–global divide. In this context, the nation state plays a critical role in reconciling foreign and domestic political pressures and promoting the competitiveness of national agriculture under competing global and triadic regional trade regimes. Firms respond to shifting competitive tensions via varied patterns of production reorganisation and efforts to manipulate rival agro-food markets politically and economically. Ultimately it is rural populations which are most deeply affected by restructuring processes which are simultaneously undermining existing agrarian livelihoods and reinforcing rural job insecurity and the political and economic marginality of rural workers.

ACKNOWLEDGEMENTS

I would like to thank Lourdes Gouveia and the editors for their helpful comments on this chapter. Support for this research was provided by the Office of Equal Opportunity, Colorado State University.

NOTES

1 For a selection of contributions to this extensive literature see *The Journal of Peasant Studies*; for excellent reviews, see Goodman and Redclift (1982) and Roseberry (1993).

2 For some of the initial statements in this broad tradition, see Aglietta (1979), Piore and Sabel (1984), and Lipietz (1987). For insightful reviews, see Jessop (1990) and Kotz *et al.* (1994).

3 This discussion is based on recent interviews with the managers of forty-four non-traditional agricultural firms in the Dominican Republic as well as on corporate and government documents.
4 Piore and Sabel (1984) argue that 'post-Fordist' production is based on flexible specialisation – small-firm production oriented toward niche markets – in contrast to 'Fordist' production which is based on mass production for mass markets.
5 While Piore and Sabel (1984) focus on the positive implications of flexible specialisation for workers, numerous studies recognise that for the majority of workers flexibility means job insecurity and low wages (e.g. Standing 1989).
6 This discussion is based on a review of government and corporate documents and recent interviews in the Dominican Republic with 216 workers in non-traditional agriculture.

BIBLIOGRAPHY

Aglietta, M. (1979) *A Theory of Capitalist Regulation: The US Experience*, London: New Left Books.

Berkerman, M. and Sirlin, P. (1995) 'Trade policy and international linkages: a Latin American perspective', *CEPAL Review*, 55 (April): 65–79.

Bobbin Consulting Group (1988) *Sourcing: Caribbean Option*, Columbia, SC: Bobbin Consulting Group.

Ceara Hatton, M. (1993) 'De reactivación desordenada hacia el ajuste con liberalización y apertura', paper presented at Research Workshop on the Dominican Republic, City University of New York, New York (23 April).

CEDOPEX (Centro Dominicano de promoción de exportaciones) (1981–94) Boletin Estadísticos (various editions), Santo Domingo: CEDOPEX.

Chambron, A. (1995) 'The European banana market', *Trade Briefing* (December): 2–5.

Chayanov, A. ([1925] 1966) *On the Theory of Peasant Economy*, Homewood IL: American Economic Association.

Conroy, M., Murray, D., and Rosset, P. (1996) *A Cautionary Tale: Failed US Development Policy in Central America*, Boulder: Lynn Reinner.

de Janvry, A. (1981) *The Agrarian Question and Reformism in Latin America*, Baltimore: Johns Hopkins University Press.

Deere, C. and Melendez, E. (1992) 'When export growth isn't enough: US trade policy and Caribbean Basin economic recovery', *Caribbean Affairs* 5: 61–70.

Dole (Dole Food Corporation) (1994) *Annual Report*, Dole Food Corporation.

Duarte, I., Baez, C., Gomez, C., and Ariza, M. (1989) *Población y Condición de la Mujer en República Dominicana*, Santo Domingo: Instituto de Estudios de Población y Desarrollo.

Exportador Dominicano (1993) 'Exportaciones nacionales se consolidan en mercado Europeo', *El Exportador Dominicano* 21, 103: 15.

Garst, R. and Barry, T. (1990) *Feeding the Crisis: US Food Aid and Farm Policy in Central America*, Lincoln: University of Nebraska Press.

Goodman, D. and Redclift, M. (1982) *From Peasant to Proletarian: Capitalist Development and Agrarian Transitions*, New York: St Martin's Press.

Investment Promotion Council of the Dominican Republic (1987) 'Investing in the Dominican Republic', Santo Domingo: Investment Promotion Council.

JAD (Junta Agroempresarial Dominicana) (1994) 'Balance del sector agropecuario en el 1993', *Agroempresa* (January–June): 16–19.

Jessop, B. (1990) 'Regulation theories in retrospect and prospect', *Economy and Society* 19, 2: 153–216.

Kotz, D., McDonough, T., and Reich, M. (eds) (1994) *Social Structures of Accumulation*, Cambridge: Cambridge University Press.

Lenin, V. ([1899] 1960) *The Development of Capitalism in Russia*, London: Lawrence and Wishart.

Lewis, D. (1995) 'The Latin Caribbean and regional cooperation: a survey of challenges and opportunities', *Journal of InterAmerican Studies and World Affairs* 37, 4: 25–55.

Lipietz, A. (1987) *Miracles and Mirages: The Crisis of Global Fordism*, trans. D. Macey, London: Verso Books.

Lozano, W. (1992) 'Trabajadores, poder, y populismo', *Revista Interamericana de Sociologia* 6, 2/3: 61–94.

Mackintosh, M. (1989) *Gender, Class and Rural Transition*, London: Zed Books Ltd.

Mintz, S. (1986) *Sweetness and Power: The Place of Sugar in Modern History*, Harmondsworth: Penguin.

Murphy, M. (1989) 'Respuestas Dominicanas frente a la crisis azucarera', Santo Domingo: Fundación Friedrich Ebert.

Piore, M. and Sabel, C. (1984) *The Second Industrial Divide*, New York: Basic Books.

Purcell, T. (1993) *Banana Fallout: Class, Color, and Culture Among West Indians in Costa Rica*, Los Angeles: University of California Center for Afro-American Studies.

Raynolds, L. T. (1994a) 'The restructuring of Third World agro-exports: changing production relations in the Dominican Republic', in P. McMichael (ed.) *The Global Restructuring of Agro-food Systems*, Ithaca, NY: Cornell University Press.

—— (1994b) 'Institutionalizing flexibility: a comparative analysis of Fordist and post-Fordist models of Third World agro-export production', in G. Gereffi and M. Korzeniewicz (eds) *Commodity Chains and Global Capitalism*, Westport, CT: Praeger.

Roseberry, W. (1989) *Anthropologies and Histories: Essays in Culture, History, and Political Economy*, New Brunswick: Rutgers University Press.

—— (1993) 'Beyond the agrarian question in Latin America', in F. Cooper, A. Isaacman, F. Mallon, W. Roseberry, and S. Stern (eds) *Confronting Historical Paradigms*, Madison: University of Wisconsin Press.

Sanchez Roa, A. (1989) *Campesinos Crisis Agropecuaria e Inflación*, Santo Domingo: Editora Corripio.

Sang Ben, M. (1988) *La Crisis de Alimentos en República Dominicana*, Santo Domingo: Universidad Iberoamericana.

Sayer, A. and Walker, R. (1992) *The New Social Economy*, Cambridge MA: Basil Blackwell.

Standing, G. (1989) 'Global feminisation through flexible labour', *World Development* 17, 7: 1077–95.

Tickell, A. and Peck, J. (1995) 'Social regulation after Fordism: Regulation theory, neo-liberalism and the global-local nexus', *Economy and Society* 24, 3: 357–86.

Trouillot, M. (1988) *Peasants and Capital: Dominica in the World Economy*, Baltimore: Johns Hopkins University Press.

Tussie, D. and Glover, D. (1993) *The Developing Countries in World Trade*, Boulder: Lynn Rienner.

UEPA (Unidad de Estudios de Política Agropecuaria) (1990) *Los Subsidios en la República Dominicana*, Santo Domingo: Editorial Gente.

USDA (US Department of Agriculture) (1994) 'Agricultural Situation Report', Santo Domingo: American Embassy, USDA Office.

Walton, J. and Seddon, D. (1994) *Free Markets and Food Riots*, New York: Basil Blackwell.

6

RESTRUCTURING PORK PRODUCTION, REMAKING RURAL IOWA

Brian Page

INTRODUCTION

Our popular conceptions of the rural Midwest are of pastoral, quiescent landscapes. Here, among undulating corn fields, distant silos, and picturesque farmsteads, we imagine hard-working families that till the earth and tend to animals. Though the actual practice of farming is strange and distant to nearly every American, we nevertheless imbue this livelihood with distinctive qualities, particularly independence, stability, and worthy simplicity – qualities sorely lacking in most urban lives. In this way, farming (and rural life more generally) has come to represent a measure of collective assurance that at least some places remain untouched by the relentlessly transformative powers of modern capitalism.

A close look at rural Iowa, however, reveals the fiction of the urban imagination. The contemporary Iowa countryside is anything but tranquil; it is filled with dramatic, even epochal, social and economic changes. Of particular importance in Iowa are shifts in hog production. Agricultural historians have observed that the practice of marketing corn 'on the hoof' was the cornerstone of economic development in Iowa during the nineteenth century (Ross 1951, Gates 1960, Walsh 1982). Through most of this century, hogs were produced on small to mid-sized farms and continued to be a central factor in the survival of family farm enterprises. But today, Iowa hog production is changing. Increasingly, hogs are produced under contract by farmers who own neither the hogs nor the necessary inputs to produce them. Instead, the hogs are owned by a contracting firm which supplies young pigs and feed to a network of farmers (called 'growers') who fatten the hogs to a specified weight under direct supervision. In exchange the farmer arranges financing, supplies the buildings in which the hogs are fattened, and is paid a fee. This new model is a striking break from the past, when independent family farms raised hogs as one component of diversified farming operations, selling their livestock in open markets. Now, contractual arrangements are steadily replacing market exchanges as nominally independent growers are brought under the control of agro-industrial firms that orchestrate relationships within the entire commodity system – from livestock production through slaughtering, processing, and marketing.

These social and organisational changes in pork production have been accompanied by shifts in technological practice and the geography of production. Indeed, nearly every aspect of Iowa livestock and meat production is in tumult. If this situation does not fit the images projected by popular culture, neither does it fit the analytic models of traditional rural studies. For years, farming was perceived as a residual social and economic activity operating under dictates distinct from those governing industrial capitalism (cf. Newby 1982).[1] The lasting dualism separating the study of agriculture from the study of manufacturing has been broken only recently by research within geography, sociology, and allied fields. This new work – falling loosely under the heading of 'the political economy of agriculture' – views farming not as a pre-capitalist vestige, but as a distinct branch of industry guided by the overarching principles of capitalist production. The initial inspiration for this field came from the writings on agriculture by the classical political economists Marx, Lenin, and Kautsky. Not surprisingly, its initial concern was the classical 'agrarian question' applied to the modern family farm, an effort that focused on the transitional or persistent nature of non-capitalist social relations in US farming. For years, debates on this agrarian mode of production dominated the literature, but over time a chorus of voices acknowledged key limitations to this thinking. In particular, critics decried over-attention to social forms of production on the farm and a corresponding failure to conceptualise adequately ways in which agricultural production is bound up with a wider process of economic development and capital accumulation (e.g. Buttel 1982, Goodman and Redclift 1985, Marsden *et al.* 1986).

As this critique gained ground, researchers focused their attention on the forces shaping agriculture from beyond the farm – resulting in a re-formulation of the field's theoretical agenda.[2] Attempts to rethink the relationship between farming and manufacturing and thereby to re-position agriculture in an industrial context have led to an engagement with the growing social sciences literature on the transformation of contemporary capitalism. As part of this effort many have turned to the industrial restructuring literature. In particular, the application of regulation theory (Aglietta 1979, Lipietz 1987, Boyer 1990) to the study of agriculture is increasingly popular, leading to the development of what can be termed 'Fordist agriculture' approaches.[3] Such research makes a decisive break from traditionally insulated discourses and opens up new approaches to the study of rural economy and society, but it does so seemingly without awareness of the widespread critique of Fordist/post-Fordist frameworks in the analysis of industrial change.[4] In fact, the uncritical extension of regulation theory into the agricultural realm represents less a synthesis of the agricultural and industrial literatures and more a subsumption of the former into the latter. Ironically, in the haste to cross the divide that separates farming from manufacturing, the distinctive technological, organisational and institutional character of agriculture is all but obliterated (cf. Goodman and Watts 1994, Page and Walker 1994, Page 1996).[5]

This situation raises (at least) two sets of questions for political economic analyses of agricultural development in the US. The first concerns our sources of theory. If the regulation approach obscures more than it reveals about capitalist development in agriculture, where else should we turn for insight? What other literatures and what other approaches hold the potential to shed light on this process in ways that acknowledge and build upon agriculture's evident specificity? The second concerns the direction of theory. Should agricultural studies simply be an importer of theoretical ideas generated in other fields of inquiry? The Fordist agriculture approach follows this unidirectional logic, as its main concern is with what agriculture analysts can learn from theories of industrial development. But is the inverse also true? Can industrial analysts also learn from the careful study of agriculture and can theories of industrial restructuring be enriched by such a dialogue?

This chapter addresses these questions through an examination of the restructuring of midwestern pork production. Three chief arguments are made. First, research on agriculture should continue to look to research on industrial restructuring. The fact that regulation-inspired approaches are seriously flawed does not mean that the engagement with the industrial literature should be abandoned altogether. Rather, that engagement should be more selective. In particular, work within the new industrial geography that focuses on production dynamics, sectoral specificity, and spatial divisions of labour provides a very useful conceptual framework for grasping the trajectory of development in livestock and meat production. Second, the political economy of US agriculture should also look back from whence it came to research on agrarian transformations in the Third World. This is not to suggest that we return to an exclusive concern with the social relations of farming, but students of US agriculture were perhaps too eager to move beyond the agrarian question in the search for a better understanding of agriculture's industrial integument. In the case of pork production, another look at the agrarian studies literature provides valuable insight into the uneven geography of restructuring, highlighting the lasting importance of place-bound class relations, cultural practice, and political resistance. The third argument is that the study of agricultural industrialisation (in pork production or more generally) provides an opportunity to scrutinise industrial theory from a forgotten perspective. In fact, the study of agriculture has several important lessons to impart to industrial studies. The following sections take up each of these arguments in turn.

AGRO-INDUSTRIAL INTEGRATION IN PORK PRODUCTION

Regulation theory is unsuited to the study of agriculture because it takes all industries to be variations on a basic theme drawn from the specific histories and geographies of one or two sectors. Within industrial studies, this tendency to over-generalise exists uneasily alongside other approaches that pay close attention to the distinctive character of specific industries. Indeed, sectoral specificity was an early theme in the new industrial geography (Massey 1984, Storper and Walker 1984) and has been expanded upon by many others (e.g. Scott and Storper 1986, Florida and Kenney 1992, Storper and Walker 1989 and Angel 1994). This research demonstrates that the process of industrialisation – through which firms and industries expand capital investment and commodity output, improve production methods, implement new labour processes, multiply the division of labour, and compete vigorously – must be looked at carefully in order to uncover the workings behind geographical patterns of economic activity, and their expansion, contraction, and upheaval. It also reveals industrialisation to be highly differentiated; industries exhibit distinctive paths of development based upon critical differences in material base, technology and business organisation – as well as upon the special qualities of people and places that enable or constrain industrial growth.

This orientation resonates with a large body of research within the political economy of agriculture that focuses on the unique character of industrial development in farming. Agriculture shows the importance of sectoral studies in spades. The reason for its specificity is the landed basis of farm production, the natural qualities of farm products, and the social relations that follow close on the heels of natural difference. These circumstances make the capitalist imperative in agriculture quite distinctive. Unlike most branches of industry, farming presents constraints to industrialisation which act to limit the productivity of labour and restrict capital investment.[6] Because of these barriers to the social and technical rationalisation of the farm labour process, non-capitalist social relations of production dominated US agriculture during the nineteenth century and continue to persist today.

While it is important to affirm the viability of household producers versus capitalist producers on the farm, it is equally important to address the effects of the division of labour in allowing capitalist domination of the overall food and fibre producing system. The division of labour itself has become a central object of study among industrial geographers because of its role in the development of the forces of production and industrial organisation (Scott 1988, Sayer and Walker 1992). Here, again, there is a dovetailing with parts of the agricultural studies literature. According to Goodman *et al.* (1987), industrialisation in agriculture has advanced along two broad fronts, capturing agricultural production itself in a pincer-like grip between suppliers of farm inputs and processors and marketers of farm output. On the one hand, elements of the overall farm labour process have been gradually assimilated into factory-based industry where they have been rationalised, mechanised, and intensified beyond anything possible on the farm. This process is one in which elements of agricultural production are appropriated by manufacturers, transformed into discrete branches of industry, and then re-incorporated back into farm production as purchased inputs. On the other hand, commodity traders and manufacturers have continuously reduced farm products to more simple and controlled industrial inputs in an effort to substitute agricultural goods with industrially produced inputs.

In this way, capitalist production has encroached on agriculture from above and below within the division of labour, the extent of industrial activities linked to agriculture has expanded greatly, and industrialists and merchants have increased their effective control (directly or indirectly) over on-farm labour.[7] So, despite barriers, industrialists behave just as they would in any other branch of industry, continually attempting to revolutionise productive methods in order to extract surplus value from the labour process. This process of 'agro-industrialisation' has moved forward slowly, but over time natural constraints have been eroded and redefined as technological and organisational innovations have been introduced. Owing to enormous differences within agriculture, the pace and form of this process varies tremendously from one crop to another. In essence, the production of a given animal or plant and the transformation of that biological product into a food or fibre commodity presents a unique set of constraints to capital. Here, the concept of a commodity chain is useful because it provides a framework for understanding the distinctive process of industrialisation surrounding particular farm commodities.[8]

If research in industrial geography underscores certain approaches within the political economy of agriculture, it also adds important elements to the analysis of agricultural development. In particular, by concentrating on the dynamics of the division of labour within wider production systems (Scott 1988, Sayer and Walker 1992), research in industrial geography points to the importance of broadening the scope of agricultural studies to encompass all stages of a given commodity chain. In order to understand the formation and restructuring of commodity chains as a whole, we need a thorough knowledge of the process of industrial transformation occurring within every stage, not just farming (Page 1996). This means more of a focus on the growth and development of each of the sectors of economic activity that surround farm production – from input provision to shipping, processing and retailing – as well as a grasp of the shifting character of the relationships linking one stage to the next.

Additionally, by highlighting the inherently geographical character of capitalist restructuring (Harvey 1982, Storper and Walker 1989), industrial geography points to the importance of a spatial perspective on agriculture, calling particular attention to the link between agro-industrial innovation and shifts in the location of production.[9] The study of agriculture demands a spatially sophisticated analysis precisely because of the sector's internal richness and diversity. Each commodity chain has its own trajectory of development and there is a geography to this – a spatial division of labour. This is not to imply any sense of immutability or natural determinism, however. Nature matters to where agriculture and surrounding activities are located but such locational constraints can be overcome via the transformation of nature and the creation of new agricultural systems through irrigation, plant and animal breeding, improved production facilities, new transport modes, etc. Indeed, commodity chains are never static; rather, they are constantly being expanded and re-configured in response to competition, market shifts and persistent social, technical and political change.[10]

How do these considerations inform an analysis of the restructuring of pork production? First, they suggest close attention to multiple paths of industrialisation within the commodity chain. On the farm, the transformation of hog raising has pivoted around a series of technological changes that challenge natural impediments to the industrialisation of livestock production (cf. Goodman *et al.* 1987). Since the 1950s, advances in animal science have enabled hog producers to concentrate larger groups of animals at a single site, thereby obviating the need for an extensive land base and allowing the systematic application of capital to production. Such investment has been targeted at the development of confinement practices which utilise a wide array of purchased inputs, including boar semen, pharmaceuticals like synthetic hormones and antibiotics, manufactured animal feed, specialised buildings, mechanical systems for heating, cooling, feed delivery, waste removal, and storage, and, of course, energy (Van Arsdall and Nelson 1984, Hayenga *et al.* 1985). Compared to past practices, confinement methods reduce total labour requirements, raise labour productivity and allow the near continuous application of labour to production.[11] So, too, confinement methods have allowed farmers to intensify greatly the process of pig reproduction and growth and thereby reduce the biological time required to bring a pig from birth to slaughter (Pond 1983).

The extension and intensification of commodity production in meat packing during the postwar era has also been circumscribed by an agro-industrial problematic, revolving around attempts to overcome limitations imposed by the fact that the industry's material input is a biological entity produced by land-based enterprises and that its chief output is perishable animal flesh (Page 1996). Specifically, the development of new pork products and the creation of new methods for packaging, freezing, and canning pork represent both an effort to solve problems of circulation associated with perishability and an effort to differentiate the form and composition of animal flesh in order to expand sales of branded, high value-added commodities. So, too, the introduction of boxed pork production increased the shelf life of fresh pork by several weeks and made distribution much more efficient.[12] All of this involved significant changes in technology and labour process. The machine-driven disassembly line was extended beyond the overhead rail via the introduction of waist-high conveyors; many processing activities (e.g. bacon and sausage manufacture) were completely mechanised; and human labour in slaughtering was augmented by mechanical stunners, skinners, knives, and saws. These innovations represent efforts to overcome constraints to mechanisation associated with the irregularity and tenacity of biological architecture. Nonetheless, meat packing remains a relatively labour intensive business due the persistently difficult character of the raw material, and for this reason meat packers remain preoccupied with attempts to intensify work and lower labour costs.

Meat retailers have a similar set of problems to solve. Through the 1950s, selling meat was labour intensive for two reasons: first, just as in the packing plants, the work of retail butchering was exacting and time-consuming due to the difficulty of separating an animal carcass; and second, the work of selling fresh meat involved individualised service, handling, and wrapping, due to the problem of perishability. Labour requirements and total labour costs were drastically reduced with the arrival of self-service meat retailing via a series of innovations that extended the shelf life of meat products, including refrigerated display cases, moisture-resistant plastic tray-containers, and temperature resistant transparent films. No longer tied to customer service, workers could spend their time cutting, grinding, weighing, wrapping, and labelling cuts of meat, tasks that were increasingly mechanised in the 1960s. The labour process of meat sales was further rationalised during the 1970s and 1980s with the adoption of boxed fresh meat. Boxed pork arrives from the packing plant in standardised, vacuum-sealed portions that can be stored for weeks and can quickly be transformed into retail cuts of meat. Grocery firms have also promoted the sale of high value-added, pre-packaged fresh and processed meat items as these products generate more profit and require only the labour of stocking.[13]

A second consideration suggested by the industrial geography literature is greater attention to an overarching process of capitalist development occurring within the commodity chain as a whole. Indeed, the key dynamic behind the restructuring of pork production today is the effort to transcend the traditional boundaries separating individual stages from one another and thereby create a single, better coordinated industry. This process of agro-industrial integration has two key aspects: organisational and technological.

Organisational integration is taking two main forms. In the Midwest, changes in the meat packing industry have played the lead role. On the basis of their success in beef packing, IBP Inc., ConAgra, and Cargill each expanded into pork packing during the mid-1980s.[14] With the entry of ConAgra and Cargill, much of the pork packing industry is now housed within large and diversified enterprises that operate across a variety of commodity chains and around the world. In the US, these firms are involved in nearly every stage of pork production, including inputs to grain farming, grain transport and processing, livestock feed manufacturing, meat packing, and processed foods production. Neither firm has a significant involvement in hog production, however, and so livestock farming and meat packing continue to be linked mostly through open market transactions, although this is now accomplished through direct contact between buyer and seller rather than through the intermediary of a livestock marketing agency. Hog farming in the region continues to be dominated by mid-sized farm enterprises using family labour augmented by some hired labour (Van Arsdall and Nelson 1984), although a number of very large-scale 'corporate' hog farms have emerged. Overall, the diffusion of capital intensive techniques has led to dramatic reductions in the number of farms raising hogs and a concomitant increase in the average number of hogs produced per farm.[15]

In the South, organisational integration has taken a different route. Southern pork production has followed the model of the poultry industry. The lead role is played by 'integrating' firms which own and operate industrialised breeding and birthing facilities (farrowing) which produce large numbers of pigs in steady cycles throughout the year, using hired labour. Most of the larger southern contracting firms produce their own hog feed at company-owned and operated feed mills. Once the young pigs (feeder pigs) reach a specified weight, they are 'put out' to contract growers for fattening (finishing). Growers take delivery of the pigs and feed from the contracting firm and are paid a fee based on the ratio of weight gained to feed consumed. Veterinary and technical services are supplied by the integrating firm, while facility type and production practices are dictated by the firm's policies. Growers are responsible for financing construction, operating costs, and any loss of pigs to disease. Marketing of the finished hogs is conducted by contracting firms, who sell their hogs through forward contracting arrangements to meat packers. In this model, then, contractual relations replace market exchanges and farmers become the equivalent of wage labourers on their own land.

During the postwar era, processes of technological change in livestock production, meat packing, and meat distribution moved along different, albeit related, paths. Clearly, innovations within each of these activities reverberated upon one another, but there was rarely a coherent strategy tying these stages of production together. In the context of organisational integration, however, formerly separate paths of innovation have become intertwined and inseparable – giving rise to an increasingly unified technological trajectory from pig embryos to packaged cuts of meat.

Technological integration has increased as meat packers have turned to firm-specific measures of individual carcass quality as the basis of their direct purchases of livestock, rather than the USDA measures of grade and yield.[16] To do this, packers first used hand-held optical probes to measure the fat levels of individual carcasses. Improved evaluation technologies use both ultrasound and electromagnetic conductivity computer imaging to discern more accurately an animal's relative levels of fat, bone, and muscle, as well as its muscle tissue composition (Marbery 1993a, Lee 1994, *Science* 1994). Using this new method, meat packers are better able to identify the type of animal they want to buy – lean animals of uniform size and weight, preferably in large numbers. Livestock producers are then paid according to their ability to meet a given packing firm's quality and volume specifications.

Meat packers want uniformity in both the size and quality of the carcass because these characteristics make mechanised slaughtering and processing more time and cost efficient. Additionally, greater volume allows for more efficient input and labour scheduling. But the quality most desired by meat packers is leanness. Lean animals yield more saleable meat per pound of carcass. Perhaps more importantly, lean animals are the foundation of new pork marketing strategies focused on the sale of 'high quality' lean meat (meat with less of both inter-muscular and intra-muscular fat) to increasingly health-conscious consumers. This includes efforts to expand sales of both fresh products ('the other white meat') and processed products (a dizzying array of low-fat, pre-packaged, branded meat items).

By paying a premium for lean animals, meat packers have increased their influence on the process of technological change in livestock production through which animal physiology is being re-shaped. In the past, the main direction of livestock industrialisation was towards the development of methods that would reduce the time and cost involved in growing livestock to slaughter weight. With the advent of merit-based buying, however, this emphasis has shifted to methods that can actually alter carcass composition away from fat and towards muscle. Efforts to create leaner animals involve the use of mass-produced genetically engineered products like growth hormones (bovine and porcine somatotropins) and amino acids feed supplements, as well as ongoing attempts to standardise species genetics, livestock feeding techniques, and management practices (Pond 1983, Looker 1990, Govindasamy *et al.* 1993). In this way, meat packers push livestock producers to pursue new innovations and to increase the scale and uniformity of production – thus opening up new avenues for the appropriation of livestock agriculture by breeding firms, feed companies, equipment manufacturers, pharmaceutical firms, and the like.[17]

Technological integration is being taken to new heights within production contracting networks. Integrating firms control the genetics of all hogs within their grower network, they supply growers with standardised inputs to production, and they impose the use of standardised facilities and production practices. The end product is a lean hog of amazingly uniform size produced at a very low cost. In both the South and the Midwest, then, agro-industrial firms are pursuing strategies aimed at intensifying each stage of production while simultaneously achieving greater coordination within the overall system of production and circulation. Ultimately, these strategies are aimed at boosting both physical productivity and labour productivity while lowering the overall costs of production in order to compete more effectively with poultry in terms of both price and quality.

A third way in which research in industrial geography informs a study of pork production is by calling attention to the role of space in the process of agro-industrial restructuring. Recent organisational and technological changes have been accompanied by geographical shifts that bring meat packers and hog producers together in livestock production regions. Without doubt, the spatial integration of livestock production and meat packing derives in part from classic Weberian dynamics, but this locational structure goes well beyond the issue of transport costs. Indeed, spatial integration plays a critical role in successful strategies of agro-industrialisation. Two points make this clear.

First, the convergence of technological trajectories in livestock raising and meat packing – the driving force of change in today's meat economy – could only have occurred in the context of spatial proximity. In the Midwest, technological integration has been accomplished largely through the agency of direct livestock marketing, a system which is dependent on the development of localised personal relationships between the buying agents of meat packing firms and individual livestock producers. Packer buyers work hard to establish a working rapport with local producers in order to ensure adequate supply of quality livestock. To this end, buyers cultivate those producers most able to meet the firm's quality specification, and actively assist others in improving the quality of their livestock. In this sense, technological directives for livestock producers are dictated not only by the invisible hand of the market but also by the forceful grip of nearby packing firms. Southern production contracting, too, involves a high degree of personal contact (in this case supervision) between the grower and representatives of the integrating firm and between that firm and meat packers. Thus, while open markets and production contracting represent quite different avenues along which agro-industrial development has advanced, each is rooted in personal relationships that take place at a tight spatial compass.

Second, proximity to livestock supply has allowed meat packers to control better the flow of livestock into their plants, resulting in improved coordination between these two stages of production. In the Midwest, a certain base percentage of daily livestock requirements are forward contracted while the remainder are purchased just a day in advance of slaughter from within roughly 150 miles of the plant (Burgett 1990). In this way, meat packers have been able to improve the regularity of flow into the plant, increase capacity utilisation and avoid costs associated with livestock storage (weight loss, feed, water, bedding). Additionally, packers have gained more flexibility in terms of daily input scheduling, allowing them to adjust production according to the movement of the margin between livestock prices and wholesale meat prices. In order to take full advantage of such flexibility, however, packing firms need a labour force that allows them to conduct slaughtering operations when animals are available and when market conditions are most favourable. This control over work scheduling has been achieved by attacking the union position throughout the industry and by fiercely resisting the unionisation of rural plants. In this sense, spatial integration has been a key element in a broad strategy aimed not only at improving scheduling flexibility but also at keeping labour costs low and intensifying work in meat packing (Perry and Kegley 1989, Page 1997).

At the regional scale, the key geographic shift is the expansion of the southern contracting system into the Midwest. Indeed, production contracting is growing more rapidly in the Midwest than in any other part of the country (Rhodes 1990, Rhodes and Grimes 1992). Behind the expansion of contract production lies an undeniable competitive force: integrated producers surpass traditional midwestern producers in terms both of physical productivity (e.g. the number of pigs weaned per litter, the average age at which pigs are weaned, loss to disease, growth rates, etc.) and labour productivity, as well as the uniformity and quality of output (Kliebenstein 1988). While contracting networks have emerged in Iowa and Illinois, the southern model has taken root most firmly on the fringes of the traditional corn belt. Tyson (an agro-industrial giant with roots in integrated poultry production) recently established a pork packing plant in Missouri that draws on company-operated hog contracting networks in Missouri, Oklahoma, and Arkansas (Marbery 1993b). Seaboard, another integrated poultry producer, purchased a pork packing plant in Oklahoma and will slaughter hogs raised through its contracting system there.

American agro-industrial firms are also faced with competition from foreign counterparts. A company called IPC – a joint venture of Central Soya (a subsidiary of the Italian firm Ferruzzi) and Mitsubishi of Japan – has recently constructed a pork packing plant in Indiana that slaughters hogs raised through a company-owned contracting network. This plant also buys hogs from a nearby contracting network run by Continental Grain, one of the nation's largest cattle feeders. The IPC plant produces branded retail cuts of meat that are vacuum sealed using modified-atmosphere technology to further extend shelf life. These meat products are then distributed to markets in Japan and other Pacific Rim nations, as well as to markets in the US, Canada, Mexico, and Europe (Morris 1992). The degree to which such developments indicate a trend towards the globalisation of pork production is unclear, however. In terms of production–consumption linkages, the vast majority of US pork production continues to be consumed domestically, while imports are insignificant. Thus, while firms like ConAgra and Cargill have undeniable global interests, their involvement in pork production retains a distinctly national character with respect to every aspect of production save labour. In fact, in many midwestern pork packing plants a significant percentage of the workers are Mexican citizens working legally in the United States.[18]

Overall, these locational shifts reveal the geographical expansionism typically associated with new and dynamic forms of production (cf. Storper and Walker 1989). They also illustrate the way in which space is continually refashioned through the ongoing transformation of nature under capitalism. With this newest wave of revolutions in agro-industrial practice have come new spatial divisions of labour, characterised by the integration of livestock raising and meat packing in rural areas under the auspices of large agro-business firms: from the division of labour to rural form, if you like (cf. Scott 1988). Though tied into global networks of trade and finance, these new spaces of rural production are isolated pockets where low wage and hazardous industrial employment is proliferating and the social relations of farming are being re-made.

CONTESTED LANDSCAPES OF RESTRUCTURING

Recent research on agrarian change in Asia and Africa comprises a second body of research that sheds light on the multiple trajectories of capitalist development in US agriculture. This work derived from a concern with the widespread restructuring of Third World agriculture and rural society in the postwar era. It was in this context that classical debates over the role of the peasantry in capitalist development were first revised and elaborated before being extended to the study of agriculture in the developed world.

According to Watts (1996) the agrarian studies literature takes its cue from the pioneering, even 'prescient' work of Kautsky ([1899] 1988). In his analysis of the transformation of European agriculture in the late nineteenth century, Kautsky focused on the relationship between specific agrarian situations and general tendencies, including the internationalisation of production and the concentration of capital in agriculture. He argued that despite these tendencies, German petty commodity producers not only persisted but expanded their role in national agricultural production owing to an effective political voice and to the persistence of limits to the spread of wage labour in farming – limits that derived both from the distinctive character of agricultural production and from the capacity of peasants for self-exploitation. Thus, Kautsky's analysis highlighted the fact that capitalist development proceeded not through the displacement of peasant households by capitalist enterprises but through the increasing articulation of these farm households with agro-industrial capital.

Building upon this work, contemporary analyses of Third World agrarian change have come to focus centrally on situating peasant households with respect to both the state and broader circuits of capital. The great strength of this literature is its attention to the ways in which these relationships vary from place to place, culture to culture. In part, this focus on the multiple paths of agrarian restructuring derives from an understanding of the heterogeneity of state–society relations in developing nations (Hart *et al.* 1989, Timmer 1992). However, it also derives from an increasing awareness of the importance of understanding both the specific character of localised agrarian social relations and the specific forms of articulation through which these local relations are linked to wider political and economic processes (cf. Neumann 1992). In particular, recent studies emerging from the field of political ecology highlight the importance of the micro-politics of peasant struggles over access to productive resources. This research demonstrates that social context and the exercise of social power at the household and inter-household scales – revolving around class, gender, and generational relations – plays a critical role in shaping the course of agrarian change in particular places (e.g. Watts 1983, Hart 1991, Carney 1993, Moore 1993, Schroeder 1993). It also moves beyond overly structural accounts wherein agrarian dynamics are driven by monolithic state action or global economic forces, to explanations that concentrate on the process of negotiation, contest and resistance in and across multiple social arenas at multiple geographic scales, from the household to the state.

These considerations inform an analysis of US restructuring in pork production by calling attention to the fact that new divisions of labour in agriculture are not simply superimposed upon the rural landscape; to the contrary, they emerge through the complex workings of power within social networks at the local scale and between the local and non-local scales.[19] Moreover, they point to the fact that the struggles over resources, labour, and technological practice that stand at the centre of the restructuring process are simultaneously struggles over culturally constructed meanings and identities (see especially Moore 1993).[20]

For example, in Iowa – the heart of traditional hog production – a new era is being ushered in as two very different production systems collide. Yet this is no automatic transition; changes in production are being channelled through a dense web of existing social relations that are themselves constantly in flux. In this process, changes are unfolding through innumerable acts of social negotiation and resistance occurring at a variety of scales. While a full discussion of this process cannot be undertaken in the context of this chapter, a brief consideration of key arenas of transformation can shed light on the complex and contested routes through which restructuring is taking place.

One arena of transformation is the farm. Contract production gained a foothold in Iowa during the farm crisis of the early 1980s (Marbery 1993c). Large-scale southern contracting firms like Murphy Farms entered Iowa hog farming at that time, but the rise of contracting was in no way a simple invasion from afar. These firms were joined by regional agro-business co-operatives like Farmland Industries and Land O'Lakes as well as by local feed companies and independent hog farmers who could not finish all of the young pigs that they produced (Rummens *et al.* 1991). Contracting firms discovered that their chief obstacle was a strong sense of independence among family labour hog farmers, who were reluctant to sign on, difficult to supervise, and very unlikely to continue with the contracting arrangement. Contractors thus compete not only against each other for new grower recruits, but also against the region's strong agrarian ideology. Some contractors target growers with no previous experience in hog farming, but most of Iowa contract growers were at one time independent farmers who switched to contract production because they lacked the financial resources to remain independent (Rummens *et al.* 1991).

Contracting both exploits and exacerbates existing lines of differentiation within the Iowa farm community. Despite the fact that Iowa farmers are bound together by a remarkably strong collective identity, there is little unanimity in their response to the rise of contract hog production. Iowa producers see that vertical integration will restrict their available market options. Yet, as a body, they are torn between wanting to protect independent hog production and not wanting to place limits on economic enterprise. This issue has become a central point of dispute within farm organisations whose members range from farmer-contractors to growers to adamant anti-contracting activists. For these organisations, the result is the emergence of sharp internal divisions, a fragmented political voice and an uncertain future direction. Ultimately, this discourse is about more than policy. It is a contest of representation revolving around the question of which groups will define the meaning of 'independent' production. The central issue is whether or not contract growers can be accommodated within the prevailing agrarian ideology. If they can, then a significant barrier to vertical integration will be removed.[21]

A second arena of transformation is the network of relationships linking farmers to their input suppliers. Farmers are not the only actors within the established commodity chain with contradictory impulses towards contract production. Some local feed companies have integrated forward into contracting in order to expand their market, but other local material and service suppliers stand to lose business, given that integrating firms favour a pattern of centralised, non-local acquisition. This problem also extends to the system of informational inputs. Many larger contracting firms are internalising this function by doing their own proprietary research on genetics, nutrition, and health (Lawrence 1992). The role of land grant universities and agricultural extension services as the providers of research and technical advice could be usurped if independent farmers are replaced by contract growers. In Iowa, state-sponsored economists are actively shaping the process of agricultural restructuring by working to provide independent producers with the means to compete with the southern-style system (e.g. Hilburn *et al.* 1988, Kliebenstein 1988). Yet the future of such efforts is unclear because of a wider discourse on agricultural policy that could lead to cutbacks in public funds.

A third arena of transformation, then, is the state. At the local level, the battle to limit vertical integration emerged in the form of farmer–community coalitions that successfully stopped the development of large-scale farrowing units and large-scale farrow-to-finish facilities, using county nuisance ordinances as well as private lawsuits based on objections to odour and to water pollution. A coalition of livestock business interests (including the Iowa Farm Bureau) countered these efforts by supporting state legislation that would establish 'agricultural enterprise zones' designed to exempt large-scale hog facilities from nuisance suits (Roos 1993). This legislation failed, as have other efforts to repeal the state's existing ban on packer ownership of livestock and involvement in contracting. Political resistance to the restructuring of hog production is exerting a profound influence on the geography of the hog–pork commodity chain. Large-scale producers and integrating firms have responded by establishing their facilities and grower networks just outside Iowa, where they still have access to low grain prices and an established meat packing industry. Some large-scale hog producers have pursued a strategy of intra-regional specialisation wherein pigs are farrowed in areas with little developed resistance to corporate livestock production but are then shipped back to Iowa for finishing and slaughter. And, as mentioned above, a new group of integrated hog production/meat packing firms are avoiding Iowa altogether and developing their operations in other parts of the greater Midwest.

It is interesting to note, however, that most of the resistance to restructuring in Iowa has been directed towards the farrow-to-finish facilities of corporate hog farms or the farrowing houses associated with integrating firms. In both cases, there is a huge and visible shift in the scale of production, and crucially, in the volume of waste output, the amount of odour generation, and in the potential for water contamination. In contrast, grower networks associated with integrating firms receive much less attention because they are so much less a visible symbol of change. The scale of facilities employed by the average contract grower closely resembles that of a mid-sized independent hog farmer; moreover, the grower himself/herself is likely to be a farmer or farm family member with long-standing ties to the local community. For these reasons, grower networks are better able to blend into the Iowa cultural landscape as they expand.

A fourth arena of transformation concerns the meat packing industry and its relationship to hog farmers and integrating firms. It is not at all clear how the nation's dominant pork packing firms (IBP, ConAgra, Cargill) will respond to the development of hog contracting in the Midwest or to the emergence of new integrated competition in meat packing. These firms have enormous investments in Iowa hog slaughtering and have established reliable networks of independent producers. In fact, through the agency of direct marketing, meat packers have been able to push farmers to produce more lean and uniform hogs, which, in turn, has stimulated a commitment to modern input intensive practices leading to an increase in the size of hog farms. Still, meat packers are playing both sides of the fence in Iowa; ConAgra and Cargill have eagerly purchased the output of emergent contracting networks because these producers are able to provide packing firms with high quality, high volume inputs (Burgett 1990). It is also quite likely that the established pork packers will try to match companies like Tyson and become fully integrated themselves, either by attempting to remove restrictions on packer-owned livestock in Iowa, or by relocating – a move that would bring incredible political pressure to bear in the Hawkeye state. Although neither ConAgra nor Cargill is directly integrated backward into hog production in the central Midwest, they are well positioned to do so. Both firms have considerable experience with integrated production in the poultry industry. Moreover, Cargill already engages in contract hog production in Arkansas (Marbery 1988). Meanwhile, IBP may be missing the competitive boat in livestock procurement. The company has no experience in the poultry industry and does not engage in contract pork production. IBP, however, recently purchased a plant near a cluster of integrated producers in Indiana and is also attempting to establish a plant in North Carolina.

Pork production in Iowa remains extremely unsettled. Production contracting has introduced new relationships throughout the commodity chain, setting into motion a dynamic and contested process of restructuring occurring between and within various arenas of transformation.

CONCLUSION: LEARNING FROM AGRICULTURE

By way of conclusion, let me turn to the last of the questions raised at the outset: can industrial analysts learn from the careful study of agriculture and can theories of industrial restructuring be enriched by such a dialogue? The answer to this question must be an emphatic yes. While the story of agro-industrial restructuring in pork production or in other agro-food commodity chains must be told and understood in their own right, they also have several important lessons to impart to industrial studies.

First, the study of agriculture demonstrates the importance of paying greater attention to the distinctive social relations, technical conditions, and divisions of labour across all sectors of industry. The food and fibre producing system exhibits a mix of social relations of production, an incredible breadth of technological development and a wide range of forms of industrial organisation. Such variation – deriving from differences among commodity chains rooted in the landed basis of production and the natural circumstances of plant and animal growth – suggests a remarkable openness in the evolution of production systems under capitalism and a multiplicity of possible paths of industrialisation. This is not to argue that generalisation is impossible, but to point out that it must be based on abstractions from variegated concrete instances, rather than imposed on a flattened landscape by over-generalisation from one or two sectors' specific histories and geographies.

Second, the study of agriculture highlights the articulation of the local and global scales within the broad arch of world capitalism. Localities are not static remnants of past rounds of investment and social activity; they are distinctive congeries of social practices that actively shape industrial development. This can be contested furiously as localities are formed and transformed through processes of industrialisation that tear apart existing social relations and precipitate struggles over their reconstitution. In this case, recent analyses of agrarian change provide a re-affirmation of a long-standing research theme in industrial geography. After all, interest in locality was in large measure sparked by Massey's (1984) pioneering analysis of the ways in which contemporary economic restructuring is shaped by the accumulated sediments of regional and local history. But, again, research on industrial restructuring has perhaps been too quick to move beyond this early message on its way to questions concerning epochal changes to and from Fordism or Toyotaism.

Finally, it is vital that industrial studies probe the historical processes through which industries and places are mutually constructed. It is not that industrial studies have ignored geographical specificity; yet local difference is too often taken to be the direct expression of universal processes without sufficient attention to the ways in which class relations, technical advance, business culture, political dynamics, etc. emerge from local circumstances. As evidence of a failure to pursue historical analyses seriously, one need look no further than the eager grasping for an errant schematic history that has swept through the field. Students of industrial restructuring would do well to question the very thin histories drawn by the regulation and flexible specialisation schools and instead follow the cue of agrarian studies by turning towards a careful consideration of the deeply embedded sources of local difference.

NOTES

1 The study of agriculture in isolation from the dynamics of industrial society has characterised research in both geography and sociology. Agricultural geography largely eschewed the issue of farm production, focusing instead on regional classification, farm structure, and the diffusion of innovations (Symons 1967, Pacione 1986). Meanwhile, rural sociology virtually abandoned the study of agriculture in favour of descriptive research on rural communities (cf. Friedland 1982).

2 For recent summaries of the history of the field, see Mann (1990), McMichael and Buttel (1990), and Buttel *et al.* (1990).

3 'Fordist agriculture' approaches have taken three main forms. One systematically adopts the central concepts and periodisations of the regulationists, using this framework as the primary lens through which to interpret the development of US farming (Kenney *et al.* 1989, 1991, Kim and Curry 1993). A second turns to the broad institutional environment of agriculture, focusing on the ways in which state-regulated international 'food regimes' govern the structure of food production and consumption and thereby shape the process of agricultural transformation (Friedmann and McMichael 1989, Friedmann 1993). And a third lets regulation theory serve as a broad political-economic backdrop to stories of agricultural change and other forms of rural transformation (Goodman and Redclift 1991, Marsden 1992, Marsden *et al.* 1993).

4 For critiques of the Fordist/post-Fordist literature see Williams *et al.* 1987, Gertler 1988, Sayer 1989, and Walker 1995.

5 Several problems with the regulation framework render it singularly unsuited to an analysis of capitalist development surrounding agriculture. First is an oversimplified and clearly inaccurate interpretation of historical American economic development (cf. Brenner and Glick 1991). As a result, the historical relationship between agriculture and industry is badly misconstrued within the Fordist agriculture model (cf. Page and Walker 1991). Second is a reductionist tendency to collapse all industrial development into the binary opposites of Fordism and post-Fordism (cf. Pollert 1988; Sayer 1989). Yet, as Goodman and Watts (1994) demonstrate, farming never has fit this rigid typology in terms of production technology, labour process, firm organisation, or competitive structure; forcing farming into either mould only obscures its great complexity and diversity. Third is a tendency to overemphasise institutional coherence within a mode of regulation in the explanation of capitalist development (cf. Walker 1995). Accordingly, Fordist agriculture approaches grant Fordist institutions the central role in the process of agricultural restructuring and substitute an over-generalised technical-organisational model for any detailed analysis of industrial structure in agriculture (cf. Goodman and Watts 1994).

6 These constraints take a variety of forms. Of crucial importance is the role of biology in plant and animal growth. There are no industrial substitutes for soil or sunlight, and the biological conversion of energy in plant development and animal gestation cannot easily be accelerated or standardised as in manufacturing (Goodman *et al.* 1987). The fact that biological time dominates these processes has other critical implications. The excess of production time over labour time in farming means that labour cannot be applied constantly to production – thereby limiting surplus extraction; while the seasonality of production greatly slows the circulation of capital (Mann and Dickinson 1978, FitzSimmons 1986, Mann 1990). Moreover, the land-dependent character of farming itself poses several constraints. Unlike a factory, capital cannot be applied to most agricultural labour processes at a single site where production can be expanded or intensified. Instead, increased production requires a spatial extension (and conversely decreased production requires a spatial contraction); yet, because land is a fixed and limited resource, and because land markets are deeply coloured by localised social conditions, farmers cannot easily or quickly adjust their investment in land (Marsden *et al.* 1986, Goodman *et al.* 1987).

7 At times, farmers have actively resisted corporate incursions into the farm economy. But, for the most part, they have been active participants in the process, eagerly absorbing the manufactured seeds, chemicals and machines that raised labour productivity while welcoming any reduction in input cost

due to improvements in industrial methods. With each wave of new technological inputs, however, farmers have watched their collective position deteriorate in the face of individual competition, as the waterwheel of farm failure and consolidation has passed through yet another cycle (Cochrane 1979). Within this broad historical current, agro-industry has therefore widened into a torrent while family labour farming has been reduced to a trickle.

8 Sources of divergence among commodity chains are many. On the farm, crops differ from one another in many key respects. Each plant or animal species has its own biological rhythm of reproduction, growth, and development; and each yields a farm product with a unique configuration of traits such as size, shape, weight, hardness, or perishability. In turn, each type of crop exerts its own requirements upon agricultural practice and thereby puts an unmistakable stamp on the direction of technological change, the farm labour process, and farm enterprise organisation (Friedland *et al.* 1981, Friedland 1984). In similar fashion, different sets of relationships link farmers to upstream suppliers and downstream processors, distributors and consumers, depending upon the particular farm product involved. And though embedded in a common political milieu, different farm commodities are governed by often quite specific regulatory frameworks (Gertler 1991, Dupuis 1993).

9 At the most abstract level, Harvey (1982) situates geographical shifts at the very heart of the capitalist dynamic of growth and crisis. On the one hand, it is a necessary condition for capital to be immobilised differentially within the landscape so as to produce surplus value from the labour process. On the other, space needs to be created for new patterns of accumulation to allow for the valorisation of all productive activities and to exploit technological innovations fully so as to maximise surplus value. Driven by the dynamic of growth and crises, capital continually see-saws in search of a 'spatial fix', and in this process the geography of capitalism is periodically refashioned. Likewise, Storper and Walker (1989) argue that industrial development unfolds in a thoroughly geographic way. Key industrial breakthroughs are both sectoral-specific and place-specific, owing to the fact that the practical mastery of technologies and business affairs is embodied in the people and organisations who learn and grow with specific industries. In the furiously dynamic process of capitalist expansion, new industries and methods repeatedly break forth, new localisations arise, and successive waves of growth are associated with rising (or renewed) industries of their time, causing great upheavals in the space-economy of capitalism. The creation of 'new industrial spaces' is therefore central to the continuing renewal of the capitalist system.

10 By focusing on the propulsive force of capitalist production as the prime determinant in the fate of regions – as opposed to allocative market functions or corporate hierarchies – industrial geographers also direct attention to the role of agro-industrialisation in broader processes of US economic development (Page and Walker 1991). For too long, agriculture has been viewed as either a nether world beyond real industry or an active break on wider capitalist growth, when in reality whole regions have grown on the basis of the expansion of the division of labour surrounding agricultural production.

11 In fact, once confinement facilities are constructed, producers are tied to a capacity utilisation logic (brought on by the debt incurred for the building and other inputs) that pushes them to keep their buildings full on a year round basis as opposed to the strongly seasonal spring–fall production pattern of the past.

12 Boxed meat production involves an extension of mechanised disassembly within the plant in which the animal carcass is broken down into primal and sub-primal cuts of meat that are then vacuum sealed and loaded into boxes. The new techniques of boning, wrapping, and packaging, pioneered in beef production and later extended into pork, were incorporated into huge and efficient state-of-the-art slaughtering/processing plants that yielded significant economies of size, based, in part, upon improved by-product recovery and sales. Boxed beef could also be shipped directly to the retail store where the final retail cuts were performed. Thus, boxed beef production revolutionised the distribution system by eliminating the need for a separate 'fabrication' stage in between the packing house and the retail store. So, too, it revolutionised retail butchering by effectively transferring the work of thousands of skilled union butchers from the fabrication warehouse to the packing house, where the tasks were de-skilled and performed by low-wage, non-union labourers (Burns 1982).

13 Changes in meat retailing track the proliferation of sales of pre-packaged 'convenience' foods of all kinds during the postwar era. In turn, this overall trend can be tied to rising national incomes as well as to the increasing participation of women in the labour force and the associated shifts in patterns of household food purchase and consumption.

14 IBP Inc. came to dominate beef packing in the 1970s by pioneering boxed beef production. Cargill entered beef production in 1979 by purchasing MBPXL, IBP's chief competitor at the time. Meanwhile, ConAgra became a major player in the industry virtually overnight by purchasing Armour in 1983; it then fashioned a meat packing giant by buying several specialised beef packing firms (including Monfort) and combining these with the remnants of Swift. These firms rolled like competitive juggernauts through the beef packing industry. Today, the 'Big Three' comprise a modern day Beef Trust that slaughters 70 per cent of the nation's fed cattle and produces 80 per cent of the boxed beef.

15 For example, the number of hog farms in Iowa declined by 53 per cent between 1970 and 1985, while the average size of the remaining farms nearly doubled (Futrell and Dhuyvetter 1986). Despite technological change and increasing levels of concentration, hog farming in the corn belt has been characterised by strong continuity in the social organisation of production due to the fact that new production practices emerged from within the ranks of traditional family labour farms.

16 The USDA grade system predicts eating quality of meat, while the yield system measures the amount of lean meat a carcass will produce. Both of these measures, however, are based upon averages and are therefore only rough estimates of quality. The reliability of the USDA system also varies given the fact that grading takes place at a pace of over 300 carcasses per hour (Lee 1994).

17 The new system of livestock buying widens the gap between the highest and lowest prices paid for animals and thereby favours those producers that have achieved greater standardisation through the adoption of input intensive practices. Those farmers who have not adopted and updated confinement systems are put at a competitive disadvantage because they cannot easily meet the size, quality and volume specifications of the packing firms. In this sense, the new directives of the meat packing industry have influenced not only the technology of hog production, but its social organisation as well.

18 This observation is based upon field work conducted by the author in June of 1996 in Columbus Junction, Iowa, site of an IBP hog slaughtering plant.

19 Recent work on agricultural change in the US and other developed nations echoes many of these themes (Marsden *et al.* 1986, 1993, Whatmore 1991, Moran *et al.* 1994, Murdoch 1994, Roberts 1994). This body of work concentrates on localised differences in farming practices, on-farm social relations, land tenure patterns, agro-industrial linkages, labour markets, and state policy, and in so doing highlights the geographical specificity of capital's penetration into the farm sector.

20 This insight echoes a growing body of research conducted within the new cultural geography mostly concerning urban restructuring processes in the US, Canada, and Europe (e.g. Cosgrove and Daniels 1988; Duncan and Ley 1993).

21 Take, for example, a brochure produced by the contracting firm Swine Graphics Enterprises, in which one of the growers working with the company states 'this relationship allows me to be my own boss and is the reason I feed for them'.

BIBLIOGRAPHY

Aglietta, M. (1979) *A Theory of Capitalist Regulation*, London: Verso Press.

Angel, D. (1994) *Restructuring for Innovation: The Remaking of the U.S. Semi-Conductor Industry*, New York and London: Guilford Press.

Boyer, R. (1990) *The Regulation School: A Critical Introduction*, New York: Columbia University Press.

Brenner, R. and Glick, M. (1991) 'The regulation approach to the history of capitalism', *New Left Review* 186: 45–119.

Burgett, D. (1990) Personal interview conducted at the Excel plant in Ottumwa, IA, 8 August.

Burns, W. (1982) 'Changing corporate structure and technology in the retail food industry', in D. Kennedy, C. Craypo, and M. Lehman (eds) *Labor and Technology: Union Responses to Changing Environments*.

Buttel, F. H. (1982) 'The political economy of agriculture in advanced industrial societies: some observations on theory and method', *Current Perspectives in Social Theory* 3: 27–55.

Buttel, F. H., Larson, D. F., and Gillespie, G. (1990) *The Sociology of Agriculture*, New York: Greenwood Press.

Carney, J. (1993) 'Converting the westlands, engendering the environment: the intersection of gender with agrarian change in the Gambia', *Economic Geography* 69: 329–48.

Cochrane, W. W. (1979) *The Development of American Agriculture: A Historical Analysis*, Minneapolis: University of Minnesota Press.

Cosgrove, D. and Daniels, S. (eds.) (1988) *The Iconography of Landscape*, Cambridge: Cambridge University Press.

Duncan, J. and Ley, D. (eds.) (1993) *Place, Culture, Representation*, London: Routledge.

Dupuis, M. (1993) 'Sub-national state institutions and the organisation of agricultural resource use: the case of the dairy industry', *Rural Sociology* 58: 440–60.

FitzSimmons, M. (1986) 'The new industrial agriculture: the regional integration of specialty crop production', *Economic Geography* 62: 334–53.

Florida, R. and Kenney, M. (1992) 'Restructuring in Place: Japanese Investment, Production, Organization, and the Geography of Steel', *Economic Geography*, 68: 146–173.

Friedland, W. H. (1982) 'The end of rural society and the future of rural sociology', *Rural Sociology* 47: 589–608.

—— (1984) 'Commodity system analysis', in H. K. Schwarzweller (ed.) *Research in Rural Sociology and Development*, Greenwich, CT: JAI Press.
—— (1993) 'The Political Economy of Food: a Global Crisis', *New Left Review* 197: 29–57.
——, Barton, A. E. and Thomas, R. J. (1981) *Manufacturing Green Gold*, New York: Cambridge University Press.
Friedmann, H. and McMichael, P. (1989) 'Agriculture and the state system', *Sociologia Ruralis* 29: 93–117.
Futrell, G. A. and Dhuyvetter, K. (1986) 'Trends in Hog Production and Production Efficiency', Working Paper No. 22, Department of Economics, Ames, IA: Iowa State University.
Gates, P. W. (1960) *The Farmer's Age: Agriculture 1815–1860*, White Plains, NY: M. E. Sharpe, Inc.
Gertler M. (1988) 'The limits to flexibility: comments on the post-Fordist vision of production and its geography', *Transactions of the Institute of British Geographers* 13: 419–32.
—— (1991) 'The institutionalisation of grower–processor relations in the vegetable industries of Ontario and New York', in W. H. Friedland, Busch, L., Buttel, F. H., and Rudy, A. P. (eds) *Toward a New Political Economy of Agriculture*, Boulder: Westview Press.
Goodman, D. and Redclift, M. (1985) 'Capitalism, petty commodity production, and the farm enterprise', *Sociologia Ruralis* 25: 231–47.
—— (1991) *Refashioning Nature: Food, Ecology and Culture*, London: Routledge.
Goodman, D. and Watts, M. (1994) 'Reconfiguring the rural or fording the divide: capitalist restructuring and the global agro-food system', *Journal of Peasant Studies* 22, 1: 1–49.
Goodman, D., Sorj, B., and Wilkinson, J. (1987) *From Farming to Biotechnology: A Theory of Agro-Industrial Development*, Oxford: Basil Blackwell.
Govindasamy, D., Liu, D. and Kliebenstein, J. (1993) 'Economic Impacts of Porcine Somatotropin on a Farrow-to-Finish Hog Farm Operation', Staff Paper No. 248, Department of Economics, Ames, IA: Iowa State University.
Hart, G. (1991) 'Engendering everyday resistance: gender, patronage, and production politics in rural Malaysia', *Journal of Peasant Studies* 19: 93–121.
Hart, G., Turton, A. and White, B. (eds) (1989) *Agrarian Transformations: Local Processes and the State in Southeast Asia*, Berkeley: University of California Press.
Harvey, D. (1982) *The Limits to Capital*, Oxford: Basil Blackwell.
Hayenga, M., Rhodes, V. J., Brandt, J. A., and Deiter, R. E. (1985) *The US Pork Sector: Changing Structure and Organisation*, Ames, IA: Iowa State University Press.
Hilburn, Kleinbenstein, J., Stevermer, E. and Trede, L. (1988) 'A Comparison of Iowa Swine Production with its Competition', Staff Paper 184, Department of Economics, Ames, IA: Iowa State University.
Kautsky, K. ([1899] 1988) *The Agrarian Question*, London: Zwan.
Kenney, M., Labao, L., Curry, J., and Goe, R. W.. (1989) 'Midwestern agriculture in US Fordism', *Sociologia Ruralis* 29: 131–48.
Kenney, M., Labao, L., Curry, J. and Goe, R. W. (1991) 'Agriculture in US Fordism: the integration of the productive consumer', in W. H. Friedland *et al.* (eds) *Toward a New Political Economy of Agriculture*, Boulder: Westview Press.
Kim, C. and Curry, J. (1993) 'Fordism, flexible specialisation and agro-industrial restructuring', *Sociologia Ruralis* 33: 61–80.
Kliebenstein, J. (ed.) (1988) *Iowa Pork Industry: Competitive Situation and Prospects*, report prepared by the Swine Task Force, College of Agriculture, Ames, IA: Iowa State University.

Lawrence, J. D. (1992) 'The US Pork Industry in Transition', Staff Paper No. 240, Department of Economics, Ames, IA: Iowa State University.

Lee, M. (1994) 'Excel: computers add accuracy to cattle pricing', *Denver Post* 14 August.

Lipietz, A. (1987) *Mirages and Miracles*, London: Verso Press.

Looker, D. (1990) 'Hog growth hormones may force changes in the pork business', *Des Moines Register* 19 August.

McMichael, P. and Buttel, F. H. (1990) 'New directions in the political economy of agriculture', *Sociological Perspectives* 33: 89–109.

Mann, S. A. (1990) *Agrarian Capitalism in Theory and Practice*, Chapel Hill: University of North Carolina Press.

—— and Dickinson, J. M. (1978) 'Obstacles to the development of a capitalist agriculture', *The Journal of Peasant Studies* 5: 466–81.

Marbery, S. (1988) 'Tyson counter-attacks', *Hog Farm Management*, March.

—— (1993a) 'Pork industry gears-up for consumer-driven market', *Feedstuffs* 22 February.

—— (1993b) 'Transition smooth at Tyson's Missouri plant', *Feedstuffs*, 8 March.

—— (1993c) 'North Carolina was Catalyst for Iowa Hog Expansion', *Feedstuffs*, 7 June.

Marsden, T. (1992) 'Exploring a rural sociology for the Fordist transition', *Sociologia Ruralis* 32: 209–30.

Marsden, T., Munton, R., Whatmore, S. and Little, J. (1986) 'Toward a political economy of capitalist agriculture: a British perspective', *International Journal of Urban and Regional Research* 10: 498–521.

Marsden, T., Murdoch, J., Lowe, P. and Flynn, A. (1993) *Constructing the Countryside*, Boulder: Westview Press.

Massey, D. (1984) *Spatial Divisions of Labor: Social Structures and the Geography of Production*, New York: Methuen.

Moore, D. S. (1993) 'Contesting terrain in Zimbabwe's eastern highlands: political ecology, ethnography, and peasant resource struggles', *Economic Geography* 69: 380–401.

Moran, Blunden, G., Workman, M., and Brady, A. (1994) 'Agro-Commodity Chains, Family Farms and Food Regimes', paper presented at the Annual Meeting of the Association of American Geographers, San Francisco, 30 March.

Morris, C. E. (1992) 'Indiana packers: American pork goes global', *Food Engineering*, November.

Murdoch, J. (1994) 'Weaving the Seamless Web: A Consideration of Network Analysis and its Potential Application to the Study of the Rural Economy', Working Paper 3, Centre for Rural Economy, University of Newcastle Upon Tyne.

Neumann, R. (1992) 'Political ecology of wildlife conservation in the Mt Meru Area of northeast Tanzania', *Land Degradation and Society* 3: 85–98.

Newby, H. (1982) 'Rural sociology in these times', *The American Sociologist* 17: 60–70.

Pacione, M. (ed.) (1986) *Progress in Agricultural Geography*, London: Croom Helm.

Page, B. (1996) 'Across the great divide: agricultural and industrial geography', *Economic Geography* (forthcoming).

—— (1997) 'Rival unionism and the geography of the meat packing industry', in A. Herod (ed.) *Geographical Perspectives on Trade Unionism*, Minneapolis: University of Minnesota Press (forthcoming).

—— and Walker, R. (1991) 'From settlement to fordism: the agro-industrial revolution in the American Midwest', *Economic Geography* 67: 281–315.

—— and Walker R. (1994) 'Staple lessons: Harold Innis, agriculture and industrial geography', unpublished paper, University of Toronto.
Perry, C. R. and Kegley, D. H. (1989) *Disintegration and Change: Labor Relations in the Meat Packing Industry*, Philadelphia: The Industrial Research Unit, Wharton School, University of Pennsylvania.
Peterson, S. (1990) Personal interview conducted at the IBP plant in Columbus Junction, IA, 24 August.
Pollert, A. (1988) 'Dismantling Flexibility', *Capital and Class* 34: 42–75.
Pond, W. G. (1983) 'Modern Pork Production', *Scientific American* 248: 96–103.
Rhodes, V. J. (1990) *US Contract Production of Hogs*, Agricultural Economics Report No. 1990–1, Department of Agricultural Economics, University of Missouri, Columbia.
—— and Grimes, G. (1992) *US Contract Production of Hogs: A 1992 Survey*, Agricultural Economics Report No. 1992–2, Department of Agricultural Economics, University of Missouri, Columbia.
Roberts, R. (1994) 'Uneven Development, Enterprise Reproduction, and Nature: Persistence of Family Farms on the Southern High Plains', paper presented at the Annual Meeting of the Association of American Geographers, San Francisco, 30 March.
Roos, J. (1993) 'Enterprise zone bill shot down in House', *Des Moines Register*, 20 April.
Ross, Earle, D. (1951) *Iowa Agriculture: An Historical Survey*, Iowa City: State Historical Society of Iowa.
Rummens, M., Kliebenstein, J., and Rhodes, V. J. (1991) 'Size and Distribution of Contract Hog Production in Iowa', Staff Paper No. 232, Department of Economics, Ames, IA: Iowa State University.
Sayer, A. (1989) 'Post-Fordism in question', *International Journal of Urban and Regional Research* 13: 666–93.
—— and Walker, R. (1992) *The New Social Economy: Re-Working the Division of Labor*, Oxford: Basil Blackwell.
Schroeder, R. A. (1993) 'Shady practice: gender and the political ecology of resource stabilisation in Gambian garden/orchards', *Economic Geography* 69: 349–65.
Science (1994) 'Designer cattle with ultrasound', 263: 327.
Scott, A. (1988) *Metropolis: From the Division of Labor to Urban Form*, Berkeley: University of California Press.
Scott, A. and Storper, M. (1986) *Production, Work, Territory: The Geographical Anatomy of Industrial Capitalism*, Boston: Allen & Unwin.
Storper, M. and Walker, R. (1984) 'The spatial division of labor: labor and the location of industry', in W. Tabb and L. Sawers (eds) *Sunbelt/Snowbelt*, New York: Oxford University Press.
—— (1989) *The Capitalist Imperative: Territory, Technology and Industrial Growth*, Oxford: Basil Blackwell.
Swine Graphics Enterprises, L. P. (1993) *Swine Graphics Enterprises Contract Finisher and Nursery Production*, Webster City, IA: Swine Graphics Enterprises.
Symons, L. (1967) *Agricultural Geography*, New York: Praeger.
Timmer, P. (ed.) (1992) *Agriculture and the State*, Ithaca, NY: Cornell University Press.
United States Department of Agriculture (USDA) (1996) 'Concentration in the red meat packing industry', *Packers and Stockyards Programs*, Grain Inspection, Packers and Stockyards Administration: Washington, D.C.

Van Arsdall, R. N. and Nelson, D. E. (1984) 'US Hog Industry', Agricultural Economic Report No. 511, Economic Research Service, United States Department of Agriculture.
Walker, R. (1995) 'Regulation and flexible specialisation as theories of capitalist development', in H. Liggett and D. Perry (eds) *Spatial Practices: Critical Explorations in Social/Spatial Theory*, Thousand Oaks, CA: Sage Publications.
Walsh, M. (1982) *The Rise of the Midwestern Meat Packing Industry*, Lexington: University of Kentucky Press.
Watts, M. (1983) *Silent Violence: Food, Famine and Peasantry in Northern Nigeria*, Berkeley: University of California Press.
—— (1996) 'Development III: the global agrofood system and late twentieth-century development (or Kautsky redux)', *Progress in Human Geography* 20: 230–45.
Whatmore, S. (1991) *Farming Women: Gender, Work and Family Enterprise*, Basingstoke: Macmillan.
Williams, K., Culter, T., Williams, J. and Haslam, C. (1987) 'The end of mass production?' *Economy and Society* 16: 405–39.

COMMENTARY ON PART II

Regions in global context? Restructuring, industry, and regional dynamics

Margaret FitzSimmons

Any restructuring of agriculture requires renegotiation and recomposition of spatial relationships at multiple geographical scales. These relationships produce and reproduce institutions that ensure:

- the organisation of the farming sector and the recruitment of its labour force;
- access to land, farming knowledge, practices, and infrastructure;
- customary commercial practices and linkages between input manufacturers, farms, processors, and marketing firms;
- capital and risk markets;
- and government policies.

All of these are intrinsically spatial and therefore constitute geographies. These geographies need not coincide: for example, the spatial domain from which farm labour is recruited may lie within the farming region itself or may be external. Connections between input producers, farmers, and processors and marketing firms are likely to reflect multiple spatial relationships at several scales and may include regional monopolies or oligopolies in input and product markets. While global capital and risk markets have now appeared, most actors in farming systems are connected to these markets through a chain of intermediaries; only the largest transnational firms access these global markets directly. Government policies may defend internal social practices and established economic alliances or, in the new era of liberalisation of agricultural trade, encourage export production of some farm commodities at the cost of relaxing trade barriers that defend other agricultural products.

What is the role of the region in this process? In the context of this multiplex and dynamic spatiality, the definition of an agricultural region becomes complex. At the simplest level, agricultural regions refer to relatively homogeneous patterns of farm (and land) ownership and farming practices. These production systems are often specialised in the production of a few agricultural commodities for national or international markets. Within a region farmers must produce those commodities for those markets or make the costly and difficult effort of building alternative linkages and pathways for other commodities. While and where existing linkages work tolerably well, farmers are likely to produce within them; only when prices fall below an acceptable minimum or an area is abandoned by marketing and processing firms as a source for a particular commodity does reorientation of the regional farm system seem worthwhile.

The ongoing process of globalisation of agricultural markets for particular commodities imposes strong pressures on some, but not all, agricultural regions. Not all major agricultural commodities are traded globally; among those that are, production in the domestic farming sectors of the core market countries may provide a substantial share of the commodities consumed, so external sources complement domestic production. World trade in wheat moves only a portion of the wheat produced; most wheat grown is consumed within the producing country. The new 'non-traditional exports' of temperate-zone fresh fruits and vegetables also complement domestic production in most markets, though differential labour costs (and resource and environmental regulations) may support the spatial reorganisation of production of these commodities. The primary markets for agricultural commodities in international trade remain in the core and a few newly-industrialising countries. Colonial and post-colonial linkages often appear in these secondary sourcing relationships, but these are subject to recomposition if national agricultures increase production.

Discussion of regional change is further complicated by the appearance of regions at both sub- and supra-national scales. The term region is generally used to reflect an organised area with an *incomplete* state, that is, an area whose political and economic institutions cannot act to regulate these activities as directly as can the nation state. Regional institutions arise from the development of local and regional customary practices, of business and social conventions; they may eventually achieve legal standing in some fashion, but they are always more vulnerable, and perhaps thus more flexible, than the formally constituted institutions of the nation state within which they are embedded or over which they become composed. Sub-national regions reflect aggregation economies and relatively local institutional strategies and linkages; supra-national regions express core–periphery relationships, regional and global geopolitical agendas, and pressures for liberalisation of agricultural policy and trade. Once established, regions tend to persist in the intermediate term, since they are overdetermined by the multiple social institutions of which they are composed. Crisis, and the weakening of these institutions, opens opportunities for restructuring.

The impetus for restructuring may be internal or external; that is, restructuring may develop from successful local innovation, local agency, which gives rise to new competitive strategies, or it may be imposed from without by powerful external actors such as national or transnational firms or the International Monetary Fund. Agricultural change is path-dependent, and both internal and external restructuring must either incorporate the political, social, and economic institutions at present in place or address the costs of destroying and recomposing these institutions. The rise of global markets for many agricultural commodities, and of institutions which coordinate and control these markets, presents a powerful pressure for restructuring which is external to many regions; however, these institutions always begin as innovations in the resolution of local problems and opportunities.

There is not a singular logic to the process of agricultural restructuring, though there are rules to the game. Restructuring occurs at multiple scales. Globalisation of markets for particular agricultural commodities constructs not a uniform terrain but a new wave of uneven development at the world scale. Globalisation modifies boundaries: it changes their meanings, it does not erase them. Globalisation searches out differences – in living standards, in the defensive strength of political institutions, in the resilient and resistant practices of people in place – to achieve the old mercantile goal of buying cheap and selling dear. In the process, globalisation creates differences – connecting the lives of rich people and poor people, of rich places and poor places, in new ways. The primary agents of globalisation in the agro-food sector are large corporations that sit at the centre of webs of relationships that link farming, processing, and marketing. Their position as coordinators and controllers of access to large and distant markets allows them to structure the production and flow of particular commodities in space and time.

The two chapters in this section present located case studies and theoretical concerns which contrast very differently constituted regions and commodity systems. The distinctions between them frame the dimensions onto which we can map the spatial relationships which restructuring encounters and constructs in place. These mappings can address the changing spatiality of institutions such as labour markets, land markets, capital markets, input markets, product markets, and risk markets in the context of the political powers of regions and nations. Laura Raynolds situates the national concerns of the Dominican Republic in the changing regional context of the Caribbean, as the United States and the European Community redefine their market realms. Brian Page examines competing strategies of regional organisation of the pork commodity chain, reviewing the implications for Iowa, the current core region, of new institutions of organisation and technological integration which subsume on-farm activities to the coordination and standardisation of the industry as a whole.

These two chapters present very different conceptions of *the regional scale*. Raynolds treats the Dominican Republic as a nation within a supra-national region, the Caribbean, addressing changes related to post-colonial, post-Cold War realignments in international trade agendas. She notes that most of the agricultural institutions internal to the Dominican Republic were initiated within the colonial period, but that these institutions have been reorganised in later periods and must now accommodate changes such as the new role of NAFTA in controlling access to the US market, which has moved the Dominican Republic into the Lome Agreement and the European Community economic realm. The spatial relationships Raynolds addresses are primarily peripheral and post-colonial; the Dominican Republic remains at the margins of geopolitical links mediated by externally determined trade policy, the international strategies of other nation states, and the agency of multinational firms. Page, on the other hand, addresses a sub-national region, analysing the new spatial conflicts which result from the competition between institutions built over time by Iowa farmers and the new pork production platforms that the corporate feed formulators and meet packers have set up in the South of the United States. Page's concern is with regions within a national market, regions within a nation state, in which farmers have built strong institutions that are now challenged by corporate initiatives.

The *process of restructuring* in agriculture involves a number of changes, including intensification, new mechanisms of coordination and control of production, and new pressures of product standardisation and just-in-time production. Restructuring in agriculture faces particular impediments, including the continuing role of private property in land and the problem of recruiting capital to agricultural activities (where the importance of nature and of natural variability creates conditions of high risk and low average rate of return to investment). Both authors note the importance of national and international marketing firms and of the substitution of contract production for open product markets. In addition Raynolds notes the role of US AID and other agencies in encouraging the production of non-traditional export crops. The primary force driving restructuring which Raynolds addresses is the gatekeeping role of the multinationals in access to markets for particular commodities.

Page gives us a richer and more analytical view of the internal workings of the process of restructuring, presenting multiple paths in different regions which reflect upon each other as a consequence of their competitive engagement within a national market. He attends to the internal structure of the commodity chain, reviewing the changes which are appearing at the farm level, in slaughter and packing, and in retailing. Page provides an impressive analysis of two paths of coordination and control: organisational change and technological integration.

Change in the structure of the industry is also an important component of regional restructuring. The industrial structure of agriculture involves relatively small numbers of input producers selling their products to a large number of farmers, who then must sell to a small number of firms that then move their products to the millions of final consumers. The system thus involves alternating moments of oligopoly and competition. There is very little vertical integration of farming itself into this system; firms that control these commodities shed the production of the commodities, and the risks involved in this production, to farmers, but control prices through oligopolistic control of markets and other strategies. Within a given region, farmers are likely to find only a few potential buyers, and to have little control over price.

Raynolds takes this structure of coordination and control through monopoly of markets somewhat for granted. She is more concerned with the way that state policy and the state's negotiations with transnational firms restructure the organisation of production in different ways: in pineapples, production follows a plantation model, replacing sugar with pineapples grown on land made available at below-market prices; banana production, in contrast, depends on production contracting, with contractees providing land and labour.

These questions of the restructuring of industrial linkages and of production practices, technologies, and standards provide the richest part of Page's contribution. He traces the shift from independent to contract production of hogs and its connection to the change from diversified to value-adding specialised hog farming. In this he pays special attention to the reorganisation of the labour process and to comparisons which contrast labour in Iowa hog farming with the battery hog farms in the South. In Iowa the foundational arrangements are those of family farming. In the new production complexes of the South, family farms are also involved but have substantially less autonomy; the system depends on different patterns of labour recruitment, capital formation, and articulation with local suppliers and processors. The distinctions are lessening as innovations being tried out in the South are then introduced to Iowa through production contracting and the increasing pressure of technical standards in production.

Both Raynolds and Page attend to the role of labour and livelihood in the regional systems with which they are concerned, and to the issues of gender, race, and nationality, which so often appear in the restructuring of agricultural labour pools and practices. In the Dominican Republic, Raynolds tells us of the redefinition of the labour pool which accompanied the closing of the sugar economy and its replacement by pineapples: the traditional use of Haitian labour was challenged and the effect of this challenge was to substitute Dominican women as a new labour force, supplemented by labour contracting. Page addresses changes in the role of labour, and in the composition of the labour force, at several points. New incorporation of hired workers, often workers distinguished by race or ethnicity, appears in the battery production systems of the New South and in the ruralisation of meat packing, where Mexican workers hired by the new rural abattoirs of the giant meat packing companies have been used to replace the strongly-unionised workers of the old trailhead and railhead plants. The use of a new work force and the relocation of slaughter and packing into production areas supports the spatial integration of the industry, an important part of the restructuring now under way.

Political institutions are an important battleground in the terrain of agricultural restructuring. Raynolds attends primarily to the role of the Dominican state within the Caribbean and the post-colonial reorganisation of provisioning the core countries, a role in which the state is a relatively powerless negotiator as TNCs remap their sourcing strategies. She points out that there are challenges to the legitimacy of the state: food struggles, the insurgence of the internal export agriculture-oriented faction against attempts to restrict agricultural subsidies, and political pressure to change the composition of agricultural labour markets. However, much of what Raynolds describes is the recomposition of externally determined institutions under the pressures of external change. Page's discussion offers much more insight about resistance to the pressures of restructuring. Farmers in Iowa have built relatively strong political institutions which now defend them against the technical and structural rationalisation being imposed by the feed and meat packing industries. Though this has offered some respite, the political culture of family farming carries forward a contradiction which restricts the way that this political struggle is framed: agrarian ideology supports the independence of farmers, but there is also a belief in economic enterprise which limits farmers' willingness to regulate each other's business choices, even where these choices may present an opening to the penetration of corporate coordination and control.

Both authors engage two distinct theoretical traditions – agrarian political economy, widely applied in the literature concerned with agrarian change in the Third World; and the new literature on industrial restructuring and regulation. The integration of these theories offers potential insights to each of these cases, but this integration will require further development: the points of entry of the two theories (class analysis in the case of the agrarian political economy literature; inter-firm linkages and the geography of transaction costs in the industrial restructuring literature) are widely separated. Regulation theory has been proposed as connecting these two points of entry, but the particular class bargain Aglietta found in the US case has only limited importance to farming and the strategies of accumulation in the agricultural sector must always engage the matter of nature in ways from which manufacturing is intrinsically relatively 'free'. In the cases addressed here, regulation theory would not offer much help: the government of the Dominican Republic is a weak post-colonial state and is unable to develop an integrated national social contract; instead it is vulnerable to alliances between particular internal interests and external powers. The situation in Iowa is too complex to be well portrayed by the relatively broad brush of regulation theory; Iowa is a region in which farmers' political struggles, in the tradition of agrarian populism, have built a set of institutions at the level of local, state, and even national government, based on the need to protect farming against the power of surrounding corporate interests, but also on the belief in the benefits of the free enterprise system. Class analysis has not been a part of this approach – indeed farmers and labour organisers saw a common anti-corporate goal in the height of the populist movement. Though Iowa farmers appear to some as propertied labourers, they see themselves as Jeffersonian freeholders (a quite different view).

CONCLUSIONS, FOR THE MOMENT

What pictures do these two chapters give us of restructuring in agriculture at the regional scale? Both authors assert the path-dependence of agricultural change, the need to recognise the importance of existing institutions and practices in the process of transformation. Both authors assert the importance of new contracting relationships, rather than vertical integration in a more traditional form, as linking farmers to the large corporations which coordinate and control particular commodity complexes and manage access to markets. But this change is not related to new processes of globalisation in these two cases: in the Dominican Republic the impact of global forces, and even the transnational corporations involved, are not new – and in the reorganisation of pork production, the forces bearing on Iowa are external to the region, and use competition between regions as a strategy for discipline and domination, but they are entirely within the nation state. The changes addressed in these two chapters treat this new phase of agricultural reorganisation as a part of the uneven development which capitalism imposes on any economic landscape – not something which is new in kind.

Is globalisation an overarching developmental trajectory for agriculture? Philip McMichael asserts that we are facing an epochal 'shift from a nationally coherent to a globally competitive economy . . . [that] the content of contemporary agrarian politics is inexorably shifting from the (nationalist) agrarian question to the (internationalist) food and green questions' (McMichael 1994: 3, 15). In contrast, Richard Le Heron suggests that the globalisation of agriculture is a political choice (Le Heron 1993). Globalisation may appear a necessary process to capital in agriculture, faced with the globalisation of manufacturing and finance capital. However, it need not be necessary to the livelihoods of farmers or consumers. The development and defence of strong institutions at the regional scale can foster institutional and technical innovation, and can help to build alliances between farmers and consumers which contain the abstraction of power over the food system to the corporations which mediate market links. But these counters must be consciously built by those who want to recapture this power.

BIBLIOGRAPHY

Le Heron, R. (1993) *Globalized Agriculture: Political Choice*, Oxford: Pergamon.
McMichael, P. (1994) *The Global Restructuring of Agro-Food Systems*, Ithaca: Cornell University Press.

PART III

GLOBALISATION, VALUE AND REGULATION IN THE COMMODITY SYSTEM

7

CREATING SPACE FOR FOOD

The distinctiveness of recent agrarian development

Terry Marsden

A PERSPECTIVE

The past century has exhibited the process of residualisation of the agricultural and the rural. Food has been taken for granted. By the end of the century this residualisation process has come to an end. For a variety of new reasons (diet, health, amenity, environment-nature concerns, reactions to modernism) the rural and the food resource defines a new centrality in people's lives. Its value is in a process of differential reconstruction.

This chapter attempts to set out some revised concepts which give more clarity to the 'post-Cold-War' development of agrarian space and the significance, more broadly, of food. The analysis is based on recent research conducted in Europe, Brazil, and the Caribbean, and specific reference is made to some of the findings in the second part of this chapter. First, it is necessary to make some general points as to my current critical standpoint. In recent papers, I and other colleagues (see Arce and Marsden 1993 and 1995) have been reconstructing an analysis of globalisation and food which provides a more sociological and geographical perspective, incorporating social agency, contingency, and uneven development into analysis. This has been done in a spirit of building upon the more macro/food regime models, and has focused specifically on attempting to understand and conceptualise the new patterns of globalisation.

In the 1990s there are new agrarian questions to be addressed, requiring a more sensitive analysis which considers the expression of contradictions and contingencies of postwar capitalist development. In this regard there is a need to move from a broad conceptualisation of *food regimes* to one which focuses upon *food networks*. In this chapter I will attempt to justify why it is necessary to reconsider earlier conceptions, and begin to point to the ways in which this can be done. The significance of this task is heightened in the contemporary period. Food networks play an increasing role in the social and political development of regions and nation states, (i.e. it's not just 'global'), in new and uneven ways. First, it is necessary here to outline some points of departure which set the scene for the development of food network approaches.

SOME RECENT DEPARTURES: FOOD NETWORKS AND VALUE

It is important to recognise some key changes in the dynamics of agro-food networks.

1 Aspects of control, power, and dependency in agro-food networks are not only based upon input-oriented corporate capital (i.e. agribusiness firms and their relationships to the farm-based sector). Increasingly, they are associated with the control and construction of *value* from the point of production. This serves to empower near-consumer agencies. This reflects the common point that 'value-adding' in food has tended to be associated downstream of the point where it is primarily produced.
2 Food networks are characterised by the significance of 'action at a distance', as well as by local and regional 'nested' links, whereby the design of foods are shaped largely by the non-farm sector.
3 Globalisation of food is paralleled by intended and unintended local and regional effects, such as the reconstitution of local diets, food scarcity in potentially agriculturally abundant places, and surges of oversupply during volatile international currency conditions. These, in their turn, shape globalisation processes.
4 Technological developments and their social effects (in terms of our established concepts of subsumption, substitutionism, and commoditisation) need to be situated in the broader context of knowledge networks, particularly surrounding the ways in which food quality is constructed. We are moving towards food supply chains which are qualitatively regulated rather than quantitatively. As a result, food 'value' becomes increasingly multi-layered, containing commodified and formerly non-commodified attributes.
5 Nation-state regulation does not always have a functional relationship with aspects of globalisation. Global processes operate largely on the basis of corporate manufacturing and retail capital; but national processes still have to maintain 'a public interest' for legitimatory reasons. Retailers (and their relationships with the nation state) become dominant actors in the reconstruction of internationalised and national food networks (see Rabobank Nederland 1994, Marsden and Wrigley 1995, 1996).
6 Food markets are becoming 're-regulated' around the demands of segments of consumers (in the North, and in the 'Westernised' parts of the South) and corporate retailers. This involves the re-naturalisation of foods as well as their continued industrialisation.
7 These processes reproduce different mosaics of regional production and

consumption spaces, new *dynamic* and *dependent* spaces providing major forces for redefining uneven development.

Three such regional spaces include new agro-industrial districts (such as the São Francisco Valley, Brazil); quality consumption and retail spaces (much of Northern Europe); and agricultural retreat and local vulnerability spaces (the Caribbean). These spaces are interlinked and cross-cut with dependent relationships and networks of power. For instance, under globalised conditions the legitimatory relationships established in one place (Northern Europe) may be significantly at odds with those in another (e.g. the Caribbean). As a result, and if food commodities move between these places, they may carry their natural features (freshness, colour) with them but leave their more embedded social legitimatory relations very much behind. Indeed, it would seem that valued consumption spaces continuously need to devalue other spaces in order to reproduce sensitive capital accumulation. Distances have to be reproduced even, indeed particularly, in a more deregulated global terrain. In this sense globalisation and deregulation tend to redistribute risks and power, making it easier in one place to extract value only at the expense of other places. This becomes a key aspect of uneven development – the parallel but contradictory construction of value – under the new globalised conditions. In this sense, globalisation is anything but a new surface eco-geometry. Rather, as this chapter attempts to show in comparative focus, it is socially insinuated into the regional organisation of food production and consumption.

These sets of principles emerge from a series of dialectical relationships and contradictions in the food sector. These concern: production/consumption; globalisation/localisation; deregulation/re-regulation; valuation/devaluation; powerful spaces/dependent spaces; cultural distinctiveness/social exclusivity. It follows that analysis of food and rural resources in the late twentieth century needs to explore how these contradictions are sourced, spatialised, and played out, and how key actors and agencies are engaged in them. This places an emphasis not only on systemic approaches. It requires a focus on how discontinuities and contradictions are *put together* in order to make things happen.

QUALITY, REGULATION, AND CONSUMPTION: KEY CONCEPTS IN THE RESHAPING OF THE NEW AGRARIAN SPACES

Value-adding has become the central concern for the post farm parts of the food chain (Rabobank Nederland 1994). However, to concentrate on the increasingly diverse ways in which corporate capital achieves this through the redesign of foods tends to restrict our focus. The construction of value of foods, and of the labour and environmental conditions to produce and convey them, requires a renewed and more integrative focus. The construction of food quality has to be seen in the context in which it is embedded in environmental value and labour value. With this in mind it becomes important to consider more broadly social theories of value and the broader definition of commoditisation. Regulation and regionalisation are inputs and outcomes of these social processes.

In empirical work in Brazil and the Caribbean a focus on this value construction in the fresh fruit and vegetable sectors has demonstrated that traditional constraints on land-based production contribute only one element in the distinctiveness of agrarian development. The value of labour and the environment are traded off against the priorities for particular conceptions of food value and quality. Risks are allocated in the networks of food supply (e.g. producers, exporters, traders, regulators) around the need to maintain different quality criteria. Under these conditions powers and risks become rearranged around actors engaged in the carrying and exchange of food goods. The externalisation of markets empowers some of the carriers and exchange actors over producers, and indeed over some national state regulators. Some networks are formally organised, some are not. Both have to deal with uncertainties translated from distant consumption spaces.

In the dominant consumption spaces in Europe, corporate retailers play a key role in translating the quality definitions of foods globally. They become key players in the allocation of risks and constraints in supply networks (Tordjman 1995). In addition, they develop their own regulatory systems which ensure a dominance over the supply of main food products. This is a highly competitive and contingent process which is vulnerable to consumer reaction. Its success depends upon the social and political maintenance of a set of power relations which have to be legitimised. In consumption spaces of the North, this is increasingly an exchange-based system, rather than one dominated by industrial relationships. Sixty-nine per cent of European consumers claim to buy their groceries from supermarkets most often, with consumers making an average of 2.14 supermarket shopping trips per week (Tordjman 1995). Moreover, on the supply side, the saturation of the quantity of foodstuffs in most developed markets has resulted in a more or less constant or declining value for the primary sector, despite rising consumer expenditure. In the US, for instance, 'farm value' has remained virtually unchanged in recent years whilst 'the marketing bill' by 1988 was almost twice what it had been a decade before. The marketing bill – incorporating the added value of the industry, trade, retail, and food services – is three times as large as farm value (Rabobank Nederland 1995).

SOME ANALYTICAL IMPLICATIONS

Some of the principles set out here justify the need to counterpoise micro-sociology with macro concerns. Both need to be brought together in new ways. This requires conceptual development whereby social actions can be explored in their different contexts of value construction, associated with globalised and regionalised food networks. New forms of institutionalisation, regulation, and spatialisation become significant in the uneven development of agrarian space (Watts 1996). This involves exploring differential processes of social translation. Global processes have actual translations at different spatial and social levels of interaction. They are constantly internalised and layered by different actors in networks of relationships. In particular, the bases of social action in globalised food networks are associated with quality, regulation and consumption. However, to recognise this does not explain how globalised food networks come into being; how they may be perpetuated or adapted; or what the differing local and regional effects of these transactions and relationships might be. In addition, we are unclear how these concerns lead to new forms of local and political consciousness which can then act to reshape local and regional spaces. In positing these questions in a rural context, we have to begin to look at both *the people* and *the foods* in the networks of supply and how these share or allocate power and responsibility, relocate risk and penalties; and begin to construct international markets and uneven forms of regional development. As Feierman (1990: 36) argues more generally, 'The wider world is not external to the local community, it is at the heart of the community's internal processes of differentiation.'

As we shall see from the case examples, a focus upon the ways in which globalisation becomes incorporated into the social life and action of food networks – in different spatial formations – raises important questions concerning the ways that food networks themselves are situated and embedded in localities. With reference to the case studies below, we can begin to illustrate how the relative significance of food and food networks in and through spaces is variable, according to the main activities in those networks, and the relative power to capture the social value of food products.

In addition, and as we have argued elsewhere (see Marsden *et al.* 1993, Murdoch and Marsden 1994) food networks as a whole are cross-cut by 'lateral' processes of restructuring which have traditionally had little to do with the food sub-sector. Nevertheless, these boundaries of economic, social, and political activity are in a process of transformation and *reconnection* in both the North and the South. A focus upon the situational position of food networks in different spaces begins therefore to raise important questions concerning the elements involved in the processes of uneven development. Food networks thus hold vertical and horizontal connections with the spaces in which they are situated. The social sums of both begin to reshape rural space both 'from the inside' and externally *with other spaces*. Socially and politically defined dynamisms and dependencies are formed and reinforced. Dependent as well as dominant spaces are created.

In what follows we can illustrate these features which are creating uneven agrarian spaces by reference to different *but related* case examples. These provide a basis for developing a more effective comparative analysis of globalisation and regulation, focusing specifically upon the differential and spatialised construction of value of foods.

THE DEVELOPMENT OF NEW AGRARIAN DISTRICTS: THE SÃO FRANCISCO VALLEY, BRAZIL

The São Francisco Valley development represents the largest irrigated agricultural development in Latin America. It is based upon the development of export agriculture, particularly mangoes, grapes, and tomatoes. By the 1990s, the realities of significantly increasing exports, partly through the expansion of irrigated lands, but more recently through productivity increases, have to be underpinned and conditioned by constructing and maintaining specific and situated quality conditions of the fruits (i.e. mangoes, tomatoes, grapes and, more recently, acerola). The nature of these quality designs and conditions extend from planting through to the point of consumption. They are not simply associated with achieving the criteria set out by either the domestic Ministry of Agriculture (Embrapa) or those of importing nations and retailers (such as USDA). Even though these agencies do have increasingly strict criteria, involving site inspections and continuous monitoring, it is important to recognise that surrounding these legal and formal values lie a series of conditions which are set by other local and external agencies. While the origin of these primarily deals with the dynamic maintenance of the quality of the food product, this has to deal with setting parameters for the specific labour and environmental conditions as well. The construction of these quality conditions, their implementation, and their ability to allocate risks and responsibilities to different parts of the food network, means that they play a significant role in constructing and maintaining the fruit supply networks themselves. In the highly regional and local competitive conditions, where overall aggregate supply becomes less of a concern to importing nations than those concerning specific design and quality, producers and exporters have to be geared to providing a consistent 'quality product'. This may vary between importing countries and it demands different production, packing, and quality control systems. This concern for quality and design increasingly emanates from the strategies of the corporate retailers situated at the consumption end of internationalised food networks. For instance, in a rather economistic and aggregated way a recent trend report on the global food trade argued:

> The retail chain makes strong demands on security of delivery and product quality. The need for large scale and security of delivery leads retail chains into backward integration. This is becoming more easy for the retail chain to effect through the increased scale of supply and

> the liberalisation of world trade. This may lead to the exclusion of the importer. Exclusion takes place earlier in the case of simple products like citrus. The importer will have to continue to manifest his added value for the retail chains. In addition to supply and guarantee of quality and logistics, storage and transhipment are important functions, through which delivery can be on demand. The transport function can be partly contracted out.
>
> (Rabobank Nederland 1993)

At the local level those actors and agencies who attempt to control these diverse quality forms begin to play a pivotal 'social carrier' role in the agro-industrial region. Petrolina, the main service centre, becomes a focus for merchants, trade, and communications which act to solidify and extend local producer and exporter networks. An agro-industrial district begins to be formed. Some of these key actors and agencies become the conduits of globalised knowledges, and internally, the gatekeepers to globalised market entry for the numerous and variable producer sector. Agreements and contracts have to be socially based and continually reviewed. This places a considerable pressure on the maintenance of quality management and its articulation through food networks.

At the heart of these processes of food network construction is the differential creation of value through the coming together of different actors to produce and shape nature. This produces a major social process. In the valley, the gradual divestment of state responsibility for the agricultural development based on irrigation (under CODEVASF) over recent years has led to the development of large agricultural production enterprises that have developed an export function. Fruit exports can be sold through farm cooperative organisations (such as VALEXPORT), but the increasing scale of development in the region has spawned new private enterprises which focus specifically on promoting and gaining entry to export markets (e.g. FRUITFORT, MAPEL). While most of these have benefited from state supported development projects (e.g. NILO COELHO) in terms of infrastructure and initial technical assistance, they are effectively privatised firms which control extensive areas of production and recruit other smaller producers in order to export their products. These firms are located in pivotal positions in the globalised food networks. They can place strict quality control criteria on smaller family producers, and they can balance the proportion of fruit which comes from them vis-à-vis their own estates. In addition, because of their modern packing house facilities they are well placed to translate globalised knowledges about retailer and consumer demands concerning specific colour, size, shape, and content of the crops. This information is translated to local producers who may have agreed a temporary export contract with the exporter enterprise. As one of the largest export firms (with control of 1,015 hectares of irrigated land largely devoted to mangoes and grapes) argued:

> Quality control is very severe. We have to follow the product process. We go onto the farm and inspect regularly. Nevertheless, possibilities are wonderful. Conditions are good. I cannot see any bottlenecks to this production. We have the most qualified and technological status. In São Paulo they are being eradicated.

This company is currently arranging an export programme to Japan, with most of current production going to the US and Europe. By 1994 it exported 1 million boxes of fruit. By 1998 it expects the volume to have increased to 2.5 million.

As well as the degree of arm's-length control these enterprises place on their enrolled producers (see Friedland 1994), they have to take particular care in the harvesting and handling of the fruit from the trees to the packing houses. Also, the timing of this process has to be modulated to fit the gaps in market supply of the importing countries and the maturation of the fruit such that it reaches the consumer's table in the 'correct' condition. This distinctive process of time synchronisation not only to market entry but also in the managed maturation and quality of the fruit is crucial to the continuation of the food network itself. For producers and exporters alike, a particularly vulnerable *time–quality episode* is that concerning the multiple and embedded handling and packaging of the fruit. The quality and its risks can change with the mobility and handling of the product. Also, large differences can emerge between family producers and large enterprises, leading to the rejection of fruits for the market. One exporting firm had ceased taking the fruits of small producers because of this variability in control and the effects it had on the delicate appearance of the mango. It is necessary to achieve an estimated time of twenty-three days between harvest and actual consumption of the mango. The appearance and quality can severely decline after thirty days. Most of the exporters are responsible for the fruits until they are landed in the host country. Also, they may not know who the buyers are until they have left Brazilian shores. Dealing with these global contingencies but at the same time maintaining quality surveillance of the products is the responsibility and major concern of the export enterprise. Moreover, the precise links in the networks are not necessarily complete or clear during critical time–quality episodes. The whole network does not always exist at the one time. Rather, different time–quality episodes are progressed from farm production, through harvesting, packaging, transporting, and distributing products. This contingency, rather than being a random or residual event, becomes a major social context within which risks and responsibilities are allocated.

In the production–harvesting–packaging process, the inevitable handling and carrying of products comes with its own risks. This implicates and promotes a gendered set of labour relations. Within the protocols concerning the limits on pesticide and herbicide uses the problems of the biological risks to the fruits is a real one. American importers, for instance, have become concerned with the incidence of the fruit fly. This has resulted in a more stringent quality control procedure being instituted in the local packing houses. This involves the drenching of mangoes in water and petroleum beeswax, with the grading process being followed by USDA officials who are rotated at monthly intervals. They have to be paid for by the exporter firm. For the European market and packing process there is a more rudimentary process, with little checking or external monitoring. It is argued by producers and exporters that drenching changes the inherent quality of the mango, but that the effect of communicating and signifying safety, standardisation, and aesthetic pleasure (by adding an externalised value to the product) ensures the maintenance of the American market.

This example of how one of the importer countries' concerns for quality gets *fixed* into the operation of the network demonstrates how quality parameters and fruits and their labour processes are malleable to different exporter and importer concerns. The constraints inherent in the disparities in production and labour time in agriculture vis-à-vis industry (see Mann and Dickinson 1980, Mooney 1985, etc.) are only one set of social constraints which give fruiticulture a distinctive position in modern capitalist agrarian development. The malleability and the sensitivity of many of the products (in this case particularly grapes and mangoes) coupled with the growing external demands for specific quality conditions, means that the management in the food networks of the specific time–quality episodes from harvesting to point-of-sale provide a significant and distinctive basis for social action and the reconstruction of value in food networks. In such regions as São Francisco, where irrigated systems can establish a harvesting pattern which continues for eight months of the year at least, this constant process of managing and coordinating time–quality episodes is a major and dominant feature of the agro-industrial region and the networks of food which flow through it. Those actors and agencies who are closest to the definition and implementation of quality conditions begin to accumulate power in the food network. This leads to a growing social and economic differentiation in the region, with smaller producers, not exclusively, but prone to exclusion in globalised food networks.

GLOBAL AND REGIONAL DIMENSIONS: REGULATING VULNERABILITY IN THE CARIBBEAN

Agriculture and environmental resource use in the Caribbean and in Barbados specifically has to be seen within the broader structural context which partly defines it. It is necessary to realise that the power to define resource use predominantly lies with the logics of global forces even if these can be adjusted and resisted at the local level. Of particular significance in understanding the variable deterioration of environmental conditions in many of the small Caribbean islands (Watts 1996) is the need to appreciate how agriculture is both externally and internally defined by particularly powerful actors and agencies. Largely, this definition sees agriculture as a predominantly export commodity system capable, to varying degrees, of generating foreign exchange and easing nationally based debts and balance of payments conditions. This logic defines notions of efficiency on the one hand and marginality on the other. Small farmers, once seen as marginal by the plantocracy are now marginal to the external commodity markets. Agriculture is largely stripped of its ecological and human significance (see Polanyi 1944, Fitzsimmons 1989). Under these conditions environmental vulnerability also becomes marginal to the priorities of production. As a result, and despite serious efforts to improve agricultural systems, this region is characterised by the progressive devaluation of food and agriculture.

In this context it also has to be recognised that agriculture and food have a particular position and definition in macro-economic planning and development. For most islands agriculture represents a declining economic sector whereby policies need to be established to arrest it for the purpose of maintaining some level of viability and national contribution to the debt crisis. In islands such as Antigua this argument has all but been lost. In Barbados and St Vincent plantation and small-scale agriculture still have a recognised role, but the legitimacy of the sector vis-à-vis tourism, residential and recreational development, the services, industry, and off-shore banking is declining. Nevertheless, it is important to view agriculture and food provision more generally as a significant part of regional macro-economic regulation. Its 'management' in this context allows other forms of development and accumulation to occur. Its marginalisation allows for the selective development of foreign capital, and a growing importation of processed foodstuffs to meet domestic consumption.

Two developments concerning these trends produce increased vulnerability for the Caribbean. One is the pressure for the cessation of preferential access of sugar and banana exports to the European markets (Agra-Europe 1994) not least from retailers and traders in Europe. This gradual process means output and competitive price reductions for producer countries. It can also exacerbate the volatility of balance of payments problems which then can restrict other forms of export growth. In Barbados, for instance, the sugar industry has been in deepening crisis (Drummond and Marsden 1995). In 1980, sugar accounted for 6.3 per cent of total GNP, by 1992 this was reduced to 1.7 per cent. A second vulnerable factor concerns the outcome associated with the GATT Uruguay round. The likelihood is that the markets of developing countries will assume greater significance to the export performance of the developed countries. Pressured by the transnational food sector, the latter are likely to insist on reductions on tariff and non-tariff barriers. The gains for the small developing countries are likely to be few. Indeed, while liberalisation is the strongest and most irresistible trend in global trading, its advantages seem to have been captured by those economic interests which can more easily transfer operations to areas where new markets for food are emerging. For instance, the development of Northern retailers and food manufacturers in countries such as Barbados is likely to further constrain local systems of food provisioning; or at the very least, marginalise their sub-economies. In this context, and particularly concerning the food sector, the model of development adopted is one which aids the interpenetration of foreign capital such that it provides new markets and opportunities for certain segments of Caribbean society over others. The most powerful interests are strategically shifting capital out of domestic agricultural production into retailing services, food processing, banking, and tourism.

It is important to recognise the interlinked nature of some of these trends, and to see how their implications are being played out in different national settings. For many of the entrepreneurial, merchant, and private professional groups in Barbados, these trends are seen not only as inevitable but as necessary for the continued development of the island. Further externalisation, built now not so much on the traditional export of sugar, is seen as the best way forward in the context of multinational trade negotiations and the bilateral trade and economic assistance agreements. Moreover, these conditions and logics are reinforced by the conditions of development aid and balance of trade support programmes financed by the multilateral financial institutions. Debt servicing and the need to attract foreign currency earnings reinforce the tendency of seeing agricultural production as if it were like other sectors of production and services. Consider the following quotation, for instance, from the Chairman of the Barbados Development Bank:

> When we talk about agriculture we talk about sugar agriculture, separating from non-sugar agriculture. What is the optimum acreage we should have in cultivation? The indications would be that the acreages have been on the decline over the years. We used to talk of 60,000, now it is more around 30,000. We would like to raise this to 70,000–80,000 tonnes. But the way things are now, fluctuating. In 1992 we had a bad crop, we came back a little in 1993 with just over 43,000 tonnes. We know for the coming year [January 1995 reaping] it is going to be way down on this year.
>
> Why I raise all this is the question of the preferential treatment, the agricultural protocol under Lome. What are going to be the terms and the conditions? Again the indications would be that the preferences will be minimised and eventually eliminated. So that's going to throw us out into the open market. So it brings into question the issue of efficiency, and can we produce at sufficiently low cost, efficiently.
>
> Now as a sort of hiatus you are looking at diversification and there

> are certain areas coming to mind. Cut flowers is one. There are certain other crops that we can concentrate on, but again we have to look at the perishability of these items, and the transport system to get these to market as quickly as possible.
>
> Foreign exchange earnings from agriculture are from all aspects a key factor as far as we are concerned. The way I see it, you know, when you do your economics in the early years you learn about the factors of production – land, labour and capital. But now one of the most crucial factors is foreign exchange. Foreign currency this is a fact.

It is worth considering how these so-called structural conditions are promoting a diversity of local responses. They are far from uniform in themselves, and they demonstrate a key feature of the significance of uneven development in the South. Trade liberalisation and deregulation do not lead to any form of level playing field. Rather, they are responsible for the selective devaluation of certain domestic resources so as to encourage the penetration of external capital and expertise. Endogenous capabilities and resource structures are further 'hollowed out'. In this sense they set the new rules for uneven development and the lack of endogenous cooperation between the islands.

Such processes of externalisation and the exposure of national-based diversification schemes to global market pressures tend to exacerbate the ephemerality of new nationally based agricultural initiatives. Meanwhile, imports of food products increase both for consumption by the local population and for the burgeoning tourist and hotel trade. This in turn increases the pressure by national governments to organise export drives of 'quality' products. As a Barbadian trade commissioner argued:

> Our imports are about ten times to one. . . . Our export figures are not substantial because we are competing with other countries and we have to look at when summer comes in and or winter comes in and then we get our products on and so on. . . . What I have been doing is trying to get them under contract. So that they [the farmers] are committed by contract to supply at a specific price so they will have consistency of pricing throughout and they are demanded to supply. If you are caught doing those things which are in breach then you can be put out on the limb. The contract should be between the farmer and exporter. Because there is no consistency of supply the exporter has to take it when they get it, so there is a big problem in terms of the consistency of supply. That is where I come in. I believe in order to stay in export we must have consistency of supply.

In Barbados over the past decade there have been a variety of attempts to reorientate the agricultural economy around the production of vegetables and fruits for both the domestic and export markets. This has been a major source of internal conflict and competition between different sectors of agriculture (e.g. small farmers and the sugar plantations wishing to replace sugar with other lucrative crops; and older and younger farmers who variably wish to retain a domestic system of provision built upon vegetable staples such as sweet potatoes, yams, cassava, carrots, cabbages). Diversifying the agricultural base in this way has several justifications which are often in contradictory relationships with each other. For the national and international trade and ministry groups it is a strategic opportunity to capture foreign exchange in the context of the debt crises facing sugar. Some of our respondents (small farmers, ministry officials) complained that while it had been exclusively the fault of the plantocracy and the large landowners that this debt crisis had arisen, government and other elites were expecting the small non-sugar sector to deliver a panacea to this long-running problem.

By 1994 there were approximately seventy-two small farmers growing hot peppers mainly for the export market and being encouraged by the Ministry and the marketing organisations. The top quality peppers go to North America and Europe. The rest are exported as sauce or mince, or sold domestically, again according to their quality. As another Ministry official, who is responsible for exports of these goods, argued in 1994:

> We have increased the acreage in hot peppers. We have increased the export of hot peppers, but because of this increase we found people want more and more in different niches in Canada, Holland, the UK. The requests are more. For all things, yams, okras, hot peppers, sweet potatoes. You find that requests are more per year. So although we increased production it does not go on a par with demand. I am hoping that by the end of next year that we can be up more than 50 per cent on supplying on all these requests.

Larger producers become a threat to the opportunities for the use of these markets in enhancing the small farm sector, and in realising some of the state investment placed over the decades in rural development schemes, irrigation and cooperative marketing systems. Many small farmers are sceptical as to the commitment of the state to their maintenance over the medium term. Moreover, the need to provide high quality and regularity of supply and to do so with the minimum of risk means that the larger producers can potentially capture the market advantages. The 'situation of exchange' (Appadurai 1986) in this sense is both highly multidimensional (in that it involves different groups of actors and agencies often for different crops) and competitive (in the sense that the prevailing agricultural conditions are such that small and large farmers are competing for the same export bounty). The networks of production and supply are thus anything but rationally organised or structured and regulated in a coherent way.

While many in government departments and the national agricultural society (which provides extra extension services to small producers) see the need to protect the markets of the small producers and the opportunities vegetables provide for intensive small scale production, the major marketing agencies and trade officials emphasise quality and regularity of supply by the development of far more regulated systems.

The larger producers and the processing interests are also seeing the development of the new export crops as one of the only options for the maintenance of a viable agricultural sector. One key actor and carrier of these processes is the director of the largest privatised dairy, who also holds the largest dairy and beef enterprise on the island (as well as many other industrial interests). He argues:

> I think that Barbados needs to diversify out of sugar to the extent that sugar only goes to 80–85,000 tonnes. I think the other lands should be put up in producing exotic tropical fruit, some beef on the land that cannot be mechanised, vegetables that are practicable. This is practical and I think small farmers and larger farmers should pay a lot more attention to exporting things like breadfruit, yams, sweet potatoes, and that sort of thing, which were grown here before in very large quantities. Going into it is going to necessitate capital. And agricultural capital has to be cheap capital and it must have a moratorium. Bankers on this island are totally committed to the buy and sell trade. Where you buy it this week and sell it next and pay, that sort of thing. Agriculture is not that sort of thing. I mean you can start an agricultural project and have a year like this where there is disaster all around you and don't pay back anything. It must have a moratorium. . . . The IDB, IMF or World Bank needs to put something together quick or this island is going to look like Antigua – devastated. It is just an island of bush. I see it coming here.

The prospect of good returns from the sale of fresh vegetables in domestic and export markets provides a major stimulus for agricultural change amongst the small farm sector of Barbados. This puts it increasingly into competition with the larger farmers and their counterparts on neighbouring islands who are following similar strategies. At the national level the markets are becoming more diverse, with contracts and relationships being developed directly with supermarkets, hotels, and public services (e.g. schools and hospitals). Competition to gain access to these markets becomes fierce amongst the farmers. Quality and regularity of supply has to be defined both by private processors and marketeers, but also increasingly by government. This process of defining quality confers power upon certain actors and agencies over others. This all tends to constrain moves towards cooperative activity. Where this does occur it is difficult to sustain when 'once and for all' price gains can be obtained from other markets. In addition, the existence of these relatively new markets increases the volatility in the prices in local markets. Rapid price variation induces more uncertainty for the small producer.

At the level of production the farmer is faced with attempting to balance these contradictory forces over production cycles. The inadequacies of the local supply of knowledge and materials makes the recourse to the North increasingly likely. Many farmers and extension officials were purchasing their supplies and gaining their knowledge from the United States. The provision of water and its escalating costs, particularly when abstracted from the domestic supply, increased the financial risks. Carrying out intense searches for potable water sources was a major factor in the work of agricultural extension officers. To survive under these circumstances increasingly requires the recourse to alternative sources of income (for instance, through development gain, when and where it can be achieved). Where this is unavailable it is difficult to maintain the level of investment required without government assistance.

The attempts by international and national agencies to define agriculture and food as commodities like any other (as commodities divorced from their material and ecological base and value) has tended to reinforce the growing scarcity and marginalisation of environmental and agricultural resources on small islands.

RETAILERS' REGULATORY TERRAIN: MAINTAINING LEGITIMACY IN THE UK

In an earlier paper which set out a broader research agenda called 'The Social Construction of International Food', we argued at the international level that it was now necessary to examine the

> socially mediated processes of food production and consumption of commodities, interfaces of different values in commodities, actors' internalisation of new cultural perceptions of health and the environment, and the unintended social consequences in those localities both producing *and* consuming international food commodities.
>
> (Arce and Marsden 1993: 309)

An important point was that actors from both the North and the South, through their social practices and interactions with local environments and cultures, create fragile bridges across international value networks. They unequally exchange values internationally. Implicit in the international and national food value exchange are the different types and spatial scales of regulation which influence the corporate food sectors and producers both within and between regions and, for a time at least, formalise the social values associated with food goods. Value exchange between regions and between different agents and actors in the food supply networks has to be constructed and mediated. Retailers play a key role in encouraging forms of exotic and globalised consumption which in turn influence the value parameters present in the production spaces discussed above. While their relative social and political power varies considerably across the advanced world, they have become major players in the social definition of the foods and the images and identities of food. In addition, their international influence in food sourcing far outweighs their cross-national selling power, even though the latter is likely to grow over the next decade.

It is worth examining the British retail sector in its global context. Whilst it is well recognised that food markets are fairly stable, producing inelasticities which contain sales, both in the North and the South evidence suggests encouragement for the corporate retail sector in particular. In the North, while food demand may have stabilised, the food market is expected to grow at about 2 per cent annually. Improvements are largely through the value-adding process, which often means no more than effectively 'de-industrialising' former 'Fordist' foods (Arce and Marsden 1993). In the South demand is increasing, especially in Southeast Asia. As incomes rise for some, pre-packed products of high quality and products which have undergone further processing begin to replace some staple foods, offering retailers significant opportunities. In Southeast Asia the food market is expected to grow by 4–7 per cent per annum. These trends are not just associated with Southern regions but also significantly include the 'new dependent spaces' which are currently outside but contiguous to the regional trading blocks of Europe (former COMECON) and North America (particularly Mexico and the Caribbean). Moving along a trajectory of larger scale outlets which reduce operating costs and encourage one-stop shopping, retailing worldwide would seem to be in the ascendance (see Agra-Europe 1994, Rabobank Nederland 1994). Unlike other sectors, however, such as manufacturing, these global trends are not necessarily built upon the global interconnectedness of retail capital. While there is evidence of retail buying alliances and some cross-national activity, the key to globalised forms of retailing lie in the potential for convergence of national and international regulatory conditions. While these may principally be seen to be the preserve of the national state, retailers through their international and national political and regulatory activities are increasingly influencing the reconstitution of those states. Of particular significance here is the increased national flexibility for retail concentration. The ten largest chains in both the EU and USA account for roughly 30 per cent of total grocery sales; in individual US states and EU member states (most notably, Denmark, the Netherlands, France and the UK) the three largest retail chains account for 40–60 per cent of the grocery market. This not only gives them a strong national base from which to explore transnational expansion – for example, Tesco's announcement to build twenty new stores in Hungary (Buckley 1995) and Sainsbury's incursion into the US – but it also enhances their powers over the food processing and wholesaling sectors, which increasingly have to answer to them on quality and value grounds.

In the British case the rising economic and market significance of the corporate retailers is well recognised (see Flynn *et al.* 1994 and Marsden and Wrigley 1995). While global pre-tax margins are relatively low (around 0.5–3 per cent), with gross margins ranging from 20–30 per cent, improvements in efficiency and/or the reductions in costs of goods sold (and/or increases in the turnover of sales) can have a potentially large influence on net margins and profits. Although British retailers have faced pressure recently (see Wrigley 1995) they have maintained and expanded exceptionally high profit margins – 5 per cent in 1980 to 9 per cent in 1990. These exceptional margins are usually interpreted in economically rational ways. These realities, however, are outcomes of a contested process of market maintenance and expansion which has to be deeply embedded in the social and political apparatus of the nation state, and increasingly the EU. As with any type of economic activity, and especially in the case of retailing, given its 'near market' location and traditional reliance upon creating markets under highly competitive and inelastic conditions, it has to rely upon and begin to remould the social and political conditions in which it finds itself.

On the face of it British retailing has been typified by high gross incomes relative to its international competitors through developing efficiencies on sourcing and the rapid development of large stores. This has allowed the efficiencies to be captured 'in house' rather than passing them directly to the consumer in the form of lower prices. In addition, looking internationally it can be argued that, for instance, German and North American counterparts have traditionally faced more competition from the discount sector and that British retailers have been more effective at developing distribution technologies.

Probably more significant in terms of these features of the uneven development of international retailing is the speed at which food commodities are *rotated*. Retailers are not interested in backward or forward integration. Indeed they prefer, increasingly, as we see in the São Francisco case, to reallocate the considerable risks in food procurement and quality maintenance to the other actors and agencies involved. Their prime focus is to increase the speed of turnover as well as to demonstrate product innovation and new value-added quality. In the US, for instance, the late 1980s and early 1990s have witnessed a rapid increase in product development. In 1991, 16,143 new products were introduced, with over 77 per cent of these being new types of food. This represented a 100 per cent increase over the previous decade.

Even taking a strictly economic approach this suggests that there is more at stake than margins and outputs. British retailers have given considerable emphasis to reducing the number of days stock is held and in balancing off supplier and banking debts by cash cycles which maximise their control of funds and stock. Here, in parallel to the time–space constraints associated with our production and supply spaces outlined above, it is the speed towards the point of sale which is all important. Efficiencies in the days stock is held has allowed the shortening of the time of internal funds rotation. In Europe, average days stock is held tends to vary between twenty-five and thirty days. This has been reduced significantly in the UK and Japan. In addition this factor in uneven retail development works not only internationally, but also intra-sectorally. It increases the efficiency differences between the 'super-league' of retailers and the rest, particularly the ailing independent grocery retailers. In the UK the total number of retail businesses has declined from 142,000 in 1976 to 86,800 in 1990. This has largely been at the expense of the specialist independent sector (i.e. the small multiples) and the specialist retailers (butchers, bakers, grocers, and dairies). While this trend may be most evident at this point in time in the UK, it is necessary to recognise that it is set in motion elsewhere. In France retail outlets have declined from 200,000 in 1966 to 150,000 in 1990; in 1992, hypermarkets and supermarkets increased shares by 20 per cent and 40 per cent respectively; while they accounted for only 6 per cent of food stores they held a combined market share of 40.5 per cent. Italy, stereotypically perceived as a very fragmented retailing nation, witnessed a doubling of the number of supermarkets between 1985 and 1992, with the number of hypermarkets increasing six-fold. The number of retail outlets dropped by 15 per cent between 1985 and 1993. These international contextual changes and the largely economic indicators which underpin them raise some interesting questions about the degree of transnational convergence or uneven development in retailing and consumption that regions like Europe are likely to experience over the next decade.

In the British case the continuing dominance of a small group of retailers and the absorption of food markets by them at the cost of independent and specialist retailers is an outcome of social and political factors as much as economic criteria. Moreover, their particular path to success, in relation and sometimes in contrast with counterparts in other parts of the advanced world, has relied upon the particular shaping of the diverse regulatory conditions that surround them. Regulation in this sense is seen as the process by which power relations (whether within government or beyond) come to be codified and expressed, and through which contested actors and agencies align in order to progress courses of action (see Hancher and Moran 1989, Clark 1992, Marsden 1992, Christopherson 1993). In the British case the forms of regulation which have been significant have developed at two levels, at least. These can, in summary, be labelled macro and micro regulation. Both have heavily involved more than simply state agencies, for they have embodied complex networks of public and private actors and agencies. At the macro level we have outlined elsewhere (see Marsden and Wrigley 1995) how the development and potentially the continuity of the neoconservative state has embodied notions of privatised consumption through the re-privatisation of former state assets (see Saunders and Harris 1990). In this sense, we argue that in the Britain of the 1980s and 1990s, retailers have at the very least been able to capitalise on the reorganisation of expendable incomes brought forth by the breaking of collective working-class interests, and the development of privatised consumer culture at the expense of a growing, and, in consumer terms, disenfranchised underclass. In addition they have absorbed relatively cheap labour out of the Fordist restructuring of the 1980s which has (not coincidentally) promoted the devaluation of their labour costs relative to foreign counterparts.

In short, the development of the neo-conservative state, initially in the UK and the US, but now increasingly elsewhere, has provided broad macro-economic and social conditions which have been conducive to the restructuring of the retail sector along corporate and economies-of-scale lines. Underneath this macro-economic and political shift has been the development of particular forms of 're-regulation' (concerning food law, private systems of quality control). These have provided the basis for specific nationally based retail capital to further shape the food supply and provision systems. Food consumption has not been homogenised by this process. Rather, it has become more reliant upon the differential development of nationally based retailers and a redefined state regulatory role. Within the broader raft of deregulatory neo-conservatism therefore, we have witnessed the development of new spatially uneven regulatory domains which begin to foster and promote the corporate retail sectors and the modes of supply and consumption they construct. During the 1990s these domains are far from stable or predetermined. The combination of food safety scares, devaluation of property portfolios, price wars with the new discounters, as well as the onset of corporatist regulation from Brussels, has tended to threaten the relatively 'benign' regulatory environment retailers experienced for much of the previous two decades. Nevertheless, what is critical is to consider the ways in which retailers, particularly in relation to the nation state, have managed to maintain their market power and, moreover, their public legitimacy in the face of potential moral and political crises associated with the gradual reconnectedness of food, health, environment and consumption concerns.

By the early 1990s in Britain, following the passage of legislation such as the Food Safety Act (1990) and the Health of the Nation White Paper (1991), which began to shift responsibility for food matters to the retailers, the major retailers have clearly become significant actors in the promotion and implementation of 'public policy'. As a result, in a very real sense, they act on behalf of the state in delivering consumer rights and choices. Reciprocally, the regulatory state has in turn become critically dependent upon the continued economic dominance of the retailers in their role as the major providers of quality food goods. The state has shifted from its postwar Keynesian position of regulating structures of provision through macro-corporatist arrangements with producers and manufacturers. Instead, more nuanced micro-corporatist relations, particularly with the service and retail sectors, provides opportunities to reorientate and develop food markets and consumption, establishing more *inclusive* rather than *universalistic* provision systems.

The 1990s has witnessed, then, a new period of retail-regulatory activity which is broadly designed to maintain markets for the retailers, support sourcing strategies based on new free-trade principles and provide a legitimate mode of consumption for the state and the public interest. In the international context this begins to demonstrate considerable advantages for British retailers over their foreign competitors, and it goes a considerable way towards explaining their growing national and international market power. These sets of regulatory conditions also help to shroud the real globalised points of connection and value established between production and consumption spaces by highlighting certain food values parameters (such as authenticity, exoticism, and freshness) but downgrading others (e.g. real social and environmental production value of products). Retailers are thus important agents both in energising global networks of production and consumption, and in culturally and economically distancing some of the particular social and spatial origins of foods.

CONCLUSIONS

The three examples cited here demonstrate different mosaics of regional production and consumption space. They illustrate how the themes of quality, regulation, and consumption operate as active concepts in creating and maintaining food networks, on the one hand, and different *spatial situations* for those networks on the other. These key themes become influential in creating uneven agrarian spaces which express the inherent contradictory nature of recent agrarian development. Food increasingly provides a basis for power and vulnerability across and within nation states. In this sense it has enhanced its distinctive contribution to social and political life through its uneven but progressive appropriation of social, natural, as well as economic value.

The cases outlined here – 'super' quality production in São Francisco; regulating vulnerability in the Caribbean; and the power of retailers in the UK – begin to give an indication of how the more regional and deregulated global food sector is reconstituting rural space. The agricultural and food accumulation process becomes more distinctive rather than less, with the value of foods increasingly constructed in the post-farm parts of food networks. Much more comparative empirical work needs to be conducted on examining the role of social actors and institutions at different 'sites' or nodes in the food networks. In this chapter the emphasis has been on beginning to question their effects on rural space and how they *create* new patterns of dominant and dependent forms of uneven development.

One implication or paradox of the growing reconnections and centrality associated with food and rural space in consumption and quality terms in the North is the way this is built upon a growing spatial inequality in terms of Southern production spaces. The revaluation of food in the North does little to encourage local value capture in Southern producer regions. Indeed, it tends to increasingly define agriculture as a somewhat detached commodity system, disengaged from its deep ecological and social base. The processes of globalisation and externalisation now impacting differentially on Southern regions suggest not so much a 'deregulated' system of control and market development as a more variable and asymmetrical set of spatially specific power relations, absorbing some local and regional agricultures over others. Hence, the power relations running (in a vertical fashion) through food supply networks come to dominate more lateral and locally situated networks. To develop strategies for coping with these external pressures it will be necessary for agricultures and food supply to be reintegrated into their local and regional settings.

In analytical terms, therefore, scholars need to begin to see agriculture and food not so much as a highly commodified and commodity based sector *sensu stricto*. Rather, we need to focus upon ways in which agriculture and food are and can be socially embedded in local and regional spaces, as part of local systems and networks. While this is beginning to occur in the North, with a growth of concern to reintegrate agriculture and food with broader aspects of living, so far it is based upon global inequalities in space and value. The paradox of new patterns of agrarian 'globalisation' is that it seemingly reproduces highly uneven production and consumption spaces based upon different value parameters. Food travels over long physical and social distances and it remains a highly distinctive economic sector.

ACKNOWLEDGEMENTS

The evidence on which this chapter is based was collected in undertaking three research projects over the period 1993–97. The Brazilian research was funded by the British Council and the Brazilian Social and Economic Research council (CNPQ) in collaboration with Drs Salete Cavalcanti and José Ferreira Irmao. The Caribbean work formed part of the UK Economic and Social Research Council (ESRC) 'Global environmental research programme', funded under the title of 'Land degradation, population density, and the water resource on small tropical islands' (with Drs David Watts and Susan Andreatta). The retailing work is an ongoing project funded by the UK ESRC 'Nation's diet' programme and is being conducted with Drs Michelle Harrison and Andrew Flynn. I would like to acknowledge the significant institutional support, as well as the support of research colleagues in the UK, Brazil, and the Caribbean.

BIBLIOGRAPHY

Agra-Europe (1994) 'Domination of food retailing by supermarkets: a world wide trend', *Food Policy International* 3, 2: 2–4.

Apparadurai, A. (1986) 'Introduction: commodities and the politics of value' (chapter 1, p. 364) in Apparadurai, A. (ed.) *The Social Life of Things: Commodities in Cultural Perspective*, Cambridge: Cambridge University Press.

Arce, A. and Marsden, T. K. (1993) 'The social construction of international food: a research agenda', *Economic Geography* 69: 293–312.

—— (1995) 'Constructing quality: emerging food networks in the rural transition', *Environment and Planning A* 27: 1261–79.

Buckley, N. (1995) 'Tesco to open 20 stores in Hungary after profits rise', *Financial Times*, 12 April.

Christopherson, S. (1993) 'Market rules and territorial outcomes: the case of the United States', *International Journal of Urban and Rural Research* 17: 174–88.

Clark, G. (1992) '"Real" regulation: the administrative state', *Environment and Planning A* 24: 615–27

Compton Bourne & Associates (1994) *Macro Economic and Trade Policies and Structural Adjustment in Relation to Regional Integration in the Caribbean. Interim report*, Compton Bourne & Associates for CARICOM Export Development Project, Bridgetown, Barbados.

Cornia, G. A., Jolly, R., and Stewart, F. (eds) (1988) *Adjustment with a Human Face. Vol. 1: Protecting the Vulnerable and Promoting Growth*, Oxford: Clarendon.

Drummond, I. and Marsden, T. K. (1995) 'A case study of unsustainability: the Barbados sugar industry', *Geography* 80, 4: 342–54.

Feierman, S. (1990) *Peasant Intellectuals*, Madison: University of Wisconsin Press.

FitzSimmons, M. (1989) 'The matter of nature', *Antipode* 21: 106–20

Flynn, A., Marsden, T. K., and Ward, N. (1994) 'Retailing, the food system and the regulatory state', in P. Lowe, T. K. Marsden and S. Whatmore (eds) *Regulating Agriculture: Critical Perspectives on Rural Change*, London: Wiley.

Friedland, W. H. (1994) 'The New Globalisation: the case of fresh produce' (chapter 10, pp. 210–232), in Bonnano, A., Busch, L., Friedland, W., Gouveia, L., and Mingone, E. (eds), *From Columbus to ConAgra: the globalisation of agriculture and food*, Lawrence: University of Kansas Press.

Hancher, L. and Moran, M. (1989) 'Organising regulatory space', in L. Hancher and M. Moran (eds) *Capitalism, Culture and Economic Regulation*, Oxford: Clarendon.

Harker, T. (1994) 'Caribbean economic performance in the 1990s: implications for future policy', in H. A. Watson (ed.) *The Caribbean in the Global Political Economy*, Boulder and London: Lynne Rienner.

Lewis, L. (1994) 'Restructuring and privatisation in the Caribbean', in H. A. Watson (ed.) *The Caribbean in the Global Political Economy*, Boulder and London: Lynne Rienner.

Mann, S. and Dickinson, J. (1980) 'State and agriculture in two eras of American capitalism', in F. Buttel and H. Newby (eds) *The Rural Sociology of Advanced Societies*, Montclair, NJ: Allanheld, Osmun.

Marsden, T. K. (1992) 'Exploring a rural sociology for the Fordist transition: Incorporating social relations into economic restructuring', *Sociologia Ruralis* 32: 209–31.

Marsden, T. K. and Wrigley, N. (1995) 'Regulation, retailing, and consumption', *Environment and Planning A* 27: 1899–1912.

—— (1996) 'Retailing, the food system and the regulatory state', in N. Wrigley and M. S. Lowe (eds) *Retailing, Consumption and Capital: Towards the New Retail Geography*, Harlow: Longman.

Marsden, T. K., Murdoch, J., Lowe, P., Munton, R., and Flynn, A. (1993) *Constructing the Countryside: Restructuring Rural Areas*, vol. 1, London: UCL Press.

Mooney, P. (1985) *My Own Boss? Class, Rationality, and The Family Farm*, The Rural Studies Series of the Rural Sociological Society, Boulder: Westview.

Murdoch, J. and Marsden, T. K. (1994) *Reconstituting Rurality. Class, Community and Power in the Development Process: Restructuring Rural Areas*, vol. 2, London: UCL Press.

Polanyi, K. (1944) *The Great Transformation. The Political and Economic Origins of our Time*, Boston: Beacon Hill.

Rabobank Nederland (1993) *The World Fresh Fruit Market*, Utrecht: Rabobank Nederland Agribusiness Research.

—— (1994) *The Retail Food Market. Structure, Trends and Strategies*, Utrecht: Rabobank Nederland Agribusiness Research.

—— (1995) *The International Food Industry*, Utrecht: Rabobank Nederland Agribusiness Research.

Saunders, P. and Harris, C. (1990) 'Privatisation and the consumer' *Sociology* 24: 57–75.

Thomas, C. Y. (1988) *The Poor and the Powerless. Economic Policy and Change in the Caribbean*, New York: Monthly Review Press.

Tordjman, A. (1995) *Trends in Europe. Consumer Attitudes and the Supermarket 1995*, Washington DC: The Research Department, Food Marketing Institute.

Watts, M. (1996) 'Development III. The global agro-food system and late twentieth-century development (or Kautsky redux)', *Progress in Human Geography* 20, 2: 230–45.

Watson, H. A. (ed.) (1994) *The Caribbean in the Global Political Economy*, Boulder and London: Lynne Rienner.

Wrigley, N. (1995) 'Retailing and the Arbitrage Economy: Market Structure, Regulatory Frameworks, Investment Regimes', and Spatial Outcomes, paper prepared for Harold Innes Centenary Conference, University of Toronto, on 'Regions, institutions and technology: reorganizing economic geography in Canada and the Anglo-American world', 23–25 September 1994; subsequently revised and updated, August 1995.

8

AGRO-INDUSTRIAL JUST-IN-TIME

The chicken industry and postwar American capitalism

William Boyd and Michael Watts

> [T]he food system is subject to two forces – that of a highly flexible demand organized according to the just in time system by modern retailing, and an industrial supply which must deal with a strongly inelastic upstream agricultural production. This inelasticity is . . . caused by the variable length of production cycles of various agricultural raw materials and by the more or less perishable nature of . . . these raw materials.
>
> (Green, Lanini, and Schaller 1994: 2)

> Given the biological phenomenon of the broiler, the process is very much a critical path, just-in-time system. . . . Schedules must be coordinated since the output of one process serves as the input for the next process. To successfully compete requires a fully integrated system of complex processes operating as a single, co-ordinated entity. In addition, considerations such as the quality level of the live broiler, the market weight . . . and skin color can be assured to a much greater degree through central management. . . . Market signals and retailing regarding what the end products should be, as indicated by consumers . . . can [also] be much more effectively transmitted back through the processing and production stages under vertical [integration].
>
> (Roenigk 1991: 6)

While it passed relatively unnoticed, 1990 was something of a watershed in the history of American dietary culture. For the first time, Americans consumed more chicken per capita than beef, an astonishing turnaround for a society which, with the exception of Argentina, has been the most beef obsessed (Gordon 1996). In fact, the growth of chicken output in the postwar period has been nothing short of explosive. In 1928 when Herbert Hoover promised 'a chicken in every pot', each American consumed, on average, half a pound of chicken a year; by 1945, the figure stood at roughly 5 pounds per capita each year, and fifty years later the figure had leapt to almost 70 pounds (roughly half a chicken a week). All expectations are that the rate of increase (while *decreasing* in the 1990s) will continue to elevate the absolute quantities of chicken consumed per head to over 80 pounds per annum by the turn of the millennium (Table 8.1).

Embedded within the tale of the rousing defeat of American beef by the lowly chicken is, of course, a much larger story which in a curious way sheds

Table 8.1 US broiler production and consumption, 1934–94

Year	*US Production (million heads)*	*Increase over previous period (per cent)*	*Price per Pound (dollars) live*	*Price per Pound (cents) r-t-c**	*Average Liveweight (pounds)*	*Per Capita Consumption (pounds) r-t-c**	*retail*
1934	34	–	0.193	–	2.84	0.7	–
1940	143	321	0.173	–	2.89	2.0	–
1945	366	156	0.295	–	3.03	5.0	–
1950	631	72	0.274	–	3.08	8.7	–
1955	1,092	73	0.252	–	3.07	13.8	–
1960	1,795	63	0.169	–	3.35	23.6	–
1965	2,334	30	0.150	26.5	3.48	29.9	–
1970	2,987	28	0.135	26.4	3.62	36.8	–
1975	2,950	-1	0.262	45.1	3.76	36.8	–
1980	3,963	34	0.279	49.1	3.93	46.6	46.5
1985	4,470	13	0.302	50.8	4.19	52.0	52.0
1990	5,864	31	0.324	54.8	4.37	61.0	60.9
1994	7,018	20	0.350	55.7	4.63	69.9	69.8

Note: *ready-to-cook

light on the very nature of postwar American capitalism itself. Residing in the chicken story is the rise of the massive southern integrators (which in the 1990s leads, via Tyson Foods, to the very apex of political power in the Clinton White House); the role of big science and the industrialisation of the US diet (our understanding of chicken nutrition now exceeds that of any other domestic animal including humans); small impoverished contract growers and vulnerable work forces in the processing plants (recall the terrifying fire in North Carolina in 1991 in which twenty-five non-unionised workers died); and, not least, radical reorganisation of the industry along the entire *filière* (a process synonymous with the rise of the US South as the heart of an increasingly export-oriented US industry that is global in scope – the lure of massive Asian markets is at stake – and transnationally competitive in character). A century ago steak and lobster were cheaper than chicken; today it can cost less than the potatoes that are served with it (Gordon 1996).

The postwar history of the national and global chicken or 'broiler' industry is one of extraordinary dynamism and growth. Advances in breeding, disease control, nutrition, housing, and processing have conferred upon US integrators some of the lowest production costs in the world. Since 1940, the industry's feed conversion rate has declined precipitously from three pounds of feed per pound of liveweight to under two pounds. Over the same period the average broiler liveweight has increased from 2.89 pounds to 4.63 pounds and the maturation period – the time required for a bird to reach market weight – has plummeted from over seventy days to less than fifty (Watts and Kennett 1995). All of this has been driven by a small number of large integrators who dominate the *filière*. Tyson Foods – the largest broiler integrator – currently accounts for 21.6 per cent of the US ready-to-cook market, and in 1995 claimed total sales of $5.5 billion. Tyson is a global company and part of a massive industry – total US broiler sales exceed $12 billion – which can rightly claim, in the words of its former CEO Don Tyson, to 'control the center of the plate for the American people'. Tyson Foods' chicken nuggets (one of its vast panoply of chicken products) are not of course the product of a global production system in the sense that, for example, Nike's Air Max Penny basketball shoe is manufactured from fifty-two components drawn from an integrated production process spanning five or more countries. But direct foreign investment in the industry, coupled with a fast growing export trade, new linkages between integrators, and a footloose fast food industry has revolutionised (and globalised) the poultry business in new and profound ways. Over the last decade, global poultry consumption has been increasing by 5 per cent per annum, while the total number of chicken slaughterings leapt from 6.5 billion in 1961 to 33 billion in 1995. Total international trade in poultry meat was about 4.5 million tons in 1994 – an estimated 9.2 per cent of global production – dominated on the export side by the US, France, the Netherlands, Brazil, China, and Thailand. And while the quantity of poultry meat exports almost quadrupled between 1982 and 1996, the centrality of the US in world trade remains unambiguous, currently standing at 30 per cent of world trade, at least twice as large as France, the second largest exporter. Yet in an increasingly competitive global industry, US exports, which now represent more than 15 per cent of national production, confront especially fierce competition from Brazil, Thailand, and China.

In addressing a dynamic, and now global, poultry industry – the history of which has yet to be told – our concern here is to highlight three questions focused in large measure on the US industry, and the southern production complexes in particular. In the first section we address two related questions: the first addresses the *genesis* of the modern broiler production system: why did it emerge in the 1940s and 1950s and why in the US South (and why in particular industrial districts in such locations as northwest Arkansas and north Georgia)? Our argument here turns on the centrality of the southern postwar political economy which provided the social and institutional context for the contract-based model of integration that has subsequently become the standard in the industry. The second question concerns the *character* of the regional production complex: in what ways does the integrated complex, constituted in sub-regional agro-industrial districts, exhibit the flexible and just-in-time (JIT) qualities associated since the 1960s with Japanese manufacturing? Our argument here is to suggest that just-in-time has its American counterpart in *agriculture* (or more properly agro-industry) and its history can be traced back to the 1950s. That is to say, in some sense it is coeval with the Japanese experience and long predates the adoption of flexibility and JIT attributes in American manufacture in the 1980s.

In the second section of this chapter we address the third question, namely: what is the distinctive nature and what we somewhat ambitiously refer to as the 'logic' of the regional production system? Our concern here is to explain not so much flexibility or JIT in *general* but the specific and contingent qualities of flexibility and JIT in the broiler industry which derive, in our view, from certain biological attributes of the commodity on the one hand, and the distinctive 'southernness' of the path to vertical integration on the other. We explore this through the idea, taken from French research, of quality (*qualité*) which suggests both social and institutional relations of the *filière* (*conventions*) and commodity-specific attributes of chicken biology and reproduction (Allaire and Boyer 1995) – themes which resonate with some of Ben Fine's (1994) claims about food provisioning systems. In making this latter argument we engage with many of the ideas relevant to current debates on agro-industrial restructuring (Wells 1996) – the construction of quality, the role of the organic in food provisioning systems, the trajectories of accumulation in capitalist agricultures – but we are simultaneously providing an account of the social basis of southern competitiveness in the industry, and hence the centrality of US exports in the dynamic and global poultry market of the 1990s. To that effect, we conclude the chapter with a brief discussion of the 'double-edged' nature of flexibility in the industry, the ongoing problem of overproduction and the resulting 'market question' facing the integrators, and the peculiar nature of quality in the industry.

AMERICAN JUST-IN-TIME: GENESIS AND DEVELOPMENT OF A FLEXIBLE AGRO-INDUSTRIAL PRODUCTION COMPLEX IN THE POSTWAR SOUTH

Big science and the productivity revolution

Before the 1930s, chicken meat production in the United States was largely the incidental result of egg production, located on small farms dispersed throughout the country (Figure 8.1) (Packers and Stockyards Administration 1967: 1). The early chicken meat *filière* (Figure 8.2) included independent breeders, hatcheries, farmers, feed dealers and manufacturers, truckers, live and 'New York-dressed' retail markets, slaughterhouses (located in the places of consumption), and merchants or commission agents who controlled the distribution networks of the larger metropolitan markets, most especially in New York. For most consumers, broilers, or young chickens, were considered a seasonal delicacy – byproducts of the 'spring hatch'. Nonetheless, by the early 1930s some small farmers, primarily on the

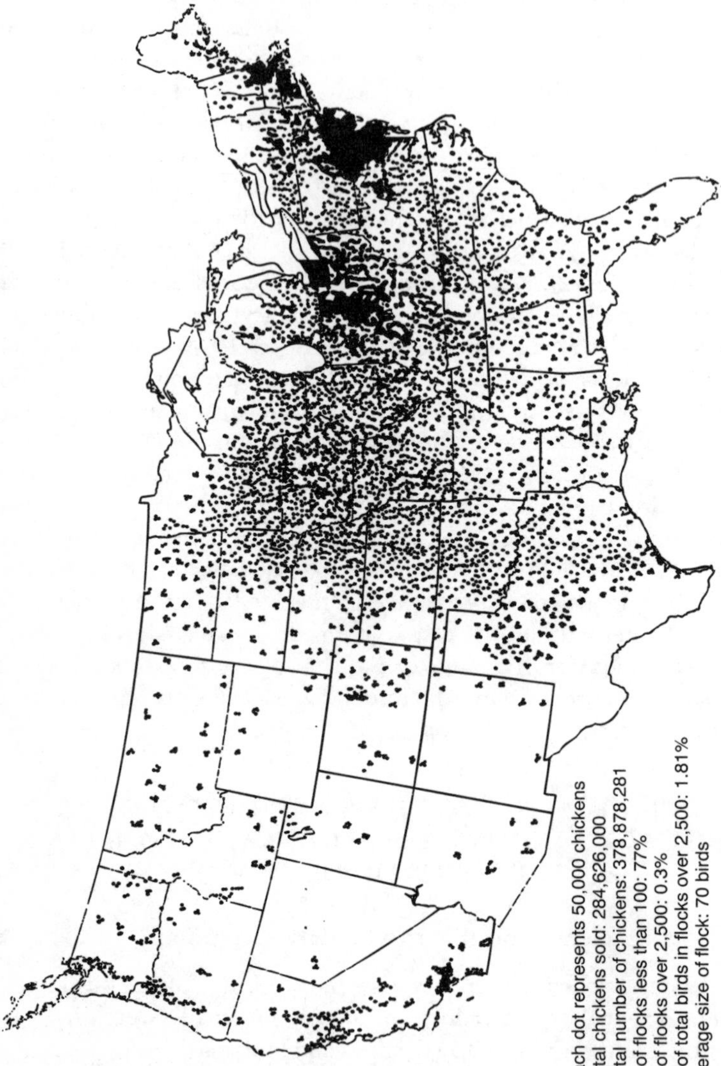

Source: Dept. of Commerce 1939, Bugos 1992, p. 149

Figure 8.1 Chickens sold, alive or dressed in 1929: number (excluding baby chicks)

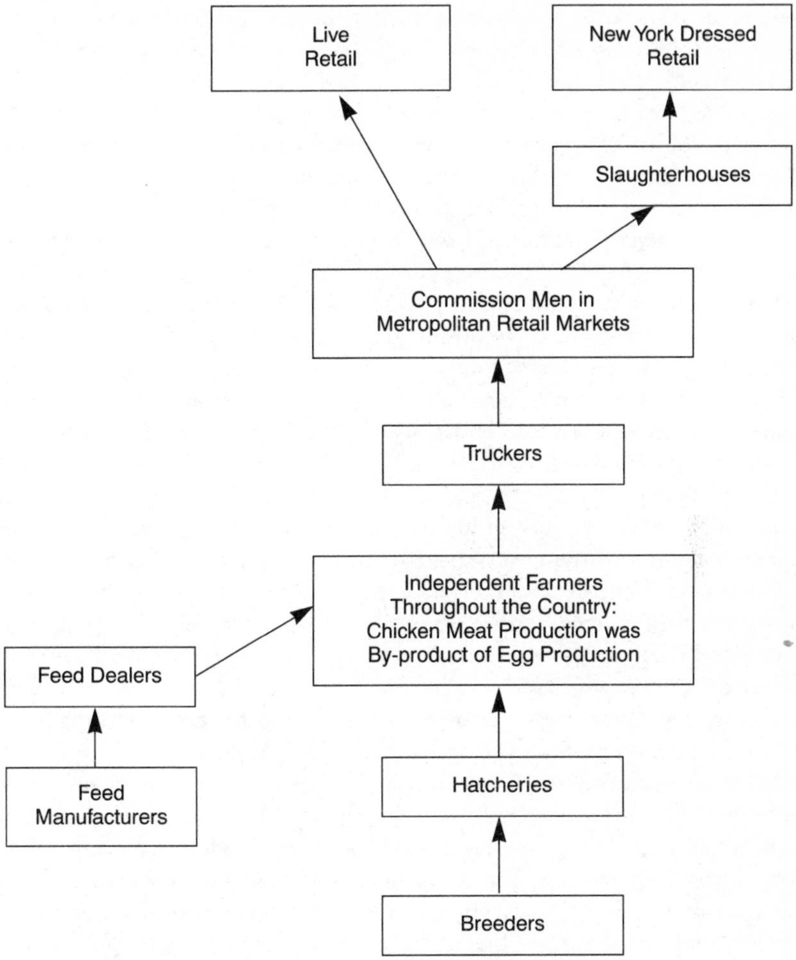

Figure 8.2 US chicken meat filière c. 1929

Delmarva peninsula, had begun to raise broilers for commercial purposes (Packers and Stockyards Administration 1967: 1, Tobin and Arthur 1964, Watts and Kennett 1995). By 1935, Delmarva accounted for some two-thirds of total US broiler production, while the average American consumed barely half a pound of broiler meat per year (Packers and Stockyards Administration 1967: 1).[1]

Two forces were central to the early development of the broiler industry. First, the move to year-round confinement, made possible by the discovery of vitamin D in 1926 (Roenigk 1991), facilitated the shift to *industrial* broiler production. Second, the federal government stepped in during the depression years to lay the foundations for the postwar productivity increases and the application of what one might call 'big science' to the industry. Although chicken breeding experiments had been established by the USDA as early as 1912, it was not until 1933, with the establishment of the National Poultry Improvement Plan (NPIP), that the government emerged as a major force in facilitating the development of new techniques in disease control, breeding, and husbandry. The NPIP, which continued on after the National Recovery Administration's poultry code was struck down along with the NRA itself in the famous Schecter brothers case[2] was critical in efforts to eliminate pullorum, typhoid, and other diseases, as well as in establishing a sustained public commitment to breeding and poultry husbandry. Several states followed by establishing their own poultry improvement associations. Gradually, mortality rates were reduced – approximately 30 per cent by the end of the 1930s (Strausberg 1995: 28, 31) – which permitted the move to more intensive confinement operations.

US involvement in the Second World War also proved to be a watershed in the still nascent broiler industry. Unlike beef, chicken was not rationed during the war, and the government set a ceiling price well above the cost of production, which, in conjunction with growing demand by both the US military and civilians, created unprecedented commercial opportunities (Tobin and Arthur 1964). In 1942, moreover, the War Food Administration placed the entire production of the Delmarva peninsula under contract for federal food programmes – creating a huge fillip for the emerging broiler areas of the American South (Bugos 1992: 148, Frazier 1995).

What proved to be decisive during the war and the immediate postwar years, however, was the revolution in primary breeding. Until the 1940s, farmers had relied on pure breeds that had been developed for egg production, with little concern for meat qualities. With increased demand for chicken meat during the war, breeders began to focus their attention on the development of specialised breeds for meat production. In 1944, the search for the broad-breasted chicken began in earnest as the A&P food chain launched a series of national breeding contests. Striking illustrations of early retailer power in product design,[3] the two 'Chicken of Tomorrow Contests' (1946 and 1951) which greatly accelerated the development of the modern broiler industry by establishing cross-breeds or hybrids as the standard throughout the industry, ushered in the age of the 'designer chicken'. Employing principles pioneered in the hybridisation of corn and other crops, primary breeders developed standard pedigrees of male and female lines that combined to produce a superior bird. By the 1950s, such cross-breeds provided the genetic basis of the modern broiler industry, and breeders focused on fine tuning their pedigrees to meet the ever more exacting demands for genetic uniformity and quality assurance. On the eve of integration in the broiler industry, moreover, the 'biological lock' of hybridisation strengthened the boundary between the primary breeders and the rest of the industry – a boundary that has persisted to this day. At the same time, major breeders developed closer networks and alliances between emerging firms in other parts of the industry – an incipient form of industrial 'conventions' (see Sylvander 1995) – and, following their lead, began to establish operations in the South, which by the 1950s had emerged as the most dynamic broiler producing area in the country (Bugos 1992, Strausberg 1995: 61–72).

Complementing these dramatic advances in breeding were substantial public investments in nutrition, disease control, and confinement technologies during the immediate postwar period, much of which was sponsored by the land grant universities and the federal government. The development of high-performance rations combined with the use of vitamin B-12 and antibiotics in feed dramatically reduced mortality and increased feed conversion efficiency (Strausberg 1995: 55).[4] At the same time, the completion of rural electrification facilitated substantial improvements in environmental control and labour productivity within confinement operations. Whereas in 1940, an average 250 'man-hours' were required to raise 1,000 birds to maturity, by 1955 the required time had dropped to 48 hours (Packers and Stockyards Administration 1967: 15). Together, better breeding, improved health and nutrition, and intensive confinement provided the scientific and technological basis for a substantial expansion in productive capacity (see Table 8.1), and, it should be emphasised, the endemic problems of overproduction that transpired during the 1950s. With the chicken of tomorrow, the stage was set for the era of high integration.

Contracting, integration, and just-in-time

In 1950, on the eve of integration, the broiler industry remained the province of a variety of specialised independent firms and small independent farmers linked through informal market-like transactions. By the end of the decade, however, the majority of production had fallen under the control of vertically integrated firms. In little more than ten years, the broiler industry evolved from an open-market system to one bound by contracting and vertical ownership. In short, broilers became one of the most tightly coordinated and institutionally dense commodity systems in US agriculture. This shift to a new model of industrial organisation – in our view a distinctively American, flexible, just-in-time production system – was matched by an equally pronounced process of spatial concentration in the American South. The modern vertically integrated broiler industry, in other words, possessed a distinctive southern accent from its inception.

At the centre of the integrated system was the contract grow-out arrangement, which emerged in the 1940s and 1950s in the American South out of efforts by local feed dealers and feed companies to protect their markets. By the mid-1950s, some 90 per cent of all broiler growers in the South received some degree of financing through such contractual arrangements (Beckler 1957: 4). Initially, the contracts were informal, open-account arrangements whereby the grower would pay after the sale of the birds for credit and supplies advanced. Most of the credit provided by the local feed dealers was in turn backed by credit from local feed mills (usually in the form of feed and feed ingredients extended on credit) and from regional commercial banks. As grow-out capacity expanded, however, increased production exerted downward pressures on prices (1954 was the first major 'price crisis' in the industry), and farmers found themselves strapped. Feed dealers, seeking to secure both their feed markets and their loans, started signing contracts with growers that ensured a certain return for the grower while retaining ownership of the birds by the feed firm. At the same time, many of the larger feed manufacturers confronted local feed dealers who were unable to meet their obligations, and subsequently moved into direct contracting with growers as a way of further protecting their own financial stake in the industry. As contracts were formalised, the feed dealers and feed companies were able to introduce better chicks and feed, more standardised management practices, and better disease control into the grow-out operations (Southern Cooperative Series 1954). In effect, this shift from open-account contracts to more formal production contracts – most of which were either share, flat fee, or feed conversion plans – marked an evolution from a simple credit arrangement to a tightly interlinked credit, input, and labour contract. It also represented the first step towards vertical integration – initiating a relentless and deepening tendency of appropriationism whereby the loci of power moved irrevocably off-farm to the integrators and suppliers (Goodman, Sorj, and Wilkinson 1987, Watts 1994a).

As with contracting, vertical integration in the broiler industry thus emerged initially as a defensive reaction intended to offer some protection from the problems of coordination and overproduction that grew out of the contract-driven expansion of productive capacity in the 1950s. Inhabiting a world of endemic gluts and volatile profit margins, many of the independent firms found themselves in a relentless cost-price squeeze. The most aggressive feed dealers and feed companies, however, began to integrate into other stages of the industry – starting with hatcheries and feed mills and later moving into processing. By the mid-1950s, a few fully integrated operations had emerged in the American South[5] but for the industry as a whole integration did not become the order of the day until the end of the decade. Meanwhile, the industry continued to experience shake-outs of smaller, independent firms throughout the 1950s – a fact that became the subject of a series of congressional hearings in 1957 and 1961 (US Congress 1957, 1961).[6]

Despite the wailing of threatened independents, integrated operations came to account for close to 90 per cent of total production by the early 1960s (Tobin and Arthur 1964). Integrators, as these firms came to be called, controlled every stage of the industry except foundational breeding. Most firms owned their own hatcheries, feed mills, and processing plants, contracting the grow-out operations to small farmers. With access to the equities market, high-volume purchases of feed, and reliable marketing outlets, such companies presented formidable barriers to entry (Strausberg 1995: 102).[7] Broilers, in other words, became synonymous with the new innovation of 'agribusiness', and by the mid-1960s large national feed companies such as Pillsbury and Ralston Purina had moved in to occupy leading positions in the industry (Strausberg 1995: 117). Feed milling capacity and business experience conferred upon these companies an early comparative advantage in an industry in which profit was derived on the basis of large volume and high throughput. During the late 1960s and early 1970s, however, as the industry was rocked by wide profit swings and periodic gluts, many of the feed companies divested themselves of their poultry operations (Marion and Arthur 1973), opening the door for regional integrators such as Tyson, Perdue, and Holly Farms to emerge as the key players of the 1970s and 1980s. In 1986, when Tyson absorbed Lane Poultry, the Arkansas firm became the largest processor in the country. Two years later, Tyson's takeover of Holly Farms (for a total cost $1.29 billion) made the firm the undisputed market leader (Strausberg 1995: 124, 127). Such high-profile takeovers, which were supplemented by a number of smaller acquisitions, have continued up until the present. In 1995, for example, Tyson's acquired Cargill's US broiler operations. By the end of that year, Tyson Foods was producing 117.2 million pounds of ready-to-cook (RTC) broiler meat every week and offering 4,600 different products with recorded total sales of $5.5 billion. (Thornton 1996a: 18). The top four companies (Table 8.2) currently account for roughly 45 per cent of total broiler output in the US; the top ten firms for over two-thirds of all RTC broiler production.

Thus, although commercial broiler production was first established on a significant scale on the Delmarva peninsula in the 1930s and 1940s, the American South of the 1950s, (and particularly Georgia, Alabama and Arkansas) proved to be the crucible in which vertical integration was forged. Since 1960, these three southern states, along with Mississippi and North Carolina, have accounted for over 60 per cent of total US production (Figure 8.3).

Table 8.2 Concentration in the broiler, breeding and egg industries: per cent accounted for by the top four firms

	1972	*1985*	*1987*	*1990*	*1994*
Broilers	7	7	na	na	42
Eggs	na	na	na	na	22
Breeders*	na	79	79	96	97

Source: Bugos 1992, Bell 1995, Watts and Kennett 1995
Note: *refers to female birds

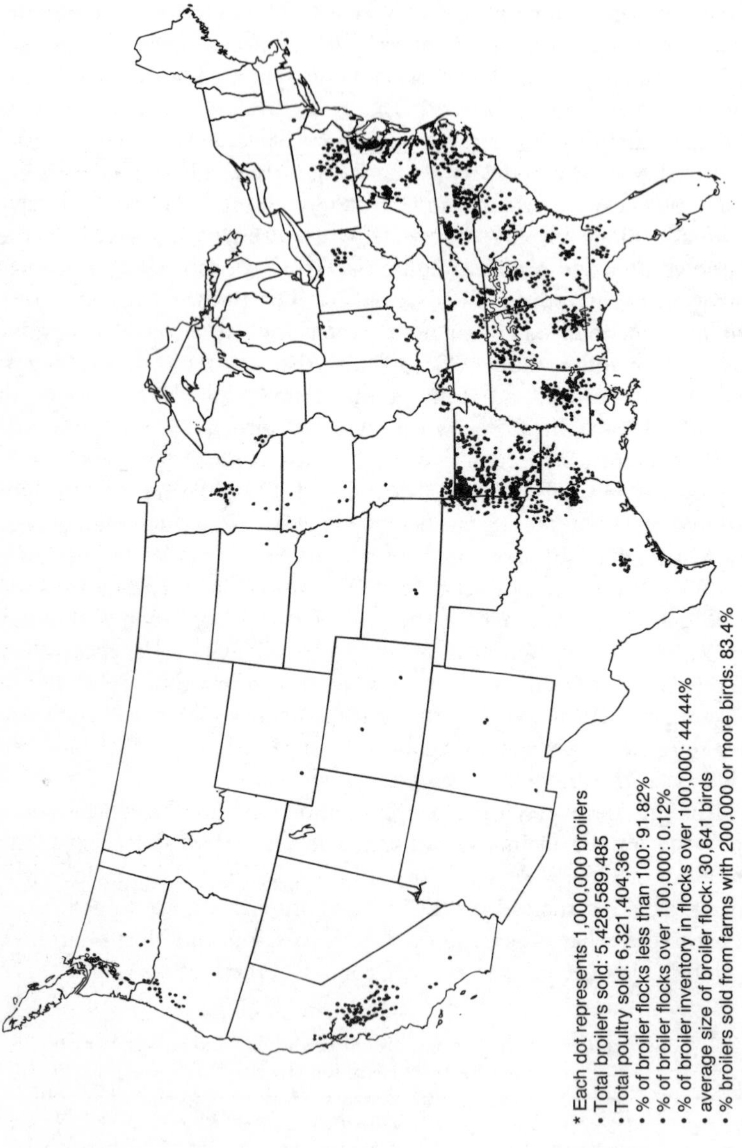

Figure 8.3 Broilers and other meat-type chickens sold, 1992

By 1965, the South Central and South Atlantic states together accounted for over 85 per cent of total US production – a share that they have roughly maintained up to the present (Perez and Christensen 1990, USDA-NASS 1995) (Table 8.3). From the beginning, moreover, broiler production has been concentrated in a few dynamic agro-industrial districts or sub-regions – most notably, northwest Arkansas, north Georgia, and north Alabama.

The rise of the southern production complex was, in short, synonymous with a radical reorganisation of the entire *filière* (Figures 8.2 and 8.4). Historically speaking, by the late-1950s – and certainly by the early 1960s – a 'critical-path just-in-time' system of industrial organisation was emerging in the broiler industry (Roenigk 1991). At the centrepiece of this new system stood the integrator complex in which the various stages of chicken production – hatcheries, breeder grow-out, feedmills, broiler grow-out, rendering, processing, and further processing – are integrated under one corporate structure. Today, Tyson's has twenty-seven such integrated broiler complexes in the South which include forty-five hatcheries, twenty-eight feed mills, fifty-five processing plants, seven distribution centres, six rendering plants, and 15,800 grow-out houses with a capacity for 35 million placements a week (Thornton 1996a: 18). The key advantage of the integrated system has been the ability to closely coordinate the various biological lags in production time so as to maximise material and time efficiencies while controlling for quality and biosecurity. Because of the metabolic sensitivity of broilers to the timing of the feed cycle, for example, some four different rations are used during the breeder phase and another three during the broiler grow-out phase (Henry and Rothwell 1995: 6). With standard processing-line speeds of 8,000 birds per hour, moreover,

Table 8.3 Leading broiler producing states (by farm value), 1993

Rank	*State*	*No. Produced (millions)*	*Liveweight (million lbs)*	*Av.Wt (lbs)*	*Price (cts/lb)*	*Total value (million $)*
1	Arkansas	1,048.8	4,614.7	4.40	36.5	1684.4
2	Georgia	960.0	4,416.0	4.60	34.0	1501.4
3	Alabama	882.2	3,969.9	4.50	34.0	1349.8
4	N. Carolina	615.2	3,137.5	5.10	32.0	1045.0
5	Mississippi	528.2	2,429.7	4.60	33.5	814.0
6	Texas	360.7	1,623.2	4.50	37.5	608.7
7	Maryland	294.7	1,326.2	4.50	32.0	424.4
8	Delaware	251.4	1,282.1	5.10	32.0	410.3
9	Virginia	244.4	1,124.2	4.60	33.0	371.0
10	California	216.0	1,101.6	5.10	32.0	352.5
	US	6,689.1	30,592.2	4.57	34.0	10,409.2

Source: USDA, *Poultry and Livestock* 1994

maintaining a steady supply of high-quality live broilers to processing facilities rests on precisely coordinated placement of 'multiplication flocks' and the capacities of the various grow-out operations (i.e. grandparent flock farms, parent stock farms, hatcheries, and broiler growing operations) (Henry and Rothwell 1995: 38–39). Coordinating the rigidities associated with broiler production times and biosecurity on the one hand with the shifting demand for the huge number of different products (4,600 for Tyson alone) on the other, requires a degree of flexibility and inventory control that is more precise and demanding than that for any other agro-food commodity.

At the level of inter-firm relations, the broiler *filière* also represents a complex production network (Figure 8.4) – another aspect of what the French group refers to as 'qualité' (see Nicholas and Valeschini 1995). Integrated operations are supported by a host of technology and input suppliers such as primary breeders, equipment suppliers, pharmaceutical and chemical firms, feed ingredient suppliers, as well as a panoply of research and technical support entities. A cursory survey of the Watt Publishing Company's electronic list of products and suppliers for the industry revealed some twenty different suppliers of vaccine products (seventeen suppliers alone for the bronchitis vaccine); thirty-seven suppliers of feed/water antibiotics; forty-five suppliers of feed pre-mixes; and over forty suppliers of processing technology.[8] Not only do the integrated firms have to coordinate their own activities, but relations between integrators and suppliers also have to be closely coordinated. Effective vaccine development, for example, depends on long-term cooperation between breeders, integrators, vaccine manufacturers, and research laboratories. As in other areas of US agriculture, this privatisation of information, technology, and service provision has important implications for the locus of power within production systems and the role of extension (Wolf and Buttel 1996). In effect, the system as a whole functions as an industrial network in which the rate of adoption of new technologies is quite rapid as suppliers are constantly working with integrators to better understand their needs and meet their criteria for improvement (Henry and Rothwell 1995: 6). Since its emergence, this remarkably advanced system of food production has been the driving force behind rapid innovation, product differentiation, and declining real prices over the last several decades.

In summary, then, vertical integration in the broiler industry proceeded initially through the signing of formal production contracts with growers, and, as coordination problems intensified, it matured into a strategy of backward and forward integration. By the early 1960s, the contract-based integrated system was highly concentrated in the American South. Its significance is two-fold, in our view. On the one hand, the centrality of contract growing to the broiler integrators raises a number of important questions originally posed a century ago by Karl Kautsky (1906) who noted

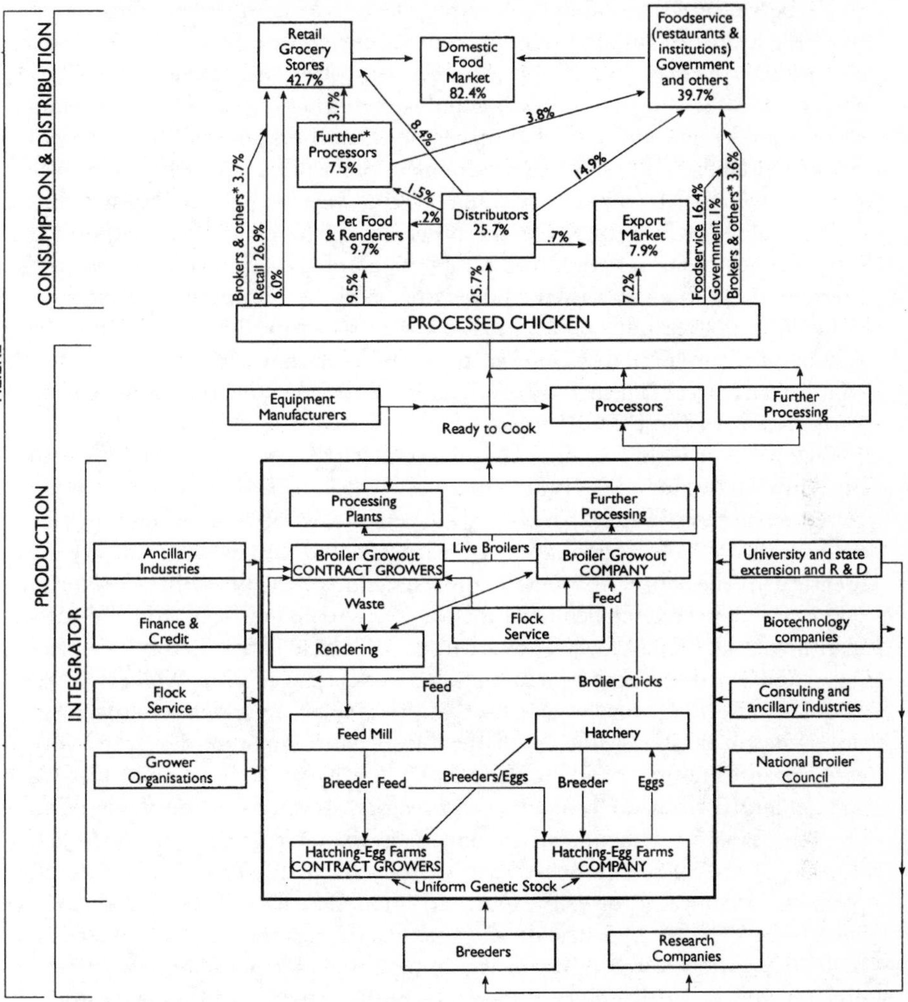

Figure 8.4 The broiler filière c. 1993

that the self-exploitative qualities of household enterprises (family farms) could be captured by capital via forms of vertical integration. This represented a form of capitalist development in agriculture in which there was *centralisation without (land) concentration*, and the dispersion of risk by agribusiness focusing on processing and sub-contracting land-based production. Kautsky pointed out of course that flexibility, and the flexibility of household labour and production systems in particular, was a key to both the competitiveness of family farms and to the ways in which industry was taking hold of agriculture. On the other hand, the genesis of the postwar integrator complex suggests that Kautsky's emphasis on flexibility has been deepened in such a way that the integrators, and their formation into territorial complexes, resemble (at a surprisingly early historical point) the flexible, just-in-time production systems customarily associated with the 'new industrial districts' (the Third Italy) and the Japanese manufacturing revolution ('Toyotaism') of the 1960s (Harrison 1994).

Of course, it would be folly to suggest that the 'geographically clustered big firm networks' (Harrison 1994: 134) of the Toyota type involving complex supplier systems, robotics, cooperative work groups, just-in-time delivery and various forms of functional, wage and numerical flexibility are identical to the southern broiler complexes. Equally, Gainesville, Georgia is not exactly a post-Confederate Third Italy. Discussion of flexibility and just-in-time in any case covers a multitude of often contradictory qualities associated with modern production networks (Sayer and Walker 1992: 170–90). Rather, our point is that forms of relational contracting, buyer–supplier relationships, and flexible networking were developed very early in agro-industry, and that these early forms of flexibility and just-in-time emerged from the biological and sectoral demands of the industry on the one hand and distinctive regional agrarian structures on the other. Broilers in this sense suggest that there is a distinctively agrarian (and perhaps 'organically driven') route to flexible, just-in-time production complexes. The complexity of the manufacturing content of broilers is, naturally, quite limited compared to automobiles; furthermore, it would be precipitous, to say the least, to represent broiler territorial complexes in the South as compelling instances of the Marshallian or Italianate districts of the Third Italy or Toyota City sort. What we wish to emphasise is how flexibility and just-in-time qualities have a *history* (and a surprisingly long history) in agro-industry – this is what interested Kautsky in 1906 – and that the broiler case provides an opportunity for thinking about the *specificity and meanings of such systems using an agrarian setting*[9] (Wells 1996) (rather than assuming that the broiler integrators simply replicate a particular model of manufacturing organisation or territorial agglomeration). All of which points to the need for a closer examination of the constraints and opportunities associated with the *biological fact* of broiler production as well as the distinctive *historical and geographical embeddedness* of this particular production system. How, in other words, has the irreducible biological character of the chicken shaped and constrained the organisation of production in the industry? And what has been the legacy of the social and geographical origins of the industry, both in the creation of low cost production systems (i.e. the social basis of competitiveness in a now global industry) and the particular sorts of flexibility (the *situatedness* of broiler JIT) characteristic of the sector? In linking these two questions we hope to shed some light on the question of the *particular* (i.e. agrarian and regional) institutional and political character of the agro-industrial production complexes themselves – to specify, in short, the quite precise ways in which flexibility and JIT are part of the peculiar dynamics and structure of the industry.

AGRO-INDUSTRIAL JUST-IN-TIME: COMMODITY LOGIC, REGIONALISM, AND THE SITUATEDNESS OF FLEXIBILITY

The biology of flexible just-in-time

Simply put, there are certain aspects of the biology of broiler production and reproduction – what Ben Fine (1994) calls the 'organic' attributes of food provisioning systems – which have important consequences for the organisation of the *filière* itself, and the 'conventions' by which quality is produced within the industry. In our view, three key areas stand out: broiler breeding; biological time lags in broiler production; and broiler microbiology. Regarding breeding, Glenn Bugos (1992) has illustrated nicely how the development of hybridisation as a form of trade protection – a sort of 'biological lock' – has contributed to the long-standing 'institutional boundary' between primary breeders and the integrators (Bugos 1992: 146). Because of the ever more exacting demands of the integrators for genetic uniformity and testing (quality) – chicks that grow evenly, without variability in body shapes, and ideally capable of meeting consumer-responsive demands (i.e. breast meat yield) without falling prey to constantly changing pathogen regimes – long-term relationships have been established between breeders, integrators, and animal health specialists. Indeed, although chicks have historically accounted for less than 3 per cent of the finished costs of the broiler, all other investments in feed, equipment, medication and so on depend on the chicken genetic endowment. In all breeding efforts, moreover, there are trade-offs between costs and performance of breeds geared to particular product mixes (Martin 1996: 26). Recent breeding efforts to increase the yield of breast meat in broilers, for example, has resulted in a higher propensity for metabolic disease and male infertility because the extra protein going to breast muscle production comes at the expense of internal organ development (Thornton 1996b: 18–22). In fact, throughout the 1980s, efforts to push the 'high-yield' envelope further have raised overall mortality rates from 5.2 per cent in 1980 to 5.65 per cent in 1990

(Scheideler 1992: 18–20) – a testament to the inevitable complexities associated with efforts to subordinate biological organisms to the dictates of industrial production.

Regarding the biological time lags facing the industry, there are still considerable lead times involved in the commitment to grow broilers to market weight despite substantial reductions in maturation time. In 1996, for example, a plant's production manager will be making decisions about the breeds to be placed in 1997 for birds to be processed in 1998 (Thornton 1995a). With the move to new product mixes, such as boneless and further processed products, the imperative of transmitting these new 'market signals' back through the production chain to breed selection is obviously critical.[10] As Dr William Rishell of Arbor Acres has noted, 'Things we're doing at the pedigree level today won't be seen by our customers in terms of broilers through the plant, for four to five years' (cited in Amey 1992: 21). Such biological lags also require precise coordination of placement schedules for the various grow-out phases in a manner that will maintain the 8,000 birds per hour required by the average processing operation. At the same time, the absence of any sort of stretch-out period for those broilers that are on feed due to rapidly mounting costs after broilers reach maturity and, conversely, the inability to speed up the process through overtime work, etc. (Tobin and Arthur 1964: 71) also create inevitable 'rigidities' in the production process.

Finally, the microbiological aspects of intensive broiler production and processing continue to exert significant influence on industry organisation and pose a number of long-term challenges. The most obvious concern is that of disease and the emergence of new pathogens that threaten the viability of intensive confinement operations. Indeed, the fact that such biological risks are greatest during the grow-out period is one of the reasons why the integrators prefer contracting to company-owned farms (i.e. the grower bears the production risk and has built-in incentives to keep his flocks healthy). A second major microbiological issue concerns product contamination and the related health risks (for workers and consumers) that are associated with the processing operations. Put simply, increasing line speed translates into increased potential for contamination (Bjerklie 1995: 46). Food safety and quality control have consequently been major regulatory issues facing the industry since 1957 when, in response to the deaths of several processing workers from bacterial contamination, the federal government passed the Poultry Products Inspection Act which mandated the inspection of all broilers sold in inter-state commerce (the act was amended in 1968 to include intra-state sales). The legislation, which went into effect in 1959 and subjected broilers to the same regulatory overview that the other meat industries had faced since 1906, required a system of organoleptic inspection of every single bird destined for human consumption. Firms in the poultry and meat industries currently consider themselves to be among the most regulated industries in the US (Bjerklie 1995). In light of the recently passed 'Food Safety Inspection Service Pathogen Reduction/Hazard Analysis Critical Control Point Program' (known in the industry as the 'Mega-Reg') which requires that the industry move from a system of organoleptic inspection of birds moving through the processing plant to one requiring systematic microbiological evaluation throughout the entire production process, such sentiments will only deepen. The final microbiological characteristic of the broiler concerns the constraints on the spatial character of the production complex posed by perishability and feeding requirements. In effect, the transport of live broilers and the multiple rations required during each grow-out phase demand that grow-out operations generally be located within a twenty-five mile radius of processing facilities (Kim and Curry 1993). Such 'production density' has often been cited as more critical than plant size in achieving economies and determining competitive position in the industry (Rice 1951, Harper 1953, Packers and Stockyards Administration 1967, Marion and Arthur 1973: 21).[11] These spatial imperatives, when added to the biological risks associated with confinement and the preference for contracts, suggest that establishing and maintaining a successful grow-out agglomeration requires precisely the type of farm structure – small marginal farms in a compact geographic area – that characterised much of the American South during the 1940s and 1950s.

Southern accents

Three key features of the economic structure of the rural South during the 1930s and 1940s underwrote the genesis of the modern broiler industry. The first was the existence of a class of small marginal farmers, located primarily on the periphery of the cotton belt, who confronted an agricultural crisis and a need for alternative forms of rural livelihood. Most of these farmers could not compete within the capital-intensive agriculture that emerged in the region during the post-Second World War period, and some clearly saw broilers as an attractive alternative. Second, the history of merchants' and finance capital in the rural South, and, in particular, the role of merchants and feed dealers in extending credit to small farmers, proved to be indispensable to the origins of the industry. By providing the institutional imperative for the contract grow-out arrangement that spread so prolifically in parts of the South during the 1950s, these feed dealers were instrumental in establishing the production base for the integrated system. Finally, the existence of a pool of surplus rural labour available to work in the processing facilities was also critical. The 'southern enclosure', as Pete Daniel (1981, 1985) has referred to the federally initiated dissolution of the southern farm tenancy system, meant that the South was awash with cheap labour. Obviously, there were other forces at work: infrastructure development (which facilitated feed grain shipments from the Midwest and lowered transportation costs for broilers), housing costs (which have historically been lower in the South) and, not least, the industrial recruitment strategies pioneered by many southern states.[12] In our view, however, they remain secondary. For us, the crux in explaining the regional character of the modern broiler industry lies in a genealogy of institutional forms and conventions – an unpacking of particular sorts of agro-industrial districts within a regional complex – rather than in some sort of shopping list of 'contributing factors'. To that effect, we will briefly explore these three 'southern accents', emphasising the distinctive character or 'situatedness' of flexibility in the industry.

Regarding the existence of a class of small marginal farmers, it should be emphasised that none of the farmers who moved into broiler production were on land that could successfully be adapted to the realities of the new capital-intensive model of crop-based agriculture emerging in the South. Broilers looked like a 'good crop on bad land'. In north Georgia and north Alabama, for example, most of the early broiler growers were small, marginal white farmers who were looking for alternatives to cotton. In northwest Arkansas, in contrast, most of the early growers saw broilers as a way out of a devastated apple and orchard economy. To say, therefore, that broiler production proliferated in the South because of the lack of alternatives for farmers – a crude index of which is the composite farm wage rate[13] does indeed capture the macro-situation, while missing the importance of sub-regional variation. Nonetheless, the fact that these farmers had so few alternatives meant that the integrators could offer lower rates of compensation than in other parts of the country (or the South for that matter). A 1967 report by the USDA noted that returns to labour for broiler growers were consistently much lower in the South than elsewhere and in fact revealed a marked *decline* throughout the 1950s and 1960s. In Georgia, for example, the report noted that net returns to labour declined from a high of 53 cents per hour in 1952 to 1 cent per hour in 1964. Between 1950 and 1965, moreover, the average return to growers in Georgia was 33 cents an hour compared with averages of 91 cents and $1.69 an hour for broiler growers in Maine and Delmarva respectively (Packers and Stockyards Administration 1967: 21).[14]

Yet, if the 'availability' of a class of small marginal farmers (not unlike the category of allotment-holding workers described by Lenin) provided the 'human capital' necessary for a successful grow-out agglomeration, it was the local feed dealers who provided the institutional and financial capital for contracting by building on the legacy of local merchant–farmer relationships that saturated southern agriculture. Several agricultural extension reports from the early 1950s claimed that such local credit arrangements were the most important overall factor in the growth of broiler production in north Georgia and other areas of the South relative to the rest of the country (Rice

1951, Harper 1953, Harper and Hester 1956: 9), and it is incontestable that early integrators were drawing on long-term, informal credit relationships with local farmers when they moved into broiler contracting. Broiler contracts emerged if not from a sort of local moral economy, then at least from a dense network of historically deep social and institutional relations. The very diversity of early production contracts – in 1953, for example, some seventeen distinct contractual arrangements with at least twenty-three different modifications were in use in north Georgia alone (Harper 1953: 15) – speaks precisely to the historical embeddedness and local specificity of these arrangements.

As contracting proceeded, however, such local diversity was soon replaced by standardised production contracts, with feed conversion 'tournaments' emerging as the most widely employed type (Knoeber and Thurman 1995). By providing performance incentives, as well as a basis for ranking growers, the standard feed conversion contract has proved to be superior to company-owned, grow-out operations utilising wage labour.[15] In short, the quality of the product can be more effectively maintained in a contracting system where growers are required to meet precise and demanding standards *and* have incentives to keep their flocks healthy. At the same time, the contract effectively absolves the integrators from providing the grower with guarantees and protections customarily accorded to normal employees.[16] Insofar as contracts are 'bunch-to-bunch', the arrangement provides the integrator with a buffer to absorb fluctuations in market conditions beyond any single grow-out phase.[17] Because grow-out facilities have few alternative uses, integrators can effectively leverage technological improvements in grow-out operations by threatening termination of contracts (Wilson 1986, Watts 1994a, 1994b). Finally, because grower investments in fixed capital represent more than half of the total investment in fixed capital in the industry[18] the contractual arrangement has provided a way of de facto freeing up capital for the integrators. In effect, the contract allows the integrator to take advantage of the chief assets of the family farm – cheap, 'docile', and flexible labour – without the burdens of equity or the costs of wage labour (both direct, i.e. actual wage rates, and indirect, i.e. surveillance, indiscipline) echoing again the flexible use of self-exploitation and risk dispersion without land concentration described by Kautsky (1906) almost a century ago in his account of vertical integration in European agricultures.

As in Kautsky's day, such 'flexibility' has had quite distinct meanings for growers and integrators. Jesse Jewell, an early integrator from north Georgia, saw integrated contracting as a flexible method for tightly regulating throughput and inventory:

> Our operation is all contracts, practically 90, 95, maybe 100 per cent. These farmers down there won't grow chickens unless they know they are not going to lose anything, so our company, in order to control

> production – and we like it that way – in order to have the regular supply of chickens coming in, we let them feed. We furnish the baby chicks and the feed. They furnish the house, the equipment, and the labour, heat, and we pay them according to the number of pounds of chickens they get out of the number of pounds of feed they use, and it is a flexible contract and as things get a little tough we can sort of twist up the contract and as they get better we have to loosen up the contract in order to hold our growers.
>
> (US Congress 1957: 217)

Several years after Jewell made his remarks, Charles R. Moultries, a grower with the Marshall County Alabama Poultry Producers Association, offered this testimony:

> I wish to impress upon the committee the fact that severe pressure and controls are being imposed on the grower by the feed companies and the processors. The role of the family-size poultry producer has been reduced to 'a cheap hired hand with a large investment'. The feed companies have a monopoly of contracts over the entire industry. The independent producer does not have a chance anymore. . . . The feed dealers set the price to be paid for growing poultry. They make just enough variation in the contracts to avoid being prosecuted for price-fixing. If the grower disagrees or objects to their demands, they refuse to deal with him, leaving him with a large debt for buildings and equipment. . . . We are a depressed people, and our only hope is through some type of legislation.
>
> (US Congress 1961: 7)

Put crudely, Jewell's contention that the contract insulated broiler growers from taking on any market risk only holds for a single grow-out period and obscures the situation of financial dependency that growers have always confronted. Despite claims by integrators and bankers that loans (which are currently on the order of several hundred thousand dollars) can be paid off within ten to fifteen years, few growers have been able to achieve financial independence. As Mary Clouse, a former grower and head of the poultry project at the Rural Advancement Fund International in North Carolina notes:

> We know of only one or two farmers who have their loans paid off. Most are kept in debt as a way to keep their 'nose to the grindstone'. All farmers are afraid that if they do not spend money on the newest equipment, their contract will be canceled.
>
> (Clouse 1995: 1)

Default rates tend to be low because loans are paid back with money from off-farm employment. Most households engaged in contract broiler production have had at least one member of the household seeking off-farm employment. A 1967 USDA report, for example, noted that almost 60 per cent of all broiler growers in Georgia sought off-farm employment (Packers and Stockyards Administration 1967: 20). Several years later, it was noted that more than 70 per cent of all broiler producers in the US required off-farm income (US Congress 1972: 128–29). More recently, a 1993 survey of Tyson's growers found that 74 per cent of the respondents had other sources of income (Clouse 1995).[19] Growers are, in other words, 'peasant-workers' in two profound senses: in one respect they have lost control over their labour process via the contract, and in another they must resort to wage labour to reproduce the household enterprise.[20]

Flexibility in its other guise appears, of course, through low-wage labour markets sustaining the broiler processing facilities[21] – the third structural feature of the postwar rural South that has been key to the growth of the broiler industry in the region. In the wake of the dissolution of farm tenancy, the postwar rural South was a reservoir of cheap labour.[22] During the late 1950s and early 1960s, as the adoption of assembly-line production techniques in the processing plants initiated a relentless effort to increase line speeds and automate, various aspects of processing, slaughtering and cut-up operations were broken into individual tasks that could be performed by low-wage labour.[23] Subsequent de-skilling and downward pressure on wages placed a premium on the maintenance of fluid, low-wage (and often highly feminised) labour markets. Up until the 1970s, surplus rural labourers displaced from southern agriculture, many of whom were African-American, provided a sufficient labour pool. Since the 1980s, however, there has been a rather significant shift to the use of immigrant labour, primarily Hispanic but also southeast Asian (Griffith 1993, 1995). By 1993, new immigrants accounted for some 25 per cent of a total work-force of over 200,000 (Hetrick 1994: 33, Griffith 1995: 147). According to David Griffith, in the face of tightening labour markets, increased worker resistance and the relentless imperative to increase productivity (i.e. line speed), the industry has had to engage in 'ever more comprehensive patterns of labour control', which is most effectively accomplished by employing immigrant labour whose presence 'serves as an incentive for native workers to submit to the terms of the plant production regimes' (Griffith 1993, 1995: 133).[24] Poultry plant managers have used kinship and friendship networks along with bonus incentives to actively recruit new immigrant labour. In Georgia, some 85 per cent of the plants visited by Griffith recruited through such channels, and 50 per cent paid bonuses to current employees who brought in new workers – many of whom found their way into poultry processing through immigrant farm labour markets. Rotating among agriculture, poultry processing, and return migration to Mexico, these immigrant workers constitute a fluid, transnational proletariat. Some plants are now surrounded by temporary trailer parks identical to the ones seen in migrant farm labour camps (Griffith 1995: 141, 143, 145) – a striking illustration of how flexible accumulation often turns on recombinative strategies of absolute and relative surplus value (informal labour markets alongside high-tech automated food processing) (Harvey 1989).

Today the line speed in a standard processing plant is 8,000 birds per hour, more than double that of the mid-1970s when line speed was governed by manual evisceration (Henry and Rothwell 1995: 40). Poultry processing workers are currently considered to be some of the most productive in all of US manufacturing and some of the most poorly paid (Hetrick 1994).[25] According to the BLS, workers in poultry plants made an average of $7.26 per hour in 1992, about 63 per cent of the average hourly wage for all manufacturing – a proportion unchanged since 1972 (Hetrick 1994).[26] Like their brethren in the grow-out system, few have enjoyed the benefits of union membership. In short, poultry processing workers confront a Taylorist work regime of unimaginable time-discipline combined with a high degree of microbiological and stress-related hazards and little recourse to collective bargaining. Worker turnover is often more than 100 per cent per year – underscoring once again the importance of maintaining 'flexible' labour pools.

CONCLUSION: FLEXIBILITY, OVERPRODUCTION, AND THE QUESTION OF QUALITY

In concluding our account of the US broiler industry and the peculiarities of agro-industrial flexibility, we will focus briefly on three key issues: namely, the double-edged nature of flexibility and the hybrid character of the broiler production complex; the endemic problem of overproduction in the industry and the perennial 'market question' facing the integrators; and the question of quality. The first issue speaks to the somewhat mixed account of flexibility, specialisation and JIT production in the southern broiler complexes. Obviously, production workers in poultry processing hardly conform to the storybook case of company unions, flexible skills, and flexible machines in the work place. As Allan Rahn, a poultry specialist at Michigan State University, put it: 'Broiler people are not used to organised labour and don't want to be' (cited in Clark 1992: 14). In the case of contract broiler growers and their relationships to the integrators, 'studied trust' is hardly adequate to the circumstances of indebted, semi-proletarianised and heavily disciplined household growers. If there is a social basis of flexibility inherent in the grow-out arrangement, it surely has been constructed more on a foundation of insecurity, rather than stability and cooperation. In a recent survey of growers for one of the major poultry firms, respondents repeatedly pointed to *fear and lack of trust* as major problems in their relations with integrators – a situation that one grower described as 'like being a one-legged man in a butt kicking contest'. At the same time, the patterns of inter-firm dependence between the integrators and the complex field of suppliers – pharmaceutical, breeders, processing equipment and so on – do indeed resemble the sorts of stable industrial districts whose attributes Amin (1994) (following Marshall) refers to as 'industrial atmosphere' and 'institutional thickness'. Moreover, the biology of broiler production – by creating certain constraints and rigidities on the production process (and the organisation of the *filière* itself) – places a premium on tight coordination of inventories and quality control at all levels. There is a sense, then, in which Roenigk's (1991) description of a critical path, just-in-time system does encapsulate some central attributes of the southern broiler production complex. But Toyotaism it is not. In reality, it is a more complex sort of hybrid in which some aspects of the *filière* resemble despotic factory regimes and coercive forms of labour control suggestive of 'bloody Taylorism' on the one hand and a veritable panopticon of contractual subservience on the other.

Second, it also needs to be said that flexible, just-in-time production – whatever its limits and character – has not resolved the endemic problem of overproduction that has plagued the industry since its inception. Periodic price depressions,[27] though perhaps less frequent than in the past, testify to the ongoing problem in spite of the stunningly successful efforts of the integrators to develop new markets and new products – to produce, in the words of Don Tyson, many chicken commodities that 'didn't even exist seven years ago'. Systemic overproduction is rooted in the fact that the broiler industry, like many industries dealing in food and natural resource commodities, operates on the basis of very low margins (USDA FSIS 1988, Bjerklie 1995). The reality of low margins means that profitability rests on 'turning volume' by expanding capacity and increasing productivity, which itself requires the successful venting of surplus production. Up until the 1980s, expanding domestic market share and product diversification were the two key strategies. The latter – encapsulated in Don Tyson's mantra 'segmentation, concentration, domination' (cited in Rodgers 1991: 51–52) – is now the backbone of all of the major firms, perhaps best illustrated by Tyson's move in the 1960s to forward contracting with the so-called institutional food market[28] (Marion and Arthur 1973: 40). Integrators themselves, through their own marketing efforts, have effectively created a mind-boggling array of chicken products as a way of venting surplus and this diversity of products is clearly linked to the sophistication and coordination inherent in the production system itself. No other agricultural commodity or agro-industry can match the capacity of the firms in the broiler industry who adjust production and develop new products with astonishing speed and flexibility, subject as we have seen to the rigidities imposed by biological time lags in breeding and the inherent trade-offs involved in maximising certain qualities such as breast meat yield.

Since the 1980s, however, the hegemony of global neo-liberalism and the passage of the 1993 GATT accords have opened up unprecedented export opportunities for low cost southern integrators (dramatically, in a number of cases, such as Hong Kong, China and Russia) just as fears of over-investment in the industry have resurfaced. US broiler exports now account for 15 per cent of national production, and exports have leapt from almost nothing in the 1950s, to 237,000 tons in 1982 to an estimated 2 million tons in 1996 (most of which has occurred since 1990, see Table 8.4).[29] US exporters have been able to take advantage of a highly segmented world chicken market that has permitted the integrators to sell breast meat at home for a premium while exporting leg quarters, feet, wings, and so on – for which domestic demand is quite low – to fulfill the culturally specific yet rapacious demand in China, Russia and east Asia. In sharp contrast to the failed export drive of the 1960s, however, the deepening of the world poultry trade in the 1990s is marked by the rise of new low-cost broiler production in the periphery – China, Thailand and especially Brazil – which renders the foreign market 'valve' a questionable strategy for US integrators even if trade barriers are overcome.[30]

Finally, there is the question of quality which so concerns the French researchers at INRA and elsewhere (Allaire and Boyer 1995). Obviously, quality has been central to the success of the postwar broiler industry in a number of ways. But the 'construction of quality', to employ Marsden's language, has in fact been integral to the industry since the 1870s (Bugos 1992). In the North American case (in contrast to France), quality concerns have focused not so much on brand name identification or the Red Label phenomenon described by Sylvander (1995), but on freshness and problems of health and disease. In this regard, quality carries its own burdens, or, perhaps more accurately, quality has proven to stand in a complex and often contradictory position with respect to the industrialisation of the production process itself. In other words, the relentless demand for uniformity and testing which undergirds the entire system can threaten its very stability through disease, overproduction, and unforeseen genetic costs. Nonetheless, Allaire and Boyer are right to suggest (1995) that the question of quality

Table 8.4 Top five destinations for US broiler meat, 1988–95 (1000 metric tonnes)

1988		*1989*		*1990*		*1991*		*1992*		*1995*	
Japan	15	Japan	101	Russia	136	Japan	125	Hong Kong	148	Russia	674
Hong Kong	46	Hong Kong	75	Japan	93	Hong Kong	105	Japan	121	Hong Kong	456
Mexico	45	Mexico	41	Hong Kong	81	Russia	83	Mexico	72	Japan	120
Jamaica	27	Canada	30	Mexico	39	Mexico	60	Canada	41	Mexico	93
Singapore	25	Singapore	26	Canada	35	Canada	34	Poland	36	Estonia	56

Source: USDA, *Foreign Agricultural Service* 1994–95

remains central and is a major axis around which competition now turns. The recent struggles over the huge California market bear this out, but in a way that forces the quality issue back into the realm of production and the 'market question'. In effect, relatively high cost California integrators can only preserve their (dwindling) market share on the basis of providing 'fresh' chicken to the local California market – unlike the frozen or 'chilled' chickens brought in from the South. In the diet- and health-conscious consumer economy in California, the retail desirability of non-frozen chicken is of course quite considerable. Due to a quirk of legal history, however, the USDA deemed that chicken chilled to zero degrees could in fact be labelled 'fresh' (in spite of the indisputable fact that the freezing point of poultry is 26 degrees Fahrenheit).[31] In a key bill (HR 203) brought to the 104th Congress in January 1995 – which was prefigured by the extraordinary spectacle of California legislators, poultry federation representatives, and none other than Wolfgang Puck bowling on the verandah of the House with purportedly 'fresh' southern chicken to make their point (see testimony by Bill Mattos, US House of Representatives, 16 June 1994) – chickens chilled to a temperature below zero were deemed 'frozen'; if they are stored at zero to 26 degrees they are 'hard chilled' and only recognised as 'fresh' if the meat has never been stored below the 26 degree Fahrenheit threshold. In this way, the likes of Zacky and Foster Farms were able to hold onto the lucrative California market in the face of what seemed to be an arcane contestation – 'discursive strategies' in the legislative sphere in poststructural parlance – over the definition of freshness. In practice this was a battle between factions of regionalised broiler capital, operating under quite different cost and profitability regimes, over market share.[32] Similarly, both the October 1996 decision of the Chinese government to ban poultry imports from ten US states on the basis of questionable disease concerns, and the failed Russian effort to block US chicken imports in the spring of 1996 on the basis of sanitation concerns, demonstrate the various ways in which 'quality concerns' can be marshalled in efforts to protect domestic industry.

Whatever its particular dimensions, therefore, the successful construction of quality seems to rest as much on the political power and influence of the integrators as it does on the ability to define, develop, and produce 'healthy' low-cost chicken products. What makes the current conjuncture so interesting and important is that the globalisation of the industry – including the export of the JIT model itself to the new agricultural countries such as Brazil, Thailand and western Europe – has rendered the world poultry market a ferociously competitive place. As worldwide demand for cheap, high-quality animal-flesh protein continues to grow, it will undoubtedly be met in large part by the chicken. The only question is whose chicken the newly enfranchised Asians, Russians and Latin Americans will be eating in the new millennium.

NOTES

1 It was not until 1952 that per capita consumption of broiler meat actually surpassed the consumption of meat from farm chickens (Watts and Kennett 1995: 7).

2 The Schecter case, which invalidated the National Industrial Recovery Act, was itself about chickens – a case that illustrated the difficulties of coming to terms with changing relations between the public and private sectors and also with the problems involved in regulating industries dealing in perishable commodities, where quality was of primary concern (Freidel 1962).

3 There is a larger point here which emphasises the extent to which retail power in the *filière*, the shaping of product design by downstream actors, and the centrality of quality – that is to say concerns seen largely as the product of the shift to flexible or post-Fordist capitalism – were all present in the golden age of US Fordism.

4 Between 1947 and 1951, the time required to grow a day-old chick into a 3.5 pound broiler was reduced by 18 per cent (Tobin and Arthur 1964: 17). Between 1940 and 1955, the feed conversion ratio for broilers declined from 4.2 pounds to 2.8 pounds (Packers and Stockyards Administration 1967: 15).

5 Considered to be the pioneer of vertical integration in the industry, Jesse D. Jewell of Gainesville, Georgia, had established a successful contract-based model of integration by 1954 (Sawyer 1971).

6 Witness the words of John Burnett, an independent processor from Haleyville, Alabama: 'These major feed mills are eagerly engaged on a wild expansion program without regard to the effects of overproduction on final marketing prices. It is the major national feed mills intention, it seems to us, to cause the supply of broilers to be so abundant that the prices will stay low until the small companies are forced from the industry. The thousands of small grower farmers will be left to whatever whim the national feed mills might have, in regard to the pay the small farmer received for the actual caring and growing of the poultry' (US Congress 1961: 106). See also the testimony of L. D. McCleskey, Jr, Georgia Feed Dealers' Association (ibid.: 32); John Bagwell, independent feed dealer, Empire, Alabama (ibid.: 37); William Harold Kitchens, Gainesville Milling Co., Gainesville, Georgia (ibid.: 95).

7 A Michigan State University study estimated in 1985 that a new fully integrated operation in the US South would require investment of roughly $85 million and would need three to four years to come on line.

8 Henry and Rothwell (1995: 3) note that 'There is little pressure to integrate at the most highly complex levels, such as genetic development and pharmaceutical and chemical inputs, or at the base commodity input levels, such as grain and proteins for feed where markets are well organised, prices are visible, and input costs of competitors can be readily assessed.'

9 It is an interesting historical irony, however, that Ohno's development of JIT for Toyota was based on his observations in the US of retailing, textiles, and agro-industry (Green, Lanini, and Schaller 1994: 4).

10 Arbor Acres, for example, has addressed the new consumer preferences in its own genetic pipeline with the development of the Arbor Acres Yield Master Male (Amey 1992: 18).

11 According to the Packers and Stockyards Administration (Packers and Stockyards Administration 1967: 13): 'Probably the most significant factor to inhibit entry of firms and growth in firm size is that of transfer costs related to the spatial density of grow-out operations.' The report goes on to note that

'economies are realised by having growers located in small procurement areas due to the added costs of delivering feed and serving distant growers' (ibid.: 23).

12 The major gap in our explanation here concerns the evolution of supplier–integrator relations in the major production areas. Whether and how this evolution was or was not distinctively 'Southern' is a question for further research.

13 When asked to describe what factors were responsible for the growth of the industry in the southern states, USDA staff economist Linley E. Jeurs quoted a 1957 USDA report that estimated that the composite farm wage rate in the southern states where broiler production had expanded most rapidly (Georgia, Alabama, and Mississippi) was less than 50 cents an hour. In contrast, he went on to note, no other region of the country had an average rate that was below 61 cents an hour and those regions with the highest average rates – the Northeast and the Pacific Coast – were also the areas where broiler production was in notable decline (US Congress 1961: 26).

14 The report defines net returns to labour as that portion of total payment that remains after all production expenses have been met and appropriate charges for capital have been made. Ray Marshall and Allen Thompson cite another USDA report from 1970 that found net returns on labour to growers in northern Alabama to be *negative* 36 cents an hour in 1969 (Marshall and Thompson 1976: 55).

15 The superiority of a contract-based system is further supported by the fact that in California, where integrators have historically been unable to shift the majority of their production to contracts, the company-owned chicken ranches are managed on a feed-conversion bonus system very similar to those used in the South. The obvious disadvantage for the California firms in this respect is the fixed capital they have tied up in the grow-out facilities. Combined with higher feed costs, this has put the California firms in a position of competitive disadvantage vis-à-vis the southern firms and was a driving force behind the fresh/frozen labelling controversy of the early 1990s.

16 With the contract, the integrator does not have to pay a minimum wage, provide workers' compensation, pay social security, or provide any other benefits. Nor does the integrator have to pay any of the property taxes or other expenses associated with the farm.

17 Those who focus on the 'risk-sharing' nature of the contract and assert that the integrators bear the bulk of the market or price risk tend to miss this point (Knoeber and Thurman 1995).

18 Today, the typical investment for a contract broiler grower is approximately $120,000 per house. Since most growers have three or four broiler houses, their total investment can approach half a million dollars. Most of the financing comes in the form of long-term loans from the farm credit system, local banks, and life insurance companies. As one might expect, moreover, most growers have to mortgage their homes and farms in order to obtain loans – all for a total net income that has been estimated to be approximately $4,000 per house per year (Clouse 1995).

19 According to Tom Smith, former senior vice president and general manager at Wayne Farms, a division of Continental Grain Co., 'The processor never saw this [contract poultry production] as a farmer's primary job. But unfortunately, and for whatever reason, poultry took over a lot of these farms. The trouble is that the way the contracts are written, if your whole income depends on raising birds for an integrator, you have to be just about the absolute best grower that integrator contracts with – at least in the top 10 per cent of growers – to make

a living, and on top of that you've got to be very frugal, prudent and conservative in your investments' (cited in Bjerklie 1994: 33).

20 Without statutory protections for bargaining, moreover, individual broiler growers are necessarily weak. Since the 1960s, efforts by contract growers to secure collective bargaining protections at the federal level – including the 1967 Agricultural Fair Practices Act – were repeatedly frustrated by the agribusiness community. To date, all efforts to strengthen the Act have failed (Frederick 1994). Despite some recent successes, recourse to the court system has proven to be very expensive and inadequate to deal with the range of issues and vulnerabilities facing growers (Roth 1995).

21 According to Mar Linder of the University of Iowa, employment in poultry slaughtering and processing has grown from 19,000 in 1934 to more than 200,000 in 1993 (the vast majority of which was in the South), a record that is second only to automobile parts in manufacturing job growth for the same period (cited in Smothers 1996). In fact, even during the 1980s, a time when employment in red meat was stagnant and the overall food industry actually lost jobs at an annual rate of 0.4 per cent, employment in poultry processing recorded a robust 4 per cent annual growth rate despite continued increases in productivity (Ahmed and Sieling 1987, Hetrick 1994).

22 Though the wage gap has certainly narrowed in the last decade or so, estimates from the late 1980s suggest that processing plants in the South during that decade could pay $1.25 per hour less than comparable plants in the Midwest (Aho and Ran cited in Kim and Curry 1993: 70). Indeed, some have argued that the South's comparative advantage in terms of low-wage employment in the processing plants has been the single most important reason why the broiler industry has proliferated in the region (Griffith 1993, Kim and Curry 1993: 76).

23 Automation started with killing, scalding, defeathering, and continuous chilling, and had advanced to automatic evisceration, deboning, and packaging by the 1980s and 1990s (Stals 1992).

24 The parallels with the strategies of labour discipline employed in the California lettuce industry and other sectors are obvious (Thomas 1985).

25 The sector has also recorded some of the highest rates of work-place injury – in 1990 the reported worker injury/illness rate for poultry processing was 26.9 incidents per 100 full-time workers (twice the average for the total manufacturing sector) (Hall 1995). Common injuries include tendonitis, carpal tunnel syndrome, respiratory ailments, and lung disease (Hagmar *et al.* 1990, Kirschberg *et al.* 1994 and Zuskin *et al.* 1994). The 1991 fire at the Imperial Foods poultry processing plant in Hamlet, North Carolina, in which twenty-five workers (out of 200) died and another fifty-six were seriously injured, was one of the worst industrial accidents in US history (US Congress 1991).

26 In 1972, when the BLS started compiling regular data on poultry processing employment, average hourly wages in the sector were approximately 62 per cent of the average hourly wage for manufacturing workers in general (Hetrick 1994, Broadway 1995: 24–5).

27 Periods of price depression include: 1954, 1959, 1961, 1967, 1970–72, and 1987 (USDA ERS 1995: 90).

28 The remarkable success of this new market, epitomised by franchises such as Kentucky Fried Chicken (KFC) and products such as the Chicken McNugget, revolutionised not only the way Americans ate but the way that poultry was marketed in the US (Strausberg 1995: 112). By the mid to late 1980s, value-added had become the obsession of the industry, and most of the integrators

were rushing headlong into mass customisation of products in response to the fierce competition over contracts with fast food chains.

29 In 1988 the top five destinations for US broilers (Hong Kong, Mexico, Jamaica, Singapore, and Japan) accounted for 158,000 tons; in 1995 Russia and Hong Kong *alone* accounted for over 1 million tons (USDA 1994–95).

30 In a recent IFC study (Henry and Rothwell 1995: 31) the cost per kilo of broiler meat RTC in Brazil was estimated to be 85.4 cents, compared to $1.05 in the US and 93 cents in China (the total operating index/cost for the three countries were respectively: 38, 46, and 46).

31 In 1993 the California legislature, under pressure from the California Poultry Industry Federation, passed law SB214 which prevents poultry from being sold as 'fresh' if it is chilled at or below 25 degrees Fahrenheit. In a ferocious legal and legislative battle in 1994 and 1995, waged at the federal level, California integrators successfully fought off a battle by the southern integrators to increase the temperature ceiling at which chilled chickens could legally be distributed and sold as 'fresh chicken' (*Fresno Bee*, 9 March 1994, *Time*, 24 July 1995, *Washington Post*, 25 August 1995).

32 In September 1995, a group of nineteen largely southeastern legislators successfully lobbied the Senate Appropriations Committee to reverse the USDA ruling. The Senate voted on 20 September by a 61–38 margin to overturn the previous federal labelling law which was to come into effect according to the Federal Register on 25 August 1996 (*Los Angeles Times*, 20 September 1995: A3).

BIBLIOGRAPHY

Ahmed, Z. Z., and Sieling, M. (1987) 'Two decades of productivity growth in poultry dressing and processing', *Monthly Labor Review* (April).

Albert, J. A. (1991) 'A history of attempts by the Department of Agriculture to reduce federal inspection of poultry processing plants – a return to the jungle', *Louisiana Law Review* 51: 1183–231.

Allaire, G. and Boyer, R. (eds) (1995) *La Grande Transformation*, Paris: INRA.

Amey, D. (1992) 'Arbor Acres Farm, Inc.: new products, changing philosophy', *Broiler Industry* (September).

Amin, A. (1994) 'The difficult transition from informal economy to Marshallian industrial district', *Area*, 26, 1: 13–24.

Beckler, R. I. (1957) 'A Summary of Recent Studies on Broiler Financing and Contracting', USDA Agricultural Marketing Service, Marketing Research Division.

Bell, D. (1995) 'Forces that have helped shape the egg industry', *The Poultry Tribune* (September): 30–43.

Bjerklie, S. (1994) 'The empire strikes back', *Meat & Poultry* (October).

—— (1995) 'On the horns of a dilemma: the US meat and poultry industry', in D. D. Stull, M. J. Broadway and D. Griffith (eds) *Any Way You Cut It: Meat Processing and Small-Town America*, Lawrence: University Press of Kansas.

Broadway, M. J. (1995) 'From city to countryside: recent changes in the structure and location of the meat- and fish-processing industries', in D. D. Stull, M. J. Broadway and D. Griffith (eds) *Any Way You Cut It: Meat Processing and Small-Town America*, Lawrence: University Press of Kansas.

Bugos, G. E. (1992) 'Intellectual property protection in the American chicken-breeding industry', *Business History Review* 66 (Spring): 127–68.

Chandler, A. D. Jr (1977) *The Visible Hand: The Managerial Revolution in American Business*, Cambridge: Belknap Press.

Clark, E. (1992) 'Poultry production returns to the Midwestern states', *Feedstuffs* 30 November.

Clouse, M. (1995) 'Farmer Net Income From Broiler Contracts', RAFI-USA.

Daniel, P. (1981) 'The transformation of the rural South: 1930 to the present', *Agricultural History* 55, 3: 231–48.

—— (1985) *Breaking the Land: The Transformation of Cotton, Tobacco, and Rice Cultures Since 1880*, Urbana: University of Illinois Press.

Dudley-Cash, W. A. (1995) 'Genetics account for nearly 80 percent of modern broiler performance increase', *Feedstuffs* 3 April.

Fine, B. (1994) 'Towards a political economy of food', *Review of International Political Economy* 1, 3: 519–45.

Fite, G. C. (1984) *Cotton Fields No More: Southern Agriculture 1965–1980*, Lexington: University Press of Kentucky.

Frazier, F. (1995) 'Demands of World War II shaped poultry industry', *Feedstuffs* 28 August.

Frederick, D. A. (1994) 'Update on the National Bargaining Law', paper presented at the 1994 Pacific Coast and National Bargaining Conference, December.

Freidel, F. (1962) 'The sick chicken case', in J. A. Garraty (ed.) *Quarrels That Have Shaped the Constitution*, New York: Harper and Row.

Gallimore, W. W. and Vertrees, J. G. (1968) 'A Comparison of Returns to Poultry Growers . . . Under Contract . . . Operating Independently', USDA Economic Research Service.

Goodman, D., Sorj, B,. and Wilkinson, J. (1987) *From Farming to Biotechnology: A Theory of Agro-Industrial Development*, Cambridge: Blackwell.

Gordon, J. S. (1996) 'The chicken story', *American Heritage* (September).

Green, R., Lanini, L., and Schaller, B. (1994) 'Technological and Organizational Innovation in the Food System', unpublished manuscript, Paris: INRA.

Griffith, D. (1990) 'Consequences of immigration reform for low-wage workers in the southeastern US: The case of the poultry industry', *Urban Anthropology* 19: 1–2.

—— (1993) *Jones's Minimal: Low-Wage Labor in the United States*, Albany: State University of New York Press.

—— (1995) 'Hay trabajo: poultry processing, rural industrialization, and the Latinization of low-wage labour', in D. D. Stull, M. J. Broadway and D. Griffith (eds) *Any Way You Cut It: Meat Processing and Small-Town America*, Lawrence: University Press of Kansas.

Hackenberg, R. A. (1995) 'Conclusion: Joe Hill died for your sins: empowering minority workers in the new industrial labour force', in D. D. Stull, M. J. Broadway and D. Griffith (eds) *Any Way You Cut It: Meat Processing and Small-Town America*, Lawrence: University Press of Kansas.

Hagmar, L., Schütz, A, Hallberg, T., and Sjöholm, A. (1990) 'Health effects of exposure to endotoxins and organic dust in poultry slaughter-house workers', *International Archives of Occupational and Environmental Health* 62: 2.

Hale, K. K. Jr, Thompson, J. C., Toledo, R. T. and White, H. D. (1973) 'An Evaluation of Poultry Processing', Georgia Experiment Station.

Hall, Bob (1995) 'The kill line: facts of life, proposals for change', in D. D. Stull, M. J. Broadway and D. Griffith (eds) *Any Way You Cut It: Meat Processing and Small-Town America*, Lawrence: University Press of Kansas.

Harper, W. W. (1953) 'Marketing Georgia Broilers', Georgia Experiment Station, University of Georgia College of Agriculture.

Harper, W. W. and Hester, O. C. (1956) 'Influence of Production Practices on Marketing of Georgia Broilers', Georgia Agricultural Experiment Station, University of Georgia College of Agriculture.

Harrison, B. (1994) *Lean and Mean: The Changing Landscape of Corporate Power in the Age of Flexibility*, New York: Basic Books.

Harvey, D. (1989) *The Condition of Postmodernity*, Cambridge: Blackwell.

Henry, R. and Rothwell, G. (1995) 'The World Poultry Industry', International Finance Corporation.

Henson, W. L. (1980) 'The US Broiler Industry: Past and Present Status, Practices, and Costs', Agricultural Experiment Station, Penn State University.

Hester, O. C. and Harper, W. W. (1953) *The Function of Feed-Dealer Suppliers in Marketing Georgia Broilers*, Georgia Experiment Station, University of Georgia College of Agriculture.

Hetrick, R. L. (1994) 'Why did employment expand in poultry processing plants?' *Monthly Labor Review* (June): 31–34.

Kautsky, K. (1906) *La Question Agrarire*, Paris: Maspero.

Kim, C. K. and Curry, James C. (1993) 'Fordism, flexible specialization, and agro-industrial restructuring: the case of the US broiler industry', *Sociologia Ruralis* 33, 1.

Kirschberg, G. F. *et al.* (1994) 'Carpal tunnel syndrome – classic symptoms and electrodiagnostic studies in poultry workers with hand, wrist, and forearm pain', *Southern Medical Journal* 87: 3.

Knoeber, C. R. and Thurman, W. N. (1995) 'Don't count your chickens . . . : risk and risk shifting in the broiler industry', *American Journal of Agricultural Economics* 77 (August).

Lasley, F. A., Jones, H. B. Jr, Easterling, E. H. and Christiansen, L. A. (1988) 'The US Broiler Industry', USDA Economic Research Service.

Marion, B. W. and Arthur, H. B. (1973) 'Dynamic Factors in Vertical Commodity Systems: A Case Study of the Broiler System', Ohio Agricultural Research and Development Center.

Marshall, R and Thompson, A. (1976) *Status and Prospects of Small Farmers in the South*, Atlanta: Southern Regional Council.

Martin, T. Jr (1996) 'Industry efficiency: a changing paradigm', *Broiler Industry* (March).

Nicholas, F. and Valeschini, E. (1995) *Agro-Alimentaire: une economie de la qualité*, Paris: INRA.

Packers and Stockyards Administration (1967) 'The Broiler Industry: An Economic Study of Structure, Practices and Problems', US Department of Agriculture, Packers and Stockyards Administration.

Perez, A. and Christensen, L. A. (1990) 'Recent Developments in the Location and Size of Broiler Growout Operations', USDA Economic Research Service.

Rice, S. T. (1951) 'Interregional Competition in the Commercial Broiler Industry', University of Delaware Agricultural Experiment Station.

Rodgers, R. T. (1991) *Broilers – Differentiating a Commodity*, unpublished manuscript.

Roenigk, W. P. (1991) *Broiler Industry Vertical Integration*, unpublished manuscript, National Broiler Council.

Roth, R. I. (1995) 'Redressing unfairness in the new agricultural labour arrangements: an overview of litigation seeking remedies for contract poultry growers', *The University of Memphis Law Review* 25: 1307–32.

Sawyer, G. (1971) *The Agribusiness Poultry Industry: A History of Its Development*, New York: Exposition Press.

Sayer, A. and Walker, R. (1992) *The New Social Economy*, Oxford: Blackwell.

Scheideler, S. E. (1992) 'Managing rapid growing broilers', *Broiler Industry* (November).

Schwartz, M. (1991) *Tyson: From Farm to Market*, Fayetteville: University of Arkansas Press.

Skinner, J. L. (ed.) (1974) *American Poultry History, 1823–1973*, Madison: American Poultry History Society.

Smothers, R. (1996) 'Unions head south to woo poultry workers', *The New York Times*, 30 January, C-22.

Southern Cooperative Series (1954) 'Financing Production and Marketing of Broilers in the South, Part I: Dealer Phase', Agricultural Experiment Stations of Southern States.

Stals, P. (1992) 'Poultry processing', *Broiler Industry* (June).

Strausberg, S. F. (1995) *From Hills and Hollers: Rise of the Poultry Industry in Arkansas*, Fayetteville: Arkansas Agricultural Experiment Station.

Sylvander, B. (1995) 'La qualité: du consommateur au producteur', *Études Recherches Systemes Agraries Developpement* 28: 27–49.

Thomas, R. (1985) *Citizenship, Gender, and Work: Social Organization of Industrial Agriculture*, Berkeley: University of California Press.

Thornton, G. (1994) 'Managing today's breeders,', *Broiler Industry* (October).

—— (1995a) 'The processing plant's most important supplier', *Broiler Industry* (May).

—— (1995b) 'Measuring the competition', *Broiler Industry* (September).

—— (1996a) 'Top 10 US broiler companies', *Broiler Industry* (January).

—— (1996b) 'High yielding broiler production: the big trade-off', *Broiler Industry* (August).

Tobin, B. F. and Arthur, H. B. (1964) *Dynamics of Adjustment in the Broiler Industry*, Boston: Harvard University Graduate School of Business Administration.

US Congress (1957) *Problems in the Poultry Industry*, Hearings before Subcommittee No. 6 of the Select Committee on Small Business, House of Representatives, 85th Congress, First Session, Washington DC: US GPO.

—— (1961) *Small Business Problems in the Poultry Industry*, Hearings before the Special Subcommittee of the House Select Committee on Small Business, House of Representatives, 87th Congress, First Session, Washington DC: US GPO.

—— (1972) *Farm Bargaining*, Hearings before the Committee on Agriculture and Forestry, Senate, 92nd Congress, First Session, Washington DC: US GPO.

—— (1991) *The Tragedy at Imperial Food Products*, Hearings before the Committee of Education and Labor, House of Representatives, 102nd Congress, First Session, Washington DC: US GPO.

USDA ERS (1995) *Poultry Yearbook*, US Department of Agriculture.

USDA (1994–95) *Foreign Agricultural Service* various issues.

USDA FSIS (1988) *Financial and Economic Dynamics of the Poultry Industry*, US Department of Agriculture, Food Safety and Inspection Service, Planning Office.

USDA NASS (1995) *Poultry Production and Value*, US Department of Agriculture, National Agricultural Statistics Service.

US ITC (1992) *Industry Trade and Summary: Poultry*, US International Trade Commission Office of Industries.

Watts, G. and Kennett, C. (1995) 'The broiler industry', *Poultry Tribune*.

Watts, M. J. (1994a) 'Life under contract: contract farming, agrarian restructuring, and flexible accumulation', in P. D. Little and M. J. Watts (eds) *Living Under Contract: Contract Farming and Agrarian Transformation in Sub-Saharan Africa*, Madison: University of Wisconsin Press.

—— (1994b) 'Contracting, social labour, and agrarian transitions', in P. D. Little and M. J. Watts (eds) *Living Under Contract: Contract Farming and Agrarian Transformation in Sub-Saharan Africa*, Madison: University of Wisconsin Press.

Wells, M. J. (1996) *Strawberry Fields: Politics and Work in American Agriculture*, Ithaca: Cornell University Press.

Wilson, J. (1986) 'The political economy of contract farming', *Review of Radical Political Economics* 18: 4.

Wolf, S. and Buttel, F. (1996) 'The Political Economy of Precision Farming', paper presented to the Agricultural Economics Association, 28–31 July, San Antonio.

Zuskin, E., Kanceljak, B., Mustajbegovic, J., Neil Schachter, E., and Stilinovic, L. (1994) 'Respiratory symptoms and immunological status in poultry food processing workers', *International Archives of Occupational and Environmental Health* 66: 5.

COMMENTARY ON PART III

'Creating space for food' and 'agro-industrial just-in-time'

William H. Friedland

The chapter by Marsden and the one by Boyd and Watts provide a veritable feast of ideas from which I want to develop three. This will involve emphasising the importance of the just-in-time (JIT) concept in agriculture and its importance in 'overcoming' nature and space; polemicising against the 'gee-whiz' character of 'flexibility' and its attendant 'flexible specialisation'; and finally, beginning to unpack globalisation, a term now endowed with a great many meanings.

OVERCOMING NATURE AND SPACE

A critical element in the establishment of the new global systems in agriculture and food requires the constant application of science and technology to the limits imposed by nature and space. An old expression in agriculture had it as 'sell it or smell it'. This has been replaced by precision agriculture, the establishment of industrial-technological conditions which control factors of production, including the delivery of the commodity to consumers.

The importance of JIT has been demonstrated by Marsden and by Boyd and Watts. For Marsden, JIT is critical for the rapid turnover of retail stocks since turnover is critical to profit-taking. The same applies to the production of broilers; as Boyd and Watts show, left to its 'natural' state, broiler output would concentrate in a spring grow-out leaving other seasons bereft of chickens.

Year-round demand for agricultural commodities must confront the reality of seasonal dislocation. This is not a problem for storable agricultural products. The storage and transportability of sugar and wheat provided the basis to feed the developing working class at the inception of capitalism. Since that time, agricultural product after product has represented a challenge to resolve the problem of annual availability.

Technologies of drying, canning, and freezing were elaborated as solutions. Beyond that, particularly in the maintenance of commodities that cannot be dried, canned, or frozen, it has been necessary to establish complex logistical systems to move commodities from production to consumption.

The battle for 'shelf-life' has been long, complex, and uneven in the multitudinous commodity inventory now demanded in modern life.

This battle has been seen as the need to overcome nature and space. In its most modern manifestation, delivering fresh fruits and vegetables from one hemisphere to another, it requires the establishment of JIT production and complex delivery systems. Getting fresh table grapes from Chile to North America and Europe within their short life-span is a remarkable feat; the same applies to salmon raised in pens in Norway for delivery fresh to locations thousands of miles away.

Agriculture seems to mimic a now well-established industrial process. In fact, agriculture has had JIT in place for many decades, ever since fresh commodities have been delivered to distant urban populations. In this respect, agriculture and industry appear to be converging, a point that should perhaps be emphasised more firmly (Pugliese 1991).

FLEXIBILITY? REALLY? FOR WHOM?

Marsden's paper raises the issue of flexibility when considering how local producers in the São Francisco valley of Brazil meet differential quality requirements in various markets, when Caribbean statelets adjust agricultural policy to fit currency exigencies, and in the shift of regulation of quality standards from state apparatuses to retail giants. Similarly, Boyd and Watts make much of flexibility in the broiler *filière*.

But just how flexible is flexibility and flexible specialisation and who benefits from flexibility? Do the actual producers – the grow-out 'farmers' – have much flexibility in their production process? Hardly, as Boyd and Watts demonstrate, for the breeds of the birds are specified contractually, as are delivery dates of chicks and grown-out broilers, the specifications of the grow-out house (square inches per bird), the hormones and antibiotics incorporated in the fodder, etc. This also applies to the iceberg lettuce grower who is contracted to a grower-shipper, or the tomato grower for processing to firms such as Heinz. Indeed, contract growing is the critical element in production and, paradoxically, a major institutional mainstay to what remains of family farming (Roy 1972, Little and Watts 1994).

The consequent subordination of such family farmers is noted by Boyd and Watts as 'peasant-workers' and by Davis (1980) as 'propertied laborers,' i.e. labourers who own the means of production but who control very few of the input factors or the methods of production which are almost all specified by the integrating contractor.

While Boyd and Watts make much of the term, their narrative admits only limited flexibility in the broiler commodity chain. They mention that 'the inability to speed up the process . . . create[s] inevitable "rigidities" in the flexible production process'. Although breeding programmes have been remarkably successful in standardising broiler chickens, the two-year period regarding breeding hardly indicates flexibility: 'In 1996, for example, a plant's production manager will be making decisions about breeds to be placed in 1997 for birds to be processed in 1998.'

So, where does this demand for the concept of 'flexibility' come from? It arises, I believe, from the dichotomisation tendencies of the social sciences – of the need to construct ideal types, of Fordism vs post-Fordism, mass production vs flexible specialisation, etc. – from the need of scholarly intellectuals to identify new developments.

Yet, just as the changes that take place are notable, so are the continuities. Rather than the sharp dichotomisations that are structurally required for ideal types, what is needed is an equivalent emphasis on continuities.

Consider Fordism-post-Fordism, which has generated a remarkable recent literature. Despite the criticism of Fordism, post-Fordist production characterised by flexible specialisation is hardly the dominant mode in production.

As is usual with ideal types, the tendency is to take exaggerated positions, emphasising the differences between types. Many researchers have critiqued the 'second industrial divide' argument by pointing out, empirically, how profound mass production/consumption remain. This also holds for agriculture (Kenney *et al.* 1989, Kim and Curry 1993). There can be little question that the mass production/consumption of wheat, corn, soybeans, beef, hogs, chickens, lettuce, tomatoes, and kiwi fruit (to mention only a few agricultural commodities) remains the dominant system of production/consumption.

At the same time that Fordism in agriculture and food must be acknowledged, it is also necessary to understand the growth of differentiated sub-systems: upscale differentiated agricultural production and food systems which cater to economically privileged strata.

How shall we characterise a system which manufactures wheat, chickens, milk, etc. on a mass production basis while, at the same time, producing artisanal stone-ground wheat, free-range and/or organic chickens, and unhomogenised and/or unpasteurised milk in glass bottles? The regulationists and the 'second industrial dividers' see only the latter as significant whereas the critics argue that the overwhelming mass of humanity in advanced capitalist countries is still characterised by the former. Both exist; there has undoubtedly been a startling growth of the latter and undoubtedly this sector of production will continue to grow. But it is extremely doubtful that it will become the dominant element in political economy.

In both production and consumption, whether in automobiles or salads, mass production continues to characterise the overwhelming bulk of economic activity and is geared at mass consumption. Rather than a dichotomised reality, there is a continuum of change. I characterise this continuum in production as Sloanism, describing a situation in which mass production is used to meet highly differentiated and specialised markets.

Even Henry Ford had to abandon Fordism. Fordism was originally embodied in two elements: the production of single model of car, the Model

T Ford, and the $5-a-day wage, i.e. the mass production of a standardised product which was sufficiently cheap that workers could afford to buy it. But Henry Ford practically destroyed his own corporation by hanging on to Fordism; he was saved by Sloanism, the system of maintaining mass production while reorganising production to grapple with differentiated markets.

Alfred Sloan of General Motors saw the automobile market as economically differentiated and built General Motors so that Chevrolet was aimed at the lowest income level and Cadillac for the top economic level, with Pontiac, Buick, Oldsmobile, and Lasalle intended for in-between strata.

But Sloan was ingenious in stratifying the automobile market even further, introducing two additional differentiating principles: the annual model change and options. By redesigning the outer shell of the automobile, he introduced planned obsolescence so that every year a new car became 'old'. By introducing options, he broadened the range of consumer possibilities to body paint, tinted glass, radios, etc. Sloan's differentiation of the market did not end mass production; it required higher levels of planning and logistics laid upon the basic system of mass production innovated by Ford.

The same approach applies in agriculture. Agricultural commodities such as lettuce and tomatoes were once handled artisanally; they were the products of 'truck farms', produced on a seasonal basis for limited geographical distribution. Iceberg lettuce became a Fordist product when refrigerated rail cars permitted its transport over thousands of miles from remote production locations to hundreds of localised markets (Friedland *et al.* 1981). Within the past two decades, we have witnessed the Fordisation of other lettuce forms: romaine, redleaf, oak, butter, etc. as well as the most recent invention, mesclun (mixed baby salad greens); all are now mass produced by firms such as Tanimura and Antle, one of the largest lettuce and vegetable producers in the world.

And just as most consumers have the option of buying mass produced chickens, it is also possible to buy artisanally produced free-range organic chickens. But what must be emphasised is that the overwhelming bulk of chicken meat sold in the US (and elsewhere) is produced in factories where chickens have been mass produced as eggs (by one set of producers) and passed on to growers-out who obtain maximal feed-to-meat ratios by scientifically determined poultry diets complete with hormones and antibiotics to minimise problems of disease in congested chicken houses.

What drives mass production in agriculture is the need for annualisation of production. Boyd and Watts make the point that chicken was once barely consumed in the US. We have gone from half a pound of chicken consumed annually in 1928 to almost seventy pounds in 1990. What permitted this change was the scientisation and technification of chicken production so that the original 'spring hatch' has now been annualised into a year-round process.

Annualisation is now found in innumerable fruits and vegetables. At the turn of the century, vegetables such as lettuce and tomatoes were available only in the summer for most Americans and Europeans. By the 1920s in the US, lettuce production became mobile and lettuce became available fifty-two weeks of the year. By the Second World War, tomatoes joined this availability (even if they were, during the winter, pallid reproductions of the original summer fruit). Since the early 1980s, table grapes, once unavailable from January to April, became annualised as Chile entered the table grape commodity system on a mass production basis.

Yet annualisation cannot occur with all agricultural products, even if new production locations are brought on line. Cherries, for example, are available in the northern hemisphere for a limited five-month period although, by 1996, a secondary season had been introduced from southern hemisphere production. Even the short northern hemisphere season is being extended as more northerly regions are brought into cherry production: Norway's cherries are now exported to Germany and mainland Europe in August when cherries are no longer available from more normal sources (Eurofruit, June 1994: 20).

As more agricultural commodities are annualised, this can only be accomplished by seasonal extensions and/or the development of new production locations. While such developments usually involve local and regional producers, the integration of production such that a just-in-time delivery system can be installed depends on large, transnational, globalised corporations such as Dole, Chiquita, Del Monte, and others. Each tendency towards annualisation produces more mass production and, at least for many of the actors involved, inflexibilities in production become increasingly specified.

UNPACKING GLOBALISATION

Globalisation is a term that many of us have clutched to our bosoms, generating a vast literature during the past two decades. But globalisation badly needs unpacking since so many variant phenomena are encompassed by it. For example, the automobile industry is recognised as one of the most globalised of commodity systems yet it is organised very differently from the apparel industries which are probably just as globalised. On any given day, after climbing out of my Brazilian pajamas, I dress in US underwear, a shirt made in India, trousers assembled in Guatemala from US components, a sweater knitted in Uruguay, and a jacket made in Korea. Women's clothing is even more globalised, coming from 'ordinary' places such as Hong Kong or India but also from such unlikely locations as Brunei, Mauritius, Moldavia, and the United Arab Emirates.

But if auto and clothing are equally globalised, each is globalised differently. In auto, producing corporations are globalised in several ways. Not only does each automobile manufacturer make cars from components from many different countries but each manufacturer is involved in joint ventures and in reciprocal ownership arrangements: General Motors and Toyota in NUMMI, Ford with Japan's Mazda, and Chrysler with Mitsubishi. Nissan, Honda, BMW, and Volkswagen now assemble in the US from components coming from the US and elsewhere.[1]

The apparel industries are globalised very differently. Producers are not globalised; rather it is the merchandisers – Esprit, The Gap, Nine West, etc. – that are. These merchandisers do not produce in their own factories; production is contracted to local, regional, and national manufacturers who produce clothing and shoes to specifications set by the merchandisers.

While US broiler production is national, as Boyd and Watts show, production is also directed to global trade as dark meat, low in demand in the US, is shipped to Russia and Asia.

Fresh fruits and vegetables are a commodity system similar to apparel; production comes from many sites in vastly separated areas but where distribution is controlled by a handful of genuinely globalised corporate distributors (Friedland 1994a, 1994b). Table grapes enter the northern hemisphere during its winter from Chile, South Africa, and India so that northerners can now consume them fifty-two weeks a year. Grapes are now joined by broccoli from Mexico and Costa Rica, kiwi fruit from New Zealand, pineapples from Mexico and Costa Rica, plums and other stone fruit from Chile, melons from El Salvador and Honduras, coloured bell peppers and on-the-vine tomatoes from the Netherlands and Belgium, to mention only a few commodities. This globalisation, however, is very different from the globalised automobile *filière*. Thus, there are commodity systems where producing corporations are globalised and others where merchandisers are globalised, although production is local, regional, or national.

At this stage, what is needed is a typology of globalisation, something which space does not permit me to develop. Rather, I simply want to note the need for developing some conceptual clarity with respect to the different forms of globalisation.

NOTE

1 The most recent and startlingly global example of auto globalisation is found in Volkswagen's new truck and bus factory (Schemo 1996). The German-based VW plant was designed by a Spanish Basque to manufacture in Brazil for markets in South America and Europe. A remarkable transnationalisation has been built into the assembly process where only 200 of the 1,000 workers in the plant are employed by VW. The others are employed by different firms that produce sub-assemblies. These include several Brazilian firms, 'local subsidiaries of transnational corporations, like Rockwell International and Cummins Engine of America and Motorenwerk of Germany. . . . The cabs . . . are outfitted by VDO do Brasil, a unit of Adolf Schindling AG Germany'. Despite the presence of many employers, employees all wear the same uniform except for the name of the

employer on the pocket patch. All employers have also agreed to a uniform wage of $374 a month, 'roughly a third as much as autoworkers in São Paulo'.

BIBLIOGRAPHY

Davis, J. E. (1980) 'Capitalist agricultural development and the exploitation of the propertied laborer', in F. H. Buttell and H. Newby (eds) *The Rural Sociology of Advanced Societies: Critical Perspectives*, Montclair, NJ: Allanheld, Osmun.

Friedland, W. H. (1994a) 'The new globalisation: the case of fresh produce', in A. Bonnano, L. Busch, W. Friedland, L. Gouveia and E. Mingione (eds) *From Columbus to Conagra: the Globalization of Agriculture and Food*, Lawrence: University Press of Kansas.

—— (1994b) 'The global fresh fruit and vegetable system: an industrial organization analysis', in P. McMichael (ed.) *Agro-Food System Restructuring on a World Scale: Toward the Twenty-First Century*, Ithaca: Cornell University Press.

——, Barton, A. E. and Thomas, R. J. (1981) *Manufacturing Green Gold: Capital, Labor, and Technology in the Lettuce Industry*, New York: Cambridge University Press.

Kenney, M., Lobao, L. M., Curry, J. and Goe, W. R. (1989) 'Midwestern agriculture in Midwest Fordism: from the new deal to economic restructuring', *Sociologia Ruralis*, 19, 2: 131–48.

Kim, C. K. and Curry, James C. (1993) 'Fordism, flexible specialization, and agro-industrial restructuring: the case of the US broiler industry, *Sociologia Ruralis* 33, 1: 61–80.

Little, P. D. and Watts, M. J. (eds) (1994) *Living Under Contract: Contract Farming and Agrarian Transformation in Sub-Saharan Africa*, Madison: University of Wisconsin Press.

Pugliese, E. (1991) 'Agriculture and the new division of labor', in W. H. Friedland, L. Busch, F. Buttel and A. P. Rudy (eds) *Towards a New Political Economy of Agriculture*, Boulder: Westview.

Roy, E. P. (1972) *Contract Farming and Economic Integration*, Danville, Ill.: Interstate Printers and Publishers.

Schemo, D. J. (1996) 'Is VW's new plant lean, or just mean?' *New York Times* 19 November: C1.

PART IV

DISCOURSE AND CLASS, NETWORKS AND ACCUMULATION

9

LEGAL DISCOURSE AND THE RESTRUCTURING OF CALIFORNIAN AGRICULTURE

Class relations at the local level

Miriam J. Wells

INTRODUCTION

While a great deal of attention has been directed in recent years to the forms of economic restructuring, much less has been devoted to its processes, especially to the ways these reflect the struggles between social classes and the changing regulatory constraints of the state. Mainstream 'flexible specialisation/flexible accumulation' scholars, in fact, underplay the involvement of social conflict in restructuring (e.g. Piore and Sabel 1984, Harvey 1989). In portraying restructuring as the necessary response of firms to an increasingly volatile and competitive global market, they imply that restructuring is inevitable and value-neutral, thereby masking its intentionality and distributional impacts. They offer us little insight into the sources and extent of industrial and local variation. Nor are we given purchase on the ways that class concerns and strategies figure in economic restructuring: how it might affect the power relations between social classes, and what initiatives class actors might take to challenge, encourage, or even effect it. Of particular and unexplored interest are the ways that new economic forms engender fundamental contests over culturally-constructed meanings and entitlements, contests that may vary decidedly across local production environments and crucially engage the apparatuses of the state.

In this chapter, I will use an ethnographic study of California's central coast[1] strawberry industry conducted between 1976 and 1989 (Wells 1996) to clarify the involvement of socio-political conflict in economic restructuring. I hope to demonstrate, first, that economic restructuring may be both a cause and a consequence of the struggles between social classes, and that its motivations may be primarily local and political rather than global and economic. Second, I will show that the meaning, processes, and extent of restructuring are heavily conditioned by the character of the industry and the immediate ethnographic contexts within which production takes place. Third, precisely because economic relations are so politicised in advanced capitalist societies, restructuring may elicit challenges which are pursued in the courts. These challenges and their responses reveal the continually evolving, agency-driven process of economic reconfiguration. Its dynamics counter both *performative* characterisations of society, which regard society's normative, economic, and political structures as the consequence of what people do and elect, and *ostensive* characterisations, which view such structures as the cause of people's views and behaviour (Latour 1986). In their stead, I propose a *reciprocal* characterisation which shows that human agency and socioeconomic structures are mutually determinant.

THE IMPACTS OF INDUSTRIAL AND LOCATIONAL SPECIFICITY

One cannot account for restructuring in this industry without attending to the ways that industrial and locational constraints shaped growers' strengths, vulnerabilities, and alternatives. After the Second World War, locational, technological, and organisational advantages concentrated market power in the hands of California berry producers, buffering them from swings in the national and global economies. Although California produced only about 4 per cent of the US strawberry crop before the war, afterwards, new cultivation methods and plant strains, and the development of transportation, handling, and quick-freezing methods, catapulted the state's strawberry industry into national market leadership. California has produced three-quarters of the nation's strawberries for the past several decades, a position achieved through its astonishing differential in productivity. California growers regularly average over 24 tons of berries an acre – almost five times the 5 tons per acre averaged by the US excluding California. The state now harvests strawberries from February through November, as compared with the one- to two-month harvest seasons for most other states. Within the state, the central coast is especially privileged. It is the largest single berry-producing region, has the longest harvest period (mid-March through November), produces the greatest proportion of high-value fresh market fruit, and has arguably the highest per-acre yields in the world.[2]

In this regional crop industry, profits did not follow the national agricultural pattern of a gradual decline between the early 1950s and 1970s, a sudden rise, and then a sharp drop in the early 1980s. Rather, real net returns have been remarkably stable and growing.[3] The vagaries of the international market have had little impact on this record, since most of the crop is sold domestically. Although increasing yields are in part responsible, it is the sustained availability of cheap and tractable workers which undergirds stable profits. Postwar technological developments and yield increases have reduced the need for non-harvest labour, but have increased labour demand within the harvest and overall. As a result, berry farms have moved away from their prewar character as family-worked enterprises, to become 'fully capitalist': that is, growers have increasingly confined themselves to the role of farm manager, and hired workers have come to perform the actual work of tending and harvesting the crop. Currently, twelve to fifteen workers are required for a very small farm of five acres at the harvest peak. Strawberries are one of the most labour intensive of all crops to produce and labour is the largest single cost, constituting about half of total production costs on the central coast. Its timing and steady availability are crucial as well, because deviation from a carefully specified timetable depresses production, and because the high value, yields, and perishability of the crop make harvest interruptions very costly. Moreover, because the fruit is so fragile, because cultural practices are so highly refined, and because the plants require such constant tending, workers' care, skill, and judgment substantially impact crop value. In fact, according to University of California researchers, the care with which workers choose, handle, and pack the crop is the single greatest determinant of market price.

The importance of harvest labour is due as well to the fact that growers cannot eliminate it and that it is virtually the only cost they can control. Mechanisation is not a viable option, as it is for producers of some other products. Existing harvest machines damage the fruit so that it cannot be sold on the higher-priced fresh market to which most central coast growers direct their fruit. The machines also destroy the plants – an untenable consequence for growers whose plants bear from seven to ninth months of the year. Currently, no one expects harvest mechanisation on the central coast. Nor is market dominance a viable means of increasing profit. Not only is the market relatively competitive, but growers cannot contain the costs of land, finance, transportation, or supplies, all of which have risen substantially since the war. Finally, because of the region's tremendous yield differentials, geographical relocation has not been an attractive means of reducing labour costs or increasing labour control. Rather, with yields from five to ten times the national average, central coast berry growers tend to cleave to the geographical resource which has yielded such rich profits over the years – the temperate, rain-free climate of these two ocean-cooled valleys. In fact, some growers liken themselves to miners in this regard. As one put it, 'On the central coast we mine climate. Whatever else happens we can't leave our mines.'[4]

POLITICAL CHANGE AND ECONOMIC RESTRUCTURING

Because of these industrial and locational constraints, central coast berry growers are particularly vulnerable to disruptions in the local context of production. It was such disturbances which, after the mid-1960s, engendered an otherwise inexplicable form of economic restructuring. That is, whereas almost 100 per cent of the workers on central coast berry farms were wage labourers from the early 1950s through the mid-1960s, at that point many growers began to shift to sharecropping. By the early 1980s, about half of the berry acreage in the region was sharecropped. At that point, however, sharecropping began to dwindle. By the early 1990s only about 10 per cent of the acreage was sharecropped, and at present it is less than that.

This shift to sharecropping is perplexing from the viewpoint of traditional neoclassical and Marxist economics, both of which posit the disappearance of sharecropping and the increased dominance of impersonal wage labour in rationalised capitalist agriculture (Mill [1848] 1915, Marx [1894] 1977). From the perspective of the economic restructuring literature, the resurgence of sharecropping might appear more explicable. Perhaps the replacement of employees by sharecroppers was a manifestation of the 'second industrial divide': a qualitative watershed in the structure of global capitalism leading to smaller-scale and more flexible technologies and organisational units, decentralised production systems, differentiated product and niche markets, and new kinds of state regulation which legitimate flexibility – all supposedly engendered by changes in the world market (Harvey 1989, Piore and Sabel 1984).

In its organisational form, certainly, strawberry sharecropping exemplified the trend in employment structure anticipated with the global watershed: that is, the replacement of permanent full-time employees with temporary and part-time workers, contractors and subcontractors, at-home workers, leased employees, the self-employed, and firms too small to be covered by protective laws. While the exact proportions of this development are unknown, observers agree that it accelerated after the mid-1970s in advanced capitalist countries and currently comprises the most rapidly growing source of work. From 25 to 30 per cent of civilian workers in the United States occupy these sorts of 'contingent' (Callaghan and Hartmann 1992) jobs, as do about 25 per cent of non-agricultural and over 30 per cent of agricultural workers in California, including sharecroppers, contractors, their employees, and many small-scale contract farmers.[5]

If one looks beyond the apparent organisational structure of strawberry sharecropping, however, to its actual functioning and causes, its attribution to a shift in the global regime is not borne out. Restructuring in this instance was not a means for growers to eliminate ponderous organisational structures, foster technological innovation, respond to changing consumer demand, or exploit new market niches so as to compete more efficiently in the world market. Rather, technologies, products, and consumer demand remained relatively constant, the world market had little impact on profits, and the extent of decentralisation and the increase in workers' flexibility and autonomy were in practice negligible. In fact, most growers said that sharecroppers were, if anything, less efficient and easy to control than wage workers.

Why, then, did producers in this regional crop industry restructure their operations after the mid-1960s, and why was this pattern of restructuring abandoned after the early 1980s? The answers to these questions have to do not with the changing structure of the global economy, but rather with changing political pressures on the local economy. In this instance, the shift from employees to independent contractors was primarily a response to three political developments: a change in border policy; the expansion of protective labour legislation; and the rise of union mobilisation. Let us briefly examine these three developments and the role of sharecropping in them.

Border policy

First, as the industry burgeoned after the war, the workers of choice were Mexican *braceros* – contract workers who were recruited and co-managed by the federal government under the provisions of a 1942 agreement with Mexico. As the years went on, however, the Bracero Program came under increasing fire for its negative impacts on the organising, wages, and working conditions of domestic workers. In 1964, an alliance between established urban labour unions, the UFW, and national civil rights activists secured the termination of the programme, cutting off this source of cheap and malleable workers and leaving recruitment networks on central coast berry farms in a shambles. Labour shortages and delays became commonplace and some observers predicted that producers would move to Mexico where labour was cheap and readily available.

These predictions did not come to pass, however, largely due to the adoption of sharecropping. Customarily, growers would invite long-time employees with large families, either former *braceros* or others, to become sharecroppers. Thus, growers got access to sharecroppers' unpaid family labour and their networks of personal acquaintance, as they displaced onto Mexican families the tasks of labour recruitment and management. Although the local labour market stabilised in a few years through the expansion of illegal immigration and the institution of a more restricted category of legal commuters called 'green carders', in the first couple of years after the end of the Bracero Program, sharecroppers helped the industry recover a stable and reliable supply of workers.

Protective legislation

Second, starting in the early 1960s and gathering momentum over the course of the 1970s, the state and federal governments became increasingly involved in regulating the treatment and establishing the rights of seasonal agricultural workers. Although the New Deal had firmly established the federal government's interventionist role in labour relations, farm workers had been explicitly excluded from most prior labour laws on the grounds that such protections were unfair and unnecessary in agriculture (Morris 1966). The tide began to turn in the mid-1960s, due in large part to pressure from the burgeoning UFW and the civil rights movement. In 1963

California instituted a minimum wage for agricultural workers, and in 1966 farm workers were included under the federal Fair Labor Standards Act (FLSA). In 1974 a minimum working age of twelve years was established for farm workers under the FLSA, and in 1975, years of UFW pressure resulted in the signing of the California Agricultural Labor Relations Act (CALRA), which gave California farm workers protections for organising and collective bargaining comparable to those granted urban workers by the National Labor Relations Act in 1935. In 1976 state unemployment insurance and workers' compensation insurance were extended to farm labourers and state overtime pay regulations were established for adult males working in agriculture.

These provisions increased the cost of deploying wage workers, circumscribed growers' managerial discretion, and gave workers legal support for organising and collective bargaining. In keeping with the intentions of the crafters of New Deal legislation, these provisions were extended to 'employees', individuals whose economic security was thought to be especially precarious in the unregulated working of the market (Marshall 1965: v–xxii, 71–134). From the start, however, three specific categories of workers were excluded from coverage under many protective laws: independent contractors, supervisors, and sharecroppers. Although coverage varied with the law being considered, strawberry sharecroppers could fit in all three categories. Thus, as evidenced by growers' accounts of their choice to use sharecroppers, the engagement of labour suppliers legally defined as 'independent contractors' promised to free growers from the costs and constraints imposed by protective legislation.

Union mobilisation

Finally and perhaps most importantly, the end of the Bracero Program facilitated the organising efforts of the UFW, which rapidly emerged as the most powerful agricultural labour union in the nation's history. By 1969 it had secured contracts covering some 20,000 jobs in California's inland table grape industry, as the result of a protracted and often violent series of strikes, boycotts, and inter-union disputes. In August 1970, it turned its attention to the large central coast lettuce industry, whose producers had just signed sweetheart contracts with the Western Conference of Teamsters in order to avert a feared UFW victory. As a result, the UFW launched a major organising drive on the central coast which culminated on 24 August in the largest strike in California's agricultural history. Some 10,000 workers walked off their jobs (Majka and Majka 1982: 203–5). Although lettuce farms were the strike's intended targets, workers in the strawberry industry were drawn into the fray as well. Production on all farms virtually halted for three weeks and central coast berry growers lost an estimated $2.2 million (Federal–State Market News Service 1972: 2, 30). Although only one contract was signed in the strawberry industry, it implemented intrusions into growers' managerial authority which berry growers considered intolerable. Moreover, the strike clearly demonstrated the dangers of union mobilisation and the unanticipated degree of solidarity among workers in different crops. Although labour–management struggles never regained the explosive level of 1970, the central coast remained the centre of union militancy into the early 1980s, keeping alive the threat of unionisation, raising the standard for regional wages, and exerting pressure for the observance of immigration and protective labour laws.

In this context of union militancy, sharecropping was advantageous to growers on several grounds. First, workers on sharecropping ranches were unlikely to join the union because they were often family members or friends. Their wages were also less responsive to union standards because they worked for themselves or for people they knew. As labour supervisors and employers, sharecroppers were not thought to be eligible for membership in the union, or for protection under the CALRA.[6] Finally, the UFW was reluctant to attempt to organise sharecroppers, or to strike the farms on which they worked, because sharecroppers were seen as successful ethnic entrepreneurs and compatriots.

All in all, then, sharecropping helped central coast berry growers mitigate some of the uncertainties, risks, and actual costs imposed by border policy, protective legislation, and union mobilisation.

THE LOCAL VARIABILITY OF RESTRUCTURING

Although sharecropping offered a refuge from changing political constraints, neither growers' motivations for sharecropping nor labour suppliers' responses to it were uniform across the regional system. To understand this unevenness of initiative and response, and also the uneven incidence and course of restructuring more generally, we must inquire more deeply into the ways that production was organised at the local level.

The micro-regional contexts of production

Strawberries were grown in three relatively distinct micro-regions within the broader central coast region: the Salinas Valley, the North Monterey Hills, and the Pajaro Valley and its proximate slopes. Over time, different production subsystems evolved in these three localities – subsystems that varied in their patterns of economic organisation and opportunity, in the composition of the class populations represented in each, and in the relations within and between social classes. Distinctive visions and conventions of workplace justice guided the relations of production in each locality as well. Although the boundaries between these systems were not firm and there was some movement across them, both growers and workers tended to concentrate their activities within a particular locality. As a result, their alternatives, perceptions, and choices were firmly based in the local context, and they experienced the impacts of regional political pressures through the filter of the micro-regional subsystem. I will lay out the basic contours of these subsystems in order to illuminate the local contingency and variability of restructuring.[7]

The Salinas Valley was the historical centre of UFW organising and of large-scale corporate farming. The state and federal offices that enforced labour and immigration laws were situated in the city of Salinas and concentrated their efforts on the Pajaro and Salinas Valley floors. Strawberry farms were largest here and absentee ownership and manager operatorship was most common. Most workers were employed by large estates operated by European Americans who had ample resources but moderate investments per acre in their crops. Growers here aimed for a larger volume of somewhat lower quality fruit than did those in the Pajaro Valley. Strawberry farms shared the micro-regional economy with a thriving vegetable industry, and many workers combined peak season work in the strawberry industry with work in vegetables before or after the berry harvest. All in all, employment opportunities were the most numerous in this locality, job access structures were the most open, and workers were the least dependent on their immediate strawberry employers. Hourly wages were relatively high, although not as high as those in the Pajaro Valley. Workers were both documented and undocumented. While on the average they had an intermediate amount of experience working in strawberries, many had been coming to the region for generations, returning to the same housing camps and apartment complexes along with other strawberry workers.

Production relations in the Salinas Valley incorporated a high degree of social distance between growers and labour suppliers and a class-based, oppositional tenor to the relations between them. The large size of ranches, the non-Mexican ethnicity of growers, and the fact that workers lived in off-the-ranch residential enclaves with considerable longevity, all weakened the personal ties between growers and workers and reinforced the solidarity among the latter. Here, the local intensity of union struggle set the standard for workplace justice. Its reference point was the United States: following the voice of the UFW, it claimed for farm workers the rights and protections granted them under the law, and it insisted that the conditions of labour should be set not by employer preference but by government standard. Workers in the Salinas Valley were the best informed about their entitlements under the law and the most aggressive in going after them. Walkouts and strikes were the most frequent here and lasted the longest.

In the North Monterey Hills, by contrast, strawberry farms were the smallest and most poorly financed and growers invested the least per acre in their crops. Crop quality was lowest here and growers directed their crops to the lower end of the fresh, and to the processed, markets. Almost three-quarters of the growers were of Mexican origin and they lacked capital, farming experience, and social connections to the local farming community. Job alternatives were the fewest in this locality and berry farms were almost the only employers. Farms were also physically isolated, tucked away in remote hollows and screened from view by pampas grass, brush, and live oak. Immigration and labour law enforcement officials rarely visited this area. Wages were lowest here and employers were the least likely to observe protective labour laws. Workers were the most dependent on their immediate strawberry employers and, at the time of my research, most were undocumented and relatively inexperienced.

Production relations in the North Monterey Hills involved a low degree of social distance between growers and labour suppliers and a patronal, deferential tenor to the relations between them. Almost all growers were owner-operators and were extensively involved in the daily operations of their ranches. Although some workers were 'walk-ons' – individuals, usually undocumented, who came to the ranch for work without prior connections to the grower – core workers were generally recruited through the growers' personal networks that extended to their Mexican villages of origin. Thus growers and workers had face-to-face relationships which were reinforced by common cultural, linguistic, geographical, and even family backgrounds. Standards of workplace justice here were firmly based on the experience of Mexico. The very act of employment, the level of wages, and the rare provision of overtime pay or workers' compensation were presented as 'privileges' – a reflection of the grower-patron's largesse and dependent upon the worker-client's personal relationship with him. Growers explicitly pointed out how much better the wages and working conditions they offered were than what workers would experience across the border. Growers were also, however, more likely to take on a diffuse obligation to care for their workers. They were more likely to help defray the cost of the *coyote* who smuggled workers across the border, to provide food or transportation (albeit for a charge), or to offer an empty van, shed, or a shovel to dig a cave for shelter. Workers here often lived on the ranch, reducing their expenses for housing but also reinforcing their dyadic ties with their employer and minimising the separate space in which they could develop solidarity with their peers. Not surprisingly, workers in this micro-region rarely contested the terms of their employment.

In the Pajaro region, strawberry farms were intermediate in size and well financed. Growers here invested the most per acre in their crops and concentrated on the top end of the fresh berry market. The largest proportion of growers were Japanese and the second largest segment were European American.[8] Growers in this area were also highly cohesive and had the most years of experience farming strawberries.[9] Job alternatives were more limited here than in the Salinas Valley, but more plentiful than in the North Monterey Hills. Although berry growers were the major employers of field workers, canneries, bush berries, apples, and vegetables also provided work. Job access structures were also the most closed, however, since most jobs flowed through networks of personal recommendation. Strawberry jobs were especially desirable because of their high wage levels (the highest of the three micro-regions and higher than most other field crops), the long harvest season, and the fact that growers were especially likely to provide legally mandated benefits and to maximise the employment periods of core workers. Thus most berry workers aspired to work the entire season with a single employer. At the time of my research, the proportion of documented and experienced workers was highest in this area.

Production relations in the Pajaro region were personalistic and patronal, as in the North Monterey Hills, but based on US standards of workplace justice, as in the Salinas Valley. Most growers were owner-operators and, on smaller ranches, were known to their employees. However, the social distance between workers and their employers was greater than in the North Monterey Hills, due to the lack of linguistic, cultural, and social commonalties, and to the fact that workers did not live on their employing ranches. Nor were workers as dependent upon or as deferential towards their employers as those in the hills: their legal immigration status and experience in the system, the greater solidarity among them, their greater contact with workers in the Salinas area, and their greater range of employment options all increased their leverage. Workers here tended to approach growers in an accommodating and negotiative – as opposed to an oppositional (Salinas) or a deferential (North Monterey) – manner. They were relatively quick to speak to their employers about wages that fell below regional UFW standards, but were also tolerant about the speed with which growers responded. For their parts, growers were especially eager to prevent the disruption of their high-value harvests and to preserve the good will of workers. As a result, walkouts were relatively brief and rare in this micro-region.

In sum, the resources and concerns of farm owners and workers, and the immediate pressures impacting them, varied significantly at the local level. This variance had a decided impact on the local incidence and meaning of sharecropping.

Growers' agency: micro-regional contrasts in the utility and adoption of sharecropping

It should not surprise us to find that sharecropping was distributed differentially across the three micro-regions and had differing utility for the growers who adopted it. Two patterns of sharecropping emerged after the mid-1960s. The organisation of production was the same in both (see Wells 1996: 234–41).

The first, which I call 'high-resource sharecropping', was primarily a response to the changed political context of production. This is the pattern with which we are already familiar. It accounted for about 90 per cent of postwar sharecropped acreage and was pursued by well-capitalised, large-scale growers in the Salinas and Pajaro Valleys. Farm owners were Japanese and European American producers with relatively high per-acre crop expenditures and a focus on the high-quality, high-priced fresh market. These producers were especially affected by shifts in the political context of production. They not only had more to lose from harvest disruptions than did growers in the other sharecropping pattern, but also more to gain by preserving a continuity of careful workers. Many were active in the California policy community and concerned to avoid slurs on their reputations. Their size and exposed locations made them preferred targets for the union and law enforcement officials. Their size also made labour recruitment a particular burden after the Bracero Program ended, so that the recruitment assistance of sharecroppers was especially appreciated. They were especially threatened by unionisation because their workers were less compliant and dependent, and also more cohesive and insistent on their rights. It was no accident that these growers shifted back to wage labour when union militancy and the legal advantages of sharecropping declined after the early 1980s.

The second pattern – 'low-resource sharecropping' – was a response to resource scarcity. It represented a minor fragment of the sharecropping resurgence and involved only several ranches on the central coast. Farm owners in this pattern were large-scale producers[10] situated in the North Monterey Hills who wanted to expand their size with limited capital and expertise. They were either Mexicans who lacked capital and farming experience and began independent farming in the 1970s, or non-Mexican investors who wanted to limit their involvement and expenses in farming. Such individuals turned to sharecropping in order to avoid labour expenditures until the harvest proceeds came in, and to capture sharecroppers' recruitment assistance and knowledge about how to tend and harvest strawberries. The changing political context was not so burdensome for them because they were marginal to the grower and policy communities and because their low per-acre crop investments and focus on lower quality, lower priced markets meant that they had less to lose from strikes. Moreover, workers in their area were highly dependent and compliant and did not participate in union activities or press for their rights under the law. Perhaps even more to the point, the union did not typically target ranches – especially Mexican ranches – in the hills, protective labour laws were unevenly observed and rarely enforced there, and these farm owners began to engage sharecroppers well after the end of the Bracero Program, when labour recruitment networks were already restabilised.

In short, strawberry sharecropping – and thus economic restructuring in the regional industry – had no single 'meaning' to the producers who adopted it. Rather its advantages and adoption reflected the economic organisation, class composition, and production relations of the particular micro-regions in which growers were engaged.

Workers' perceptions: micro-regional contrasts in sharecroppers' evaluations of sharecropping

Workers' assessments of sharecropping also varied by locality. Both those engaged by low-resource growers in the hills and those engaged by high-resource growers on the valley floors made their choices to sharecrop based on a comparison with wage labour. When asked their reasons for sharecropping, both identified two major advantages: potentially higher household income and greater independence/autonomy. They also cited one major disadvantage: the greater risk of income variation. Their weighting of these concerns varied by locality, however, because of differences in the institution of sharecropping, in standards of work place justice, and in the resources and options of labour suppliers.

Sharecroppers on low-resource ranches in the North Monterey Hills elected to sharecrop from positions of considerable personal dependency. As wage workers, most were undocumented, minimally connected to other work, and limited in their job options. For them, sharecropping was mainly a means of increasing the likelihood of repeat employment and consolidating their families on this side of the border. Sharecropping did not assure greater economic well-being because soils were sandier and ranches were steeper and more erosion-prone in this locality, and because owners often under-maintained their ranches so that market proceeds were low and variable. Sharecroppers' incomes could easily fall below those of employed wage workers. Some worked all winter for less than enough to pay back their debts, surviving only from the incomes of otherwise-employed family members. It was not uncommon for farm owners in this locality to leave farming in bad years and return to wage labour, taking with them the resource of sharecropping jobs.

Despite these risks and hardships, low-resource sharecroppers did not openly challenge the terms of their employment. Their views of their work were heavily tinged by feelings of deference and personal obligation towards the grower-patron. They saw sharecropping as a favour achieved through their special relationship with him and they had little sense of entitlement as to how they should be treated. Written contracts were rare, though sharecroppers had considerable latitude in executing their tasks because growers supervised so little. When asked about the differences between sharecroppers and wage workers, such sharecroppers said they were definitely worse off than wage labourers who worked for a 'good company' (high-resource ranch). Those who continued to sharecrop hoped for good years, and their solutions to economic hardship and disillusionment tended to be personal. The most common dream of those who aspired to something better was a steady wage labour job in lettuce or strawberries on one of the valley floors.

Sharecroppers on high-resource ranches in the Pajaro and Salinas micro-regions felt differently about their jobs. This sort of sharecropping was clearly a step up. It could move a family with available household workers into the middle class. Sharecroppers had detailed, written yearly contracts. Although they could be terminated with little ceremony, their contracts were renewed if they did their jobs well. Ranches were annually planted and well maintained, their crops sold at the top end of the market, and farm owners were continuously and securely in business. As a result, the risk of income variation was much lower. While sharecroppers' incomes were still more variable than those of wage workers, in 1985 the income of the high-resource sharecropping families interviewed ranged from $15,000 in a bad year to $45,000 in a good year, far more than the $6,300 averaged by wage workers who had no guarantee that other family members would be employed.

Such sharecroppers tended to see themselves as entrepreneurs on the way up. Because they started from positions of less dependency, and because conventions of workplace justice on the valley floors emphasised their entitlement to fair treatment under US law, they asked more from the institution of sharecropping. When they became sharecroppers they were more likely to have been documented, they worked for ranches that offered relatively good wages, their household members were more often locally resident and employed, and their own job alternatives were more numerous. For some, household incomes had been comparable as wage workers. As a result, other advantages of sharecropping sprang to the fore. Most important was the promise of greater independence and autonomy. Many hoped that sharecropping would be a stepping stone to independent owner-operatorship, a goal actually achieved by a few. Such sharecroppers felt they were more like growers than workers: they saw themselves as the patrons (*patrones*) of those they hired, as 'almost' the owners (*dueños*) of the small businesses on their parcels of land, and as the farmers (*agricultores*) who directed the tending and harvesting of the crop. Whatever the dimensions of the qualifier 'almost' for individuals, in their eyes they had a moral agreement with growers to be treated as independent contractors, an agreement legitimised by the state through their written contracts. As a result, they were frustrated to find that the actual range of their independence was very limited. In one landmark instance such frustrations led a group of sharecroppers to bring their complaints to the state. This case underscores the importance of human agency and offers an in-depth look into the ways that economic and legal structures are interconnected through the ongoing processes of class struggle.

Workers' agency: economic conflict reaches the courts

In 1975 a group of fifteen sharecroppers working for Driscoll Berry Farms in the Salinas Valley, frustrated at their supervisor's failure to respond to their demands that they be given more extensive information about the sale of their berries and be allowed to market outside the company, brought their complaints to a law firm in Salinas. Upon the sharecroppers' initial instruction, their attorney prepared a class action suit against the farm owner and the farming corporation of which he was part, Driscoll Strawberry Associates (DSA). This suit, *Real v. Driscoll Strawberry Associates, Inc.* (*Real*), was issued on behalf of the estimated 200 sharecroppers who had signed DSA contracts in the state. Its initial charges were antitrust damages, fraud, misrepresentation, and breach of contract. Thus at the outset *Real* was based on the vision of workplace justice and of sharecroppers' entitlements, which were expressed in their contracts – that they were legally 'independent contractors', small entrepreneurs illegitimately prevented by a large, monopolistic business from exercising their rights to conduct their businesses as they chose (*Real*1975: 2–9).

After he filed these first charges, however, the attorney conducted further interviews with his clients and developed a different understanding of their plight. As his clients described the working relations on their ranches, he said, it became clear to him that they were not unfairly constrained independent entrepreneurs, but rather inadequately protected and remunerated employees. He based this assessment on his discovery that, in the actual relations of contribution and control on sharecropping ranches, sharecroppers did not enjoy the defining characteristics of an independent contractor, as set out in the federal employment test designed to determine legitimate coverage and exemption under protective laws such as the FLSA (*US v. Silk* 1947: 716) (see Table 9.1).

The attorney presented this view to his clients, asking them to reflect on the discrepancy between the contractual representation of their independence and the economic reality of their dependency. Some protested

Table 9.1 Criteria of independent contractor and employee

	Independent Contractor	*Employee*
Degree of outside control over worker	lesser	greater
Opportunity for profit or loss	greater	lesser
Investment in the facilities	greater	lesser
Permanency of the relationship	lesser	greater
Skill required in the operation	greater	lesser
Extent to which work is an essential/ integrated part of the business	lesser	greater

vigorously because, as they said later, they hoped to get the court to side with them in supporting their independent contractor status. As they reflected on their experience, however, they acknowledged that in the ways identified by the federal employment test, they had more in common with employees. In the end, they decided to let their attorney file an additional cause of action charging that their contracts were a 'sham' intentionally designed to mislead them 'into not understanding their true status as employees' (*Real, Complaint* 1975: 3). This charge claimed unspecified damages for violations of the FLSA. As the case was played out, this last cause of action was the one that prevailed. When the court granted summary judgment in favour of the defendants, on the grounds that there was insufficient evidence to proceed, the sharecroppers' attorney decided to appeal the case on the FLSA charges alone. After considering the evidence, the appellate court reversed the summary judgment, ruling that there was indeed sufficient evidence to believe that these sharecroppers were employees. At that point, the defendants offered to settle out of court, the attorney and his clients accepted, and in March 1981 the case was dismissed.

Although the appellate court did not rule definitively on the merits of the case, its decision clearly supported the employee status of sharecroppers. Because this ruling was published, it could be used as a precedent in other cases. Thus *Real* altered the interpretation and application of the law by establishing a legal perspective on a type of labour supplier that did not exist when the law was enacted. The impacts of *Real* were far reaching. The case had been followed closely by central coast growers, labour suppliers, and law enforcement officials. After the ruling, local workers' compensation insurance companies informally began to require berry growers to carry insurance for their sharecroppers, as if they were employees. A group of growers brought in a labour relations expert to clarify the legal consequences of *Real* and to advise them as to how to proceed. Recognising that they could be liable for huge fines and back payments for FLSA and child labour law violations should they continue to use sharecroppers, some growers immediately shifted back to wage labour. Others revised their organisational structures to give more actual independence to subcontractors.

Elsewhere in the state and nation, *Real* became a reference point and support for labour advocates in the increasing number of legal battles waged over the legal designation and prerogatives of contemporary sharecroppers. In California, a suit was brought in 1985 against a grower near Gilroy who produced pickling cucumbers for Vlasic Pickle Company. This suit, *S. G. Borello and Sons, Inc. v. State Department of Industrial Relations*, built on the perspective developed in *Real*, charging the grower with violations of child labour and workers' compensation laws. In 1989, after a widely publicised series of exchanges, the California Supreme Court ruled that cucumber sharecroppers 'were obvious members of the broad class to which workers' compensation protection is intended to apply' (*Borello* 1989: 345). Dealing what the newspapers called a 'stunning blow' to California growers,[11] it articulated the implications of *Borello* for the coverage of such sharecroppers under other protective labour laws. Meanwhile, starting during *Real* and continuing through the late 1980s, other cases were brought, challenging the independent contractor status of pickling cucumber sharecroppers in the Midwest and Texas.[12] The sharecroppers' lawyers in these cases built their arguments in direct communication with the attorneys in the California cases. The resultant rulings established an increasingly decisive national legal stance supporting the employee status of sharecroppers in labour intensive crops.

Not surprisingly, strawberry sharecropping began to dwindle on the central coast after the early 1980s. By the end of the decade only about 10 per cent of the acreage was sharecropped. This shift was in part a result of the altered legal consequences of using sharecroppers, and in part a reflection of other changes in the conditions that had initially motivated sharecropping. The recruitment networks destabilised by the end of the Bracero Program had long since been restabilised. Not only did a steady supply of experienced strawberry workers flow into the central coast, but the supply and dependency of workers increased in the early 1980s as Mexico's economic crisis encouraged out-migration. At the same time, the state of California entered a period of conservative political leadership in which the enforcement of the CALRA seriously flagged. The UFW's presence on the central coast also diminished, in part because it claimed that bringing representation votes and unfair labour charges before the pro-grower ALRB was futile, and in part because it deemed that political lobbying and consumer boycotts were the effective tactics of the 1980s and 1990s (Wells 1996: 90–97). In short, the political context shifted to reduce the benefits and increase the costs of sharecropping, and once again economic reconfiguration ensued.

POLITICS, ECONOMIC RESTRUCTURING, AND SOCIAL CLASS

We have observed through this analysis that economic restructuring can be both a consequence and a cause of locally based social conflicts. Rather than being an automatic organisational response to technological and economic shifts in the world market, restructuring may be an intentional tactic pursued by certain sorts of producers to help them mitigate locally experienced political challenges. This evidence counters explanations of restructuring which privilege global influences over local, economic forces over socio-political, and structure over agency. It demonstrates that restructuring is much more uneven and diversely motivated than commonly assumed, and that human agency is crucial to its incidence. Moreover, it is not only the agency of the powerful that makes a difference, but also that of the less potent. In this case, workers' challenges altered the conditions that made restructuring initially advantageous to growers, thus eliciting another round of restructuring. These exchanges demonstrate the dynamic character of economic restructuring – its embeddedness in the ongoing processes of class struggle. This evidence casts doubt upon familiar apocalyptic claims that the structure of the economy has reached a sort of end point in which direct-hire jobs and empowered workers are relics of the past (Castro 1993, Morrow 1993, Bridges 1994). It suggests instead that the cards have not all been played, that the game continues, and that human initiative remains an important source of the unexpected.

The industry and economic sector in which producers are involved clearly affect the likelihood of and motivations for restructuring. In the present case, the insulation of producers from global market shifts, the centrality of hand labour to producers' profits, and the substantial advantages of producing in a particular locality all contributed to producers' choices to restructure. It may be, indeed, that agricultural producers are particularly prone to such a recourse, not only because of their necessary and varying bonds to land and climate, but because they are dependent on the not-entirely-overcome natural rhythms of biological processes. As a result, they are unevenly able to relocate to escape local social struggles or to achieve optimal pricing for production factors. Of course, agricultural industries vary among themselves in this regard, and some extractive and service industries are tied to locale as well (e.g. Hobsbawm 1964, Yarrow 1979). Thus they too may be especially motivated to develop compromises with local challenges which involve economic reconfiguration. Nonetheless, such persisting barriers to rationalisation are more common in agriculture and may be one of the most important ways in which the processes of agricultural development remain particular.

This study cautions us, as well, against assuming that economic restructuring is necessarily evidence of an 'industrial divide' – a qualitative watershed in the structure of the global economy. Here we saw that restructuring was a contextual and reversible response to the mix of national and regional political pressures that shaped labour markets at the local level. Thus it behooves us, in our assessments of restructuring, to distinguish carefully between economic reconfigurations that are part of the 'ordinary' functioning of a continuous regime, and those that are part of a transition from one regime to another. Such distinctions cannot be made from the remove of world economic structure. Rather, they must be grounded in the in-depth study of local systems.

The foregoing analysis demonstrates also the decidedly local meaning of economic restructuring and the local shaping of class interests and initiatives. Sharecropping had no single utility in the economic strategies of central coast berry growers. Nor did sharecroppers benefit from it, perceive it, or respond to it in a single way. Rather, the utility and likelihood of its election by growers, and the manner and likelihood of its contestation by sharecroppers, depended on the portion of the class population considered and the immediate ethnographic context in which those individuals operated.

Here, we are in a position to re-evaluate the interconnection between economic restructuring and the laws and apparatuses of the state. As demonstrated by the case examined here, economic restructuring not only alters the organisation of production, it can also alter, and initiate significant contestation over, the moral economy of productive relationships. However, the discourses of justice surrounding such relationships are not unitary. Not only do they vary radically at the point of production, but, because the state has so substantially intervened into economic relationships in advanced capitalist societies, its laws embody definitions of the identities, prerogatives, and obligations of the participants in production which have substantial ideological and material consequences. There are resonances and tensions between the two levels of moral discourse – the local and the legal – and the interplay between them can itself be transformative. Moreover, because the law so consequentially constrains the resources of economic actors and the shape of economic relationships, challenges to those relationships may build into challenges to the law. Such challenges initiate a process of class contestation which differs in important ways from contests played out in the fields: it abides by different conventions of discourse, it engages different institutions and allies, and it impacts on individuals situated far from the immediate arena of conflict. In the process, economic structures, legal standards, and local visions of workplace justice may all be transformed.

This analysis demonstrates that the work place norms emerging from economic restructuring are not determined simply by economic structure, as suggested by ostensive characterisations of society, nor chosen freely by individuals, as suggested by performative characterisations (Latour 1986). The former portrays a notion of social values that underplays the role of human agency and errs on the side of mechanical materialism. The latter proffers a position that underplays the structural constraints shaping individual and group choice and is overly idealistic. Both accord insufficient attention to the role of contingent historical circumstances. One of the most important findings of the present study is that economic and ideological structures evolve together, as they are engaged and challenged by the exchanges between social classes. Clearly, the initiatives of and struggles between farm owners and workers in this case shaped the organisation as well as the moral economy of work. However, their perceptions and choices were also shaped by their places in socio-economic structure – both by their relationship to the means of production and by the character of the local organisation of production.

The case of *Real* is particularly illustrative in this regard. The sharecropper-plaintiffs' senses of entitlement to bring a suit against farm owners were reinforced by the moral economy and opportunity structure of their micro-region, and by their legal characterisation in their contracts. At the same time, the processes of the legal battle highlighted the contradiction between their alleged (and initial) position in the legal order and their actual positions in the economic order. Thus, although their perceptions of work place justice were not determined *solely* by their class position, they were determined *also* by that position. In other words, the relationship between human agency and socio-economic structure – and between ideological and economic structures themselves – is revealed as thoroughly reciprocal. The two jointly and continuously shape one another and they are linked by the contingent processes of class struggle.

NOTES

1 For the purposes of this study, California's central coast consists of Santa Cruz and Monterey Counties with their urban centres of Watsonville and Salinas, respectively.

2 Some central coast growers produce over 50 tons of berries per acre, with certain varieties in certain years (see, for example, 'Cultural Practices Gave One Grower 60 Tons per Acre', *Western Fruit Grower* April 1978: 13, 48). County farm advisers point out that while southern California counties sometimes average higher per-acre yields, individual yields for adequately financed producers are highest on the central coast, where yield averages are depressed by the higher proportion of small, low-resource producers.

3 Between 1959 and 1985, real product prices fell by 41 per cent and costs rose by 216 per cent, but yields also rose by 311 per cent, so that costs per pound actually fell by 23 per cent and net returns rose by 40 per cent.

4 Growers point out that they are relatively undemanding as to soil quality and topography. While flat terrain is best, contour planting can utilise relatively steep slopes. Similarly, water and nutrients can be (and are) added. The main thing they need from the soil is good drainage and a lack of salinity. Only the climate cannot be 'improved' by human means.

5 My estimates here are based on my own research and on Berch 1985, Callaghan and Hartmann 1992, Castro 1993, Dillon 1987, Morrow 1993, and Villarejo and Runsten 1993.

6 In 1979 a case was brought in the Santa Maria Valley just to the south of our area, charging that the dismissal of a sharecropper for organising his peers was a violation of the CALRA (*Álvara v. Driscoll Strawberry Associates*). This charge was dismissed by the Oxnard regional director of the ALRB in 1981 and the dismissal was upheld upon review by the board's general counsel on the grounds that Álvara's supervisory status disqualified him from protections under the CALRA.

7 For a thorough exploration of the local organisation of production, see Wells (1996: 97–142, 188–228).

8 Almost half of the growers were Japanese. Together Japanese and European American producers accounted for 74 per cent of the growers and 88 per cent of the acreage.

9 For some European Americans, berry growing was a family tradition stretching back to the late 1800s; most Japanese Americans became independent

owner-operators after 1952, when the Alien Land Laws prohibiting foreign-born Japanese from owning agricultural land were discontinued.

10 Large-scale ranches were unusual in this micro-region, so that while the lower resource expenditures, limited grower experience, and limited exposure to changing political constraints were typical of the locality, the large size of ranches was not.

11 See, for example, 'Court Deals a Blow to Growers on Issue of Share Farmers', *Los Angeles Times*, 24 March 1989, pt I: 3, 23.

12 See, for example, *Brock v. Lauritzen*, 624 F. Supp. 966 (ED Wis. 1985); *Donovan v. Brandel*, 736 F.2d 1114 (6th Cir. 1984); *Donovan v. Gillmor*, 535 F. Supp. 154 (ND Ohio 1982); *Appeal Dismissed*, 708 F.2d 723 (1982); *Marshall v. Brandel*, No. G76–393 CA6, slip op. (WD Mich. 17 January 1983); *Sachs v. United States*, 422 F. Supp. 1092, (ND Ohio 1976); *Salinas v. United States*, No. B-82–140 (SD Tex. 1982); *Secretary of Labor, US Dept of Labor v. Lauritzen*, 835 F.2d 1529 (7th Cir. 1987).

BIBLIOGRAPHY

Álvara v. Driscoll Strawberry Associates, Inc. (Alvara), Nos. 79-CE-1-SM, 79-CE-2-SM (Cal. Agricultural Labor Relations Board 1981).

Bach, R. (1978) 'Mexican immigration and the American state', *International Migration Review* 12 (Winter): 536–58.

Bain, B. and Hoos, S. (1963) *The California Strawberry Industry: Changing Economic and Marketing Relationships*, Giannini Foundation Research Report No. 276, Berkeley: University of California.

Berch, B. (1985) 'The resurrection of out-work', *Monthly Review* 37, 6: 37–46.

S. G. Borello & Sons, Inc. v. Dept of Industrial Relations (Borello), 48 Cal. 3d 341 (1989).

Bridges, W. (1994) 'The end of the job', *Fortune* 19 September: 62–74.

Burawoy, M. (1985) *The Politics of Production: Factory Regimes under Capitalism and Socialism*, London: Verso.

Callaghan, P. and Hartmann, H. (1992) *Contingent Work: A Chart Book on Part-Time and Temporary Employment*, Washington, DC: Economic Policy Institute.

Castro, J. (1993) 'Disposable workers', *Time*, 29 March: 43–47.

Dillon, R. (1987) *The Changing Labor Market: Contingent Workers and the Self-Employed in California*, Sacramento: State of California, Senate Office of Research.

Federal–State Market News Service (1972) *Marketing California Strawberries, 1967–71*, San Francisco: US and California Departments of Agriculture.

Harvey, D. (1989) *The Condition of Post-Modernity*, Oxford: Basil Blackwell.

Hobsbawm, E. (1964) *Labouring Men: Studies in the History of Labour*, London: Weidenfeld and Nicolson.

Latour, B. (1986) 'The powers of association', in J. Law (ed.) *Power, Action, and Belief: A New Sociology of Knowledge*, London: Routledge and Kegan Paul.

Majka, L. and Majka, T. (1982) *Farm Workers, Agribusiness, and the State*, Philadelphia: Temple University Press.

Marshall, T. H. (1965) *Class, Citizenship, and Social Development*, New York: Anchor.

Marx, K. ([1894] 1977) *The Process of Capitalist Production as a Whole. Vol. 3 of Capital*, New York: International Publishers.

Mill, J. S. ([1848] 1915) *Principles of Political Economy*, London: Longmans, Green.

Morris, A. (1966) 'Agricultural labor and national labor legislation', *California Law Review* 54, 5: 1939–89.

Morrow, L. (1993) 'The temping of America', *Time*, 29 March: 40–41.
Piore, M. and Sabel, C. (1984) *The Second Industrial Divide: Possibilities for Prosperity*, New York: Basic Books.
Processing Strawberry Advisory Board (PSAB) (1989) *Annual Report, 1988*, Watsonville, Cal: Processing Strawberry Advisory Board.
Real v. Driscoll Strawberry Associates, Inc. (Real), No. C75–661-LHB (ND Cal. filed 4 April 1975).
US Department of Agriculture (USDA) (1989) *Agricultural Statistics, 1988*, Washington, DC: GPO.
US v. Silk, 331 US 704 (1947).
Villarejo, D. and Runsten, D. (1993) *California's Agricultural Dilemma: Higher Production and Lower Wages*, Davis: California Institute for Rural Studies.
Wells, M. J. (1996) *Strawberry Fields: Politics, Class, and Work in California Agriculture*, Ithaca and London: Cornell University Press.
Yarrow, M. (1979) 'The labor process in coal mining: struggle for control', in A. Zimbalist (ed.) *Case Studies on the Labor Process*, New York: Monthly Review Press.

10

FIELD-LEVEL BUREAUCRATS AND THE MAKING OF NEW MORAL DISCOURSES IN AGRI-ENVIRONMENTAL CONTROVERSIES

Philip Lowe and Neil Ward

INTRODUCTION

Farm pollution in Britain and in much of northern Europe shifted from being a 'non-issue' during the 1970s, contained within the realms of the agriculture and water policy communities and treated as little more than a technical side effect of efficient production, to a situation where it became seen as a pressing environmental problem, a key contributor to declining water quality and an issue in need of a much stronger regulatory approach on the part of governments. Most social science analyses of the issue have, to date, tended to concentrate on the 'high politics' of farm pollution, where parliamentary committees hold inquiries, official commissions make their reports and acts of Parliament are drawn up and debated. In contrast, this chapter examines the worlds of farmers and of regulatory officials 'on the ground' to provide an account of this shift in the nature of farm pollution and its regulation. In doing so, we also attempt to engage with wider theoretical questions around agriculture, pollution, and regulation. The chapter employs perspectives derived from 'actor-network theory' to examine how, through the struggles between actors, fixed dichotomous categories around agriculture and pollution are being undermined. Such an approach, we argue, can provide a constructive and complementary contribution to the study of the political economy of agro-food systems, particularly given the recent concern in the literature with accommodating contingency and heterogeneity (Goodman and Watts 1994, Whatmore 1994).

NATURE, RURALITY AND MORALITY

In industrial society, agriculture served as a major source of natural values and a key mediator of natural morality. However, this role is now being eclipsed by the environmental movement. It is members of the environmental movement who now define the moral high ground when it comes to what counts as natural and what counts as 'polluted'. Because of this change, contemporary conflicts between agriculture and the environment have a much wider significance indicating fundamental shifts in meaning and sources of authority. Farmers have traditionally seen themselves, and been seen by others, as guardians of the natural environment harnessing nature to meet society's needs for food. Yet now, in Britain and across the advanced industrial world, they are coming to be widely stigmatised as 'environmental criminals' absorbing huge amounts of public subsidy to intensify and expand production with little regard for the natural balance of the countryside. One consequence of this shift has been the imposition by government of various controls on agriculture to restrain the excesses of agricultural productivism. What these environmental regulations mean in practice is that farmers are increasingly confronted by regulatory officials armed not only with new powers but also with a new form of moral authority. In the regulation of polluting agricultural practices, and also in terms of the making of these new moral discourses, much of what is of interest to us occurs in the encounters between farmers and regulators and the exchanges that take place over nature, morality, and the law, and over what constitutes sound agricultural practice. In our research[1] we have employed an interactionist approach drawing upon 'actor-network theory' to study these issues (following Callon 1986, Latour 1993, Law 1994, Murdoch 1995). The use of actor-network theory helps reveal how the old meanings that used to underpin the traditional 'social order' of industrial society are being challenged as well as how new orders are being discursively constructed and resisted. Part and parcel of this breakdown is the renegotiation of conceptual and institutional boundaries through the struggles between different groups of actors with different ways of seeing the world. Industrial society (and its associated modernist project) was built around the maintenance of a series of interlocking dualisms: nature–society, rural–urban, agriculture–industry, for example. In this chapter we explore how, through struggles between actors, these categories are being undermined.

We are keen to be sensitive to the problems of the use of dualisms in *explaining* the social world, but are aware of the risks of 'throwing out the baby with the bathwater'. As Law writes,

> to turn away from dualisms doesn't mean that we should avoid the ordering strains *towards* dualism built into the modern project. Instead, we should seek to treat dualism as a social *project*, a sociological topic, rather than treating it as a resource. Accordingly, the argument is that modernism more or less successfully (although partially and precariously) *generates* and performs a series of such divisions.
>
> (Law 1994: 138, emphasis in the original)

Following Law, we seek to examine the collapse of old dualisms through the struggles between social actors, and their 'partial and precarious' replacement.

Actor-network theory provides, we would suggest, an epistemology and methodology that helps draw our attention to the interactions and social relations between actors. For us, the approach has also highlighted the central role of *discourse* in the construction of meanings and the way different actors see things differently. (Here a discourse is taken to mean a set of statements which provide a framework in which something – farm pollution, for example – is made meaningful.) Moreover, it is suitably 'modest' (after Law 1994) in its claims about how we characterise socio-economic and environmental change.[2]

As we have suggested above, industrial society seems riddled with dualisms. Latour (1993), for example, has argued that much of the modernist project can be understood in terms of the attempts to maintain a false separation between the 'natural' and the 'social' (see also Norgaard 1994), but besides nature and society we can see a host of other dichotomies too. Foucault suggested that industrial society sought systematically to manage a form of 'social order' through the imposition and maintenance of certain predominant categories. Fixed, dichotomous categories, indeed, seem central to class-based societies.

The dichotomies that characterised modern, industrial society did not exist in isolation from each other. In addition to nature–society, we might identify other important dichotomies as those of male and female, home and work, capital and labour, and urban and rural. Crucially, these dichotomies reinforced one another and together formed a means of systematic 'ordering', of making and maintaining a social order (Law 1994). By providing a place for everything, they helped keep everything in its place, and thus establish the 'natural' order of things. Through such interlocking dichotomies and their morally prescribed boundaries of what was natural and what unnatural, identity and social behaviour became fixed as a social structure.

It is the particular linked categories of 'nature' and 'the rural' that we would like briefly to explore here. The notion of the rural has long been closely bound up with the morality of nature (Mumford 1961, Bell 1994). In counterpoint to the rise of the industrial city, rurality was constructed as 'natural' and became a source of moral affirmation and condemnation. In contrast to the supposed innocence of rural life, the city was perceived as corrupting, not only of traditional morality and social hierarchies, but also of nature, through industrial pollution.

Agriculture, in part because of its very embeddedness in rurality, was thus projected as a key mediator of natural morality. This is not necessarily to say that farmers have continually been able to 'claim the moral high ground', but it is to say that an important discourse was created (through the representations made by politicians, agricultural leaders, poets, artists and philosophers, to name but a few) which links agriculture with natural morality. The creation of this discourse, in turn, provided a resource for agriculturalists in promoting and defending agriculture. For example, the discourse has been drawn upon to emphasise the farmers' management and stewardship of the countryside, with the implication (and in the case of many European nations, the policy outcome) that environmental concerns around agriculture are best left to self-regulation and voluntary schemes. Equally, the equation of agriculture with the rural and with the natural on the one hand, and of pollution with the industrial and the urban on the other hand, long precluded from recognition the category of agricultural pollution.

What has happened, in recent years, is that while nature continues to be a source of moral values and an inspiration to efforts to sustain or recreate a new moral order, the actors that now mobilise this notion are environmental groups and activists. It is they who have become the key mediators of natural morality. Bell (1994) has shown, for example, how the ex-urban middle classes in the British countryside reaffirm the link between nature and morality in rural life, but no longer necessarily through the medium of agriculture. In exploring the use of nature as a source of identity and moral understanding by rural dwellers, Bell also draws our attention to the mutability and variability of nature as a concept. Its variability, Bell argues, 'allows villagers to make a range of arguments for their right to stand on its moral rock, a range that they can apply to their own varied backgrounds, experiences and beliefs' (Bell 1994: 8). Yet 'nature', insofar as ecosystems and the biosphere are concerned, is strictly morally neutral. It is human values that are projected upon it. Thus, it seems to us, the task of the social scientist when it comes to these issues is to identify where claims about morality are made and to ask how is the moral order constructed and how is this order then imposed upon others. It is because of our concern to address these questions that we employ an interactionist approach and have turned to actor-network theory.

ENVIRONMENTAL MORALITY AND THE CHALLENGE TO AGRICULTURE

The discourse which equates agriculture with natural morality has been drawn upon in the past to deny the existence of pollution from farms. For example, when the Royal Commission on Environmental Pollution in Britain chose to examine the agricultural sector in the late 1970s (RCEP 1979), the response of the agricultural community was that when it came to pollution, agriculture was more sinned against than sinning. The Commission, in part as a concession to the mobilisation of this discourse, in its final report followed chapters on pollution by nitrates, pesticides, and wastes from farms with a chapter examining the *impact* of predominantly urban and industrial pollution *on agriculture*. Indeed, the Commission's report, with its neutral title *Agriculture and Pollution*, was the product of a period when farm pollution did not really figure as a politicised problem. During this 'pre-politicisation' period, in a number of countries farm pollution was actively constructed as a 'non-problem' by the agricultural policy community, seeking to fight a rearguard action against an advancing environmental morality and its calls for greater regulatory control over all forms of pollution. The Control of Pollution Act 1974 exempted British farmers from prosecution for pollution if they were following 'good agricultural practice', for example, but similar exemptions were also framed in the Netherlands, Germany, and Denmark in the 1970s.

Thus in the struggle even to define the very existence of agricultural pollution, two conflicting forces have been at work. The first can be characterised as 'environmental'. A general increase in interest in environmental issues and the NIMBY (Not-in-my-back-yard) phenomenon as rising numbers of articulate middle-class people have moved to the countryside may well account for the growing recognition and public intolerance of farm pollution (see Ward *et al.* 1995). On the other hand, the ways agricultural scientists and agricultural policy officials defined pollution and its seriousness, which in the past predominated and confined the issue to the technical realm (Lowe *et al.* 1996), still continue to have great influence. Despite increasing environmental concerns and surpluses of many agricultural products, faith in high-tech farming has not been undermined among many agricultural scientists and farmers. Their self-images are deeply entrenched. The high-tech intensive trajectory of agricultural production continues, in which pollution is still seen as an unfortunate side-effect of efficient production, albeit one which occurs in a context of heightened public concern.

In the 1980s, farm pollution shifted from being a non-issue to an issue of considerable public and political concern. This shift came about as a consequence of a number of separate developments. First, agricultural pollution problems were mounting in a physical sense helped not least by the government's encouragement over the preceding three decades of technological change, expansion, and intensification of livestock production through grants, subsidies, and advice. For example, in the UK numbers of pollution incidents were rising and the general quality of the water environment seemed to be in decline. Second, the notion of farming as a morally worthwhile activity began to be fundamentally questioned. Subsidies and surplus production in Europe undermined the productivist ethos in agriculture and, following the imposition of milk quotas in 1984, responsible farming actually came to mean producing less in some cases. Finally, the conditions came into place for the ascendancy of the new environmental morality. A crucial factor in the UK was the government's proposal to privatise the water industry, which provoked considerable popular concern over protection of the water environment and opened up the issues of water quality and pollution regulation to public and political scrutiny.

Thus agricultural pollution emerged as a politicised problem. But its very recognition transgressed established categories of meanings and thereby clearly indicated that some of the old dualisms – agriculture–industry, urban–rural, nature–society – were crumbling. Even so, the recognition of agricultural pollution came as a shock and strong legislative action followed swiftly. For example, just a few months after the government first acknowledged farm wastes as a significant source of water pollution, legislation was introduced which removed agriculture's exemption from pollution control and subjected farm effluents to much more detailed regulation. (To illustrate our point here, a similar tale of the sudden forging of a new conceptual category, with associated boundary-crossing, leading to a morally charged atmosphere and to pressure for regulatory reform could be constructed around the public recognition of marital rape).

As a result, the pollution inspectors (PIs) of the new National Rivers Authority (NRA)[3] came, by the late 1980s, to be patrolling rural Britain armed with the discourse of the new environmental morality and accompanied with new regulatory powers. The general outlook of pollution inspectors is of pollution as a form of environmental crime. As such, they are reflecting the new morality generated by the environmental movement. But in the field, pollution inspectors have to engage with and adapt to what we might term the moral economy of the farmers.

AGRICULTURE'S MORAL ECONOMY

At the time of our field study in Devon, in south west England, farm pollution was causing great controversy among livestock farmers and was certainly among the most pressing issues that dairy farmers faced nationally (Centre for Agri-food Business Studies 1991). The pollution problem, we would agree, has not been seen by farmers in general as a turning point but as yet another in a succession of crises – which have included the imposition of milk quotas, the storm damage of 1987, BSE and so on – seemingly visited on them by a capricious world. Although most farmers accept the desirability of preventing pollution of rivers by farm wastes, they often remain unaware of how to do this. Hence, farmers feel they have attempted to do what they can, in the main, to minimise pollution risks but within what might be termed their 'productivist rationale' which, at the same time, acts as a major constraint in tackling the problem of pollution.

Here in our analysis we would like to draw upon the idea of the *moral economy* of the farmers. The notion was introduced by Scott (1976) in his work on the peasantry in Southeast Asia. By the moral economy of the peasants, he meant their notions of justice and their working definitions of what is fair and what is unfair. A study of the moral economy of a particular group can tell us what is likely to alienate their sympathies and what may mobilise their support. Our concern has been to examine the moral economy of dairy farmers with respect to farm pollution and to explore how a new moral discourse is introduced, constructed and maintained. How have the agents of the new environmental morality surrounding agriculture sought to impose these values upon the farming community, and how in the exchanges between actors are new orders constructed? In studying these interactions, we have tried to adhere to the maxim in actor-network theory of 'symmetry' (see Latour 1993, Law 1994) whereby the researcher should approach actors and phenomena in the same way, and not to 'start by assuming that there are certain classes of phenomena that don't need explaining at all' (Law 1994: 10). Symmetry should apply across all the divides (nature–society, micro–macro, human–technical, etc.). Operationalising this principle in our study has involved treating the main groups – farmers, environmentalists, and pollution regulators – on an equal basis.

Important within the farmers' moral economy is the ethos of productivism. The technological nature of farming has been transformed in the postwar period, but for most farmers the moral economy of farming has remained unchanged. Society needs food and so depends upon the success of farmers in carrying out their work. This fact has provided the obvious and undeniable foundation for portraying agriculture as a vital and socially worthwhile activity. The experience of urban industrialisation and the consequent spatial and occupational separation of the bulk of the population from its own basic provisioning served to heighten this sense of dependency, particularly in the century of total war and the threat thus posed to the food security of urban populations. But for farmers the rationale of the sector was not simply to provide, but became, above all, to expand. The quest to 'make two blades of grass grow where only one once grew' has been with us as long as agriculture itself, but this maxim became elevated to a new status during the technological revolution that began in the 1930s and continued through the postwar decades. What Paul Thompson (1995) has called the philosophy of productivism, which sees not only production as good, but also more production as always better, became more firmly ingrained at the farm level, where technological changes brought profound increases in the output of individual farms. That philosophy indeed became the basis for agricultural policies across much of the Western world. Understanding productivism is essential to understanding the farmers' moral economy, how they view the environment and how the farm pollution problem poses for many of them such conceptual difficulties.

Productivism is an important ethos in farming life across the advanced industrial economies. In much of Europe and North America it resonates with a Protestant work ethic which sees sloth rather than greed as sinful, links virtue and industriousness and sees greater wealth as the just reward for hard work. In this sense, *productive* farmers are seen as *better* farmers, with greater yields displaying the virtuousness of hard work. This work ethic has become especially associated with agrarian culture, particularly since urban culture has come to celebrate virtues such as refinement, taste, and artistic achievement (virtues which bespeak of leisure time) and vices, such as unemployment, welfare dependency, and hedonism (which bespeak of idleness and dissolution).

Technological changes have also meant that a form of 'technocentric productivism' has flourished in farming. Thompson describes such a philosophy as 'the headlong and unreflective application of industrial technology for increasing production' (Thompson 1995: 70). As such, it could almost be described as an anti-environmental philosophy, save for one important fact. Productivism in agriculture is usually combined with and qualified by the duty of stewardship, the responsibility to care for nature. This notion is often expressed by farmers in terms of keeping the land 'in good heart', including a sense that sound farming involves passing on the farm to the next generation in 'a better condition'. Farmers also readily acknowledge a wider role that they play within the countryside, and often talk of their contribution to the making of the much valued rural landscape. They will refer to farming's past contribution as a rhetorical defence against contemporary accusations of their pollution of nature. However, according to Thompson, traditional agrarian stewardship 'is conceived as a duty ethically subservient to production; hence when stewardship would entail constraints on production, duties to nature seldom prevail over the productivist ethic' (Thompson 1995: 72). In effect, modern technology has driven a wedge between the farmer's interest in production and traditional agrarian stewardship, although within the moral economy of the farmer the discourse of stewardship is still drawn upon. The crucial distinction, however, between productivism and stewardship as a set of values is that the former is institutionalised in the techno-economic system within which farmers are embedded while the latter is individualised and seen as a matter of personal responsibility.

THE FARMER AND THE FIELD-LEVEL BUREAUCRAT

Actor-network theory focuses our attention on the interactions between actors, and, rather than adopting *a priori* distinctions between categories such as 'natural', 'social' and so on, compels us to ask how categories (orders, regimes – call them what you will) are made and maintained. It also helps cope with different actors 'seeing things differently'. A look at virtually any environmental issue will typically turn up a clamour of certainties, opinions, strategies, and prescriptions concerning what the real problems are, how they might be rectified, and what needs to be done to achieve this. Farm pollution is no different. We wanted to be able to approach our research in a way which allowed us to account for differences in the ways in which actors 'see' things without having to resort to the uncritical elevation of some actors' views or the equally uncritical denial of others. Our interest centred on finding out how actors' worlds are constituted and what actions they provoke; how pollution events, and their solutions, are assembled.

We can identify two other literatures helpful to our approach. The first originates in studies of the implementation of public policy and concerns the work (in exercising discretion, interpreting policy and in dealing with the complexity of the social world) of the so-called 'street-level bureaucrat'.[4] The second literature originates in development studies and rural sociology where 'actor-oriented' perspectives have been developed to examine 'encounters at the interface', typically between rural producers and representatives of states, development agencies or technology interests (see, for example, Long 1989).

One strong theme that emerged from our discussions with the farmers was that tackling pollution amounts to a problem for them because it involves them in dealing with bureaucrats – people who make and enforce the ominous 'rules and regulations' that restrict what farmers can do. Linked to the general complaint about bureaucracy, seen almost inherently as a 'bad thing', was a set of complaints about the way that rules kept changing, about National Rivers Authority (NRA) advice being vague or even contradictory, and about the NRA not understanding the day to day practical problems facing dairy farmers – the NRA not being 'practically minded people'. Finally, many farmers interviewed complained specifically about the confrontational and aggressive approach of individual NRA officials.

From the perspective of the officials, the street-level bureaucrats literature tells us that 'the decisions of street-level bureaucrats, the routines they establish, and the devices they invent to cope with uncertainties and work pressures, effectively become the public policies they carry out' (Lipsky 1980: xii). Such a perspective suggests caution over the feasibility of top-down policy change amidst the inertia of established ways of working and field-level relationships. It is through implementation 'on the street', (or in our case, in the field) that de facto regulatory policy is created (see also Ham and Hill 1993). Of crucial importance is the development by street-level bureaucrats of practices which enable them to cope with the pressures they face. According to Ham and Hill, public officials are placed in a particularly difficult position vis-à-vis their clients. They may be putting into practice political decisions with which they do not agree; they are facing a public normally without the option available to some private firms of going elsewhere if its demands are unsatisfied; and the justice of their acts is open to scrutiny, by politicians and sometimes by the courts. Given the limited means available to them, they are obliged to make judgements but to do so in a way that is not vulnerable to external challenge. Formal and informal rules are bound to play a major part in their working lives.

In dealing with these pressures, policies can become reshaped as bureaucrats seek to bring some order into their own lives. The issue becomes one not of total conformity to rules, but rather the ways in which officials make choices to enforce some rules while disregarding others. Lipsky explains that bureaucrats believe themselves to be doing the best they can under adverse circumstances and they develop techniques to salvage service and decision-making values within the limits imposed upon them by the structure of their work. They develop conceptions of their work and of their clients that narrow the gap between their personal and work limitations and the service ideal (Lipsky 1980: xii).

Given their pressurised contexts, the field-level bureaucrats we studied conformed to this model. The literature suggests their coping strategies may include choosing easy rather than difficult cases, routinisation of procedures and working methods, standardised classification of the regulated world and of client groups, and adopting a cynical attitude to ambitious goals and their replacement with more personal goals.

The literature on policy implementation identifies a spectrum of regulatory policy styles, with legislative approaches at one extreme and laissez faire (with its associated risks of policy 'capture') at the other. Between these two extremes lies the notion of 'flexible enforcement' (see Bardoch and Kagan 1982) where discretion is exercised over centrally devised rules, but where the identification of 'good apples' and 'bad apples' amongst target groups comes to be of crucial importance. We found this to be so in our study of pollution inspectors and farmers. Pollution is not a straightforward issue. 'Responsible farmers' can accidentally pollute in unfortunate circumstances. 'Irresponsible farmers' are carelessly slap-dash and should have known better.

Moreover, although pollution is measurable in terms of its impacts on aquatic life, what counts as pollution is socially defined. Therefore, just what pollution *is*, and how much is acceptable become key questions. Our study found different definitions of pollution and different understandings of the seriousness of the problem. While we found a reasonable consistency in definitions amongst environmental regulators in the field, there was a greater variety of views amongst farmers. However, when the tolerance of pollution is considered, greater disparities can be found. In general, there was a greater tolerance of diffuse rather than gross pollution, even among official agricultural advisory bodies, while the farming community tended to be tolerant of what they saw as 'accidental' pollution, where it did not seem that the individual farmer was at fault, but could even be seen by the community as a victim of circumstances – such as heavy rain, the failure of machinery or even overwhelming officialdom.

The public, meanwhile, is increasingly intolerant of pollution, perhaps in part because of the continuing gross pollution by farms. However, although water pollution is classified as a criminal offence, it is only in recent years that the view of farm pollution as 'environmental crime' has become widespread. The strongest adherents of this view are amongst environmental groups and certain sections of the public. It is also gradually becoming the orthodoxy amongst environmental regulators but is not in keeping with the perspective of most agricultural advisors and farmers. Fundamental struggles continue to centre on whose representation of the nature of the problem shall prevail. The basic conflicts are between the definition of farm pollution as a problem *for* farming, which leads to agriculturally led solutions (such as upgrading storage facilities according to the needs of the production system), and farm pollution as a problem *of* farming, which implies environmentally led solutions (water quality standards, with changes being made to production systems as appropriate).

A basic perceptual divide, running through our analysis and cutting across the various groups covered, is the one between pollution as a technical problem (a form of 'rule breaking') and pollution as something discreditable that attracts blame (a type of morally reprehensible 'environmental crime'). The crucial distinction is whether or not any moral opprobrium attaches to it. The notion of pollution as wrong-doing is a modern one – it is certainly a long way from the Victorian notion of 'where there's muck, there's money' that associated pollution with industriousness and wealth creation. More recently, it has been 'wicked industrialists' who cause pollution, and while farming continued to be seen in a naturalistic light, the social recognition of the existence of farm pollution was not possible. Significantly, even wicked industrialists have tended until recently to be treated as merely rule breakers. This is largely because pollution was regarded in relative, not absolute, terms and was considered in the context of production. Industrial pollution was the norm, and may well have been seen as the necessary concomitant of jobs and prosperity. It is only with the advent of the contemporary environmental movement that this equation has been challenged. For long, though, agricultural pollution remained a submerged issue, precisely because farming was seen as the antithesis of industry. Arguably, it needed the naturalistic image of farming to be attacked from other sources (e.g. over wildlife, habitat and landscape destruction, animal welfare problems, etc.) before agricultural pollution could even be recognised.

Amongst the farmers we interviewed there was the full spectrum of views on pollution – from perceptions of agricultural waste as 'natural', to pollution as merely a form of rule breaking, to one of environmental crime. Arguably, this represents a spectrum, from those who would emphasise the productive context in which pollution arises to those who would emphasise its environmental consequences (in other words, from an inward- to an outward-looking perspective). The former view is a discourse that farmers' leaders have helped to structure. It is one which, incidentally, happily sees farmers translated as instruments of government policy – any responsibility for pollution when farmers are conforming to policy is thus displaced.

Environmental regulators are confronted by a much more complex set of judgments. Traditionally the pollution inspectors of the water authorities simply treated pollution as a technical issue. However, the establishment of the NRA as a national regulatory body, cast by itself and the environmental lobby in the role of environmental watchdog, has injected a new, morally charged discourse into the proceedings. The pollution inspectors must thus operate day to day in a farming world (and arguably in an institutional tradition) in which pollution is predominantly considered a technical issue. The NRA, though, has increasingly adopted a stance that water pollution is unacceptable and reprehensible.

In general, pollution inspectors have an absolute notion of farm pollution. To pollute a river or stream is a wrongful act, a form of environmental crime. Moreover, it is wrong because of its effects. Usually with a training in biology, pollution inspectors have an expert understanding of the sensitivity of the water environment and the functioning of aquatic ecosystems and they are aware of the potency of farm effluents. They therefore fully appreciate the damage that farm pollution can do. Their task is to *make rivers clean* and they are reluctant to regard *any* amount of pollution as acceptable.

In contrast, farmers have a relative notion of pollution informed less by an understanding of its consequences and more by an understanding of the context in which it arises. Daily, dairy farmers have to deal with considerable volumes of farm effluent. In certain circumstances, particularly in heavy rain or when storage facilities fail, effluent escape is unavoidable, and may be mitigated by the circumstances – for example, the rapid dilution and dispersal of pollutants in a river in spate. Of course, a large spillage of effluent is regrettable but 'a little won't hurt' – after all, farm wastes are 'natural' substances. For farmers, the morality of farm pollution concerns the morality of the deed – whether the pollution was deliberate or accidental. Their sense of personal worth, responsibility, and circumstances all come into play around the morality of farming and the morality of pollution.

While the pollution inspectors are armed with new powers, it is important to recognise that the farmers are not powerless in the face of a seemingly much stricter regime. It is farmers who must be persuaded or dragooned into acting differently. Their limited scope or inclination for action is a major determinant of the pace of change. In his study of forms of resistance among peasants in rural Malaysia, Scott examined the ordinary, everyday weapons of relatively powerless groups in the face of state bureaucracy or capitalist exploitation (Scott 1985). Scott's subjects, as subordinate classes, were interested less in changing the larger structures of the state and the law than in what Hobsbawm (1973) appropriately called 'working the system . . . to their own minimum disadvantage'. In such a quest, the weapons at the peasants' disposal included foot dragging, false compliance, feigned ignorance, and so on. In similar terms, what 'weapons' do British dairy farmers use when confronted by new regulatory controls? We have identified three types of strategy. The first involves 'keeping your head down', hoping no problems arise which will draw the regulators' attention to your farm and hoping the issue of pollution just 'goes away'. The second strategy is to pursue delaying tactics. These might involve attempting to draw the regulators into negotiations in an attempt to play for time – to avoid spending on pollution control facilities or to deflect prosecution. Delay tactics might also include the farmer attempting to treat the NRA as advisors, perhaps to implicate them in decisions. Finally, delay tactics may involve the farmer trying to explain the specific difficulties posed by the farm circumstances, such as the poor financial conditions or the level of indebtedness, in order to 'educate' pollution inspectors about the particular constraints faced. A third 'weapon' of resistance is to attempt to 'blackmail' the NRA, through threats that the farm may go out of business if the pollution issue is pressed, through threats of non-cooperation, or even through threats of violence. In our interviews with farmers and our participant observation studies 'shadowing' regulatory officials in their work, we learned of examples of each of these strategies.

In response to the variety of farmers, and because of the farmers' ability to exercise sanction over regulatory power, whether through inaction or malicious action, the pollution inspectors have to develop coping strategies. Often these strategies involve the pollution inspectors developing working classifications of farmers as 'good' and 'bad' farmers (see also Bardach and Kagan 1982). These categories are relative ones involving distinctions between conscientious or 'persuadable' farmers on the one hand and 'rogue' or recalcitrant farmers on the other. Crucially, the pollution inspectors' categories resonate with those of the farming community in being based on the moral worth of the farmer rather than the consequences of the pollution.

Usually, pollution inspectors' weapons of environmental morality are thus only deployed when they correspond to the farmers' own moral values. The only farmers automatically prosecuted in the courts for causing pollution are those responsible for major, serious pollution incidents (that is, in obeisance to environmental morality). Beyond that, the only farmers to be prosecuted are those viewed essentially as 'rogue' farmers – not only by the pollution inspectors but by the farming community also. It is as if prosecution has to have either the tacit acceptance of the farming community, *or* it has to be obviously seen by the pollution inspectors to reinforce the categories held by the farmers themselves (i.e. between what is seen as responsible and what is seen as utterly irresponsible behaviour). Overall, under 10 per cent of farm pollution incidents are prosecuted in an average year. Thus, although pollution inspectors are armed with the outlook and powers of the new environmental morality, they find in their efforts practically to improve river quality that they are dependent upon the cooperation of the farming community. To secure that cooperation they have to work with the grain of what that community considers is fair and unfair, right and wrong. Punishing a farmer who deliberately emptied slurry into a river would be acceptable, but it would not be acceptable to prosecute where overflows or equipment failure had occurred in heavy rain, or where a farmer was doing his best to improve his waste management facilities, or where enforcement action might put a farmer out of business.

CONCLUSIONS

The distinction between pollution as a technical problem and pollution as environmental crime encompasses two quite separate patterns of enrolment with two competing moral discourses. Technical problems are appropriately dealt with through adjustments to farming practices, effected by advice, information, the formalisation of standards and incentives – this is the realm of farm waste regulations, capital grants for pollution control equipment, and official agricultural advice. Criminal acts, in contrast, elicit remonstration, public condemnation, and prosecution – this is the realm of pollution incidents, environmental campaigning, and the courts. In the implementation of farm pollution regulations, government agricultural advisors and the farming community have sought to enrol the NRA into a technical definition/solution of the problem, with the interpretation/ implementation of the regulations and of the Capital Grant Scheme central to the process of enrolment. On the other hand, environmental pressure groups and a wider public are seeking to enrol the NRA into treating pollution as an environmental crime, with the emphasis on the recording, investigation, reporting, and prosecution of pollution incidents. Central to the enrolment process here is the ability of a member of the public to report a pollution incident which must then be investigated. The availability of published information on incidents then gives the opportunity for environmental groups both to express alarm about the scale of the problem and to condemn what can be presented as leniency.

In this chapter we have examined a collapsing order and the efforts to impose a new order. Significantly, those efforts have only had *partial* success. As Law writes, 'the social world is complex and messy' (Law 1994: 5), but in wanting to cleave to an order, we are merely clinging to the unrealisable modernist dream which seeks to monitor, control, and legislate (Law 1994, Norgaard 1994). But structures, orders, systems and so on – the very focus of so much intellectual inquiry in studies of the political economy of agro-food systems – are best viewed as *outcomes* of social practice. In their studies, many social scientists look to society itself, expressed in the form of concepts such as class, institutions, norms, interests, hegemony and so on, to explain what is going on in the social world. They work with what Latour (1986) terms an *ostensive* definition of society. Latour contrasts the ostensive definition, in which society itself is a cause of actors' views and behaviour, with what he terms a *performative* definition, in which society is the consequence of what actors do, and in which it is the actors themselves who in practice state what society is about. The performative definition of society thus differs radically from the ostensive definition. It shifts the focus of social science analysis to how society is *made*, in contrast with the ostensive definition's concern with what society *is in essence*. Under the performative definition, the conventional idea of society is turned upside down. Social concepts (such as class, for example) often conceived as causes of social action within an ostensive definition, become outcomes within a performative definition. In practice, the performative definition changes the task of the social scientist from one of invoking abstract social concepts to explain why actors do what they do, to one of investigating how social groupings and meanings are forged and maintained through time. It does not deny that society exists, but emphasises how society is constituted. Crucially, a performative sociology has fundamentally different (although possibly complementary) ambitions compared to traditional approaches in the political economy of agro-food systems.

These two definitions of society have different implications for the way in which the empirical observation that actors may 'see things differently' should be treated. The ostensive definition requires that the observer judge one representation 'authentic' and others 'mistaken'. The performative definition, on the other hand, is concerned about how society's attributes are settled in practice. It can accommodate the reality of 'seeing things differently' in a way that does not deny particular rationalities and does not privilege particular actors. Actor-network theory, therefore, is a methodology to study a process whereby orders, structures, systems and so on are constructed and resisted. We are not necessarily calling for the wholesale adoption of actor-network theory as some sort of panacea for the study of the political economy of agro-food systems, far from it. We are, however, suggesting that the search for new, all encompassing paradigms is an insufficiently modest objective. Better to revel in theoretical and methodological diversity and be promiscuous in our interests and approaches.

We have suggested from our study that a useful empirical focus for studying the construction of new orders around agriculture and the food system is what we have called the field-level bureaucrats. We propose a blending of insights from implementation theory (on street-level bureaucrats) and development studies (on encounters at the interface, moral economies and the weapons of the weak). The approach of Long and colleagues to the study of 'encounters at the interface' between rural dwellers and public agencies helps draw our attention to the struggles, negotiations and interactions between actors occupying divergent life-worlds with different 'ways of seeing'. Long defines the 'social interface' as 'a critical point of intersection or linkage between different social systems, fields or levels of social order where structural discontinuities, based upon differences of normative value and social interest, are most likely to be found' (Long 1989: 1–2). We have tried to view these 'levels of social order' through the prism of competing moralities.

NOTES

1 Our recent work has mainly been conducted with colleagues Judy Clark and Susanne Seymour. It has examined the emergence and management of agro-environmental controversies in Europe, and specifically, the pollution of ground and surface waters in Britain with farm waste effluents from the dairy sector and pesticides from the cereals sector. The research was funded by the Economic and Social Research Council under its Joint Agriculture and Environment Programme.

2 For Law (1994), 'modest sociology' is built on four principles which are: (1) symmetry – the notion that everything deserves explanation and should be explained in the same terms; (2) non-reduction, or the notion that no *a priori* distinctions between actors should be drawn before the analysis begins; (3) social recursiveness, which implies that 'social processes' drive themselves and are not driven by what lies outside; and (4) reflexivity, which implies that the sociologist's account should not be privileged above others.

3 The National Rivers Authority was established in 1989 as part of the privatisation of the water industry. It had regulatory responsibilities for water quality and pollution control. In April 1996, the functions of the NRA were merged with those of other environmental regulatory bodies to form the Environment Agency.

4 Because our empirical interest has been in the regulation of farmers' environmental practices by local pollution inspectors, we have modified the label to one of 'field-level bureaucrats'.

BIBLIOGRAPHY

Bardach, E. and Kagan, R. (1982) *Going by the Book: The Problem of Regulatory Unreasonableness*, Philadelphia: Temple University Press.

Bell, M. M. (1994) *Childerley: Nature and Morality in a Country Village*, Chicago: Chicago University Press.

Callon, M. (1986) 'Some elements of a sociology of translation: domestication of the scallops and the fishermen of St Brieuc Bay', in J. Law (ed.) *Power, Action and Belief: A New Sociology of Knowledge?*, London: Routledge and Kegan Paul.

Centre for Agri-food Business Studies (1991) *A Survey of Dairy Farmers' Attitudes Towards MMB Reform*, Cirencester: Royal Agricultural College.

Goodman, D. and Watts, M. (1994) 'Reconfiguring the rural or fording the divide?: Capitalist restructuring and the global agro-food system', *Journal of Peasant Studies* 22: 1–49.

Ham, C. and Hill, M. (1993) *The Policy Process in the Modern Capitalist State*, (second edn), Hemel Hempstead: Harvester Wheatsheaf.

Hobsbawm, E. (1973) 'Peasants and politics', *Journal of Peasant Studies* 1: 3–22.

Latour, B. (1986) 'The powers of association', in J. Law (ed.) *Power, Action and Belief: A New Sociology of Knowledge*, London: Routledge and Kegan Paul.

—— (1987) *Science in Action*, Milton Keynes: Open University Press.

—— (1993) *We Have Never Been Modern*, Hemel Hempstead: Harvester Wheatsheaf.

Law, J. (1994) *Organizing Modernity*, Oxford: Basil Blackwell.

Lipsky, M. (1980) *Street-Level Bureaucracy*, New York: Russell Sage.

Long, N. (ed.) (1989) *Encounters at the Interface: A Perspective on Social Discontinuities in Rural Development*, Wageningen: Department of Sociology, Agricultural University.

Lowe, P., Ward, N., Seymour, S., and Clark, J. (1996) 'Farm pollution as environmental crime', *Science As Culture* 5, 4: 588–612.

MAFF/WOAD (1991) *Code of Good Agricultural Practice for the Protection of Water*, London: MAFF Publications.

Marsden, T., Murdoch, J., Lowe, P., Munton, R., and Flynn, A. (1993) *Constructing the Countryside*, London: University College London Press.

Mumford, L. (1961) *The City in History*, London: Secker and Warburg.

Murdoch, J. (1995) 'Actor-networks and the evolution of economic forms: Combining description and explanation in theories of regulation, flexible specialisation and networks', *Environment and Planning A* 27: 731–57.

National Rivers Authority (1992) *The Influence of Agriculture on the Quality of Natural Waters in England and Wales*, Water Quality Series Report No. 6, Bristol: National Rivers Authority.

Norgaard, R. (1994) *Development Betrayed: The End of Progress and a Co-Evolutionary Revisioning of the Future*, London: Routledge.

Royal Commission on Environmental Pollution (1979) *Agriculture and Pollution*, 7th Report, Cmnd 7644. London: HMSO.

Scott, J. (1976) *The Moral Economy of the Peasant: Rebellion and Subsistence in Southeast Asia*, New Haven: Yale University Press.

—— (1985) *Weapons of the Weak: Everyday Forms of Peasant Resistance*, New Haven: Yale University Press.

Thompson, P. (1995) *The Spirit of the Soil: Agriculture and Environmental Ethics*, London: Routledge.

Ward, N., Lowe, P., Seymour, S., and Clark, J. (1995) 'Rural restructuring and farm pollution regulation', *Environment and Planning A* 27: 1193–211.

Whatmore, S. (1994) 'Global agro-food complexes and the refashioning of rural Europe', in N. Thrift and A. Amin (eds) *Globalization, Institutions and Regional Development in Europe*, Oxford University Press.

Williams, R. (1976) *Keywords: A Vocabulary of Culture and Society*, London: Fontana.

COMMENTARY ON PART IV

Fields of dreams, or the best game in town

Richard A. Walker

While representing two very intriguing studies of agricultural change under the spur of social conflict, the chapters by Miriam Wells and by Philip Lowe and Neil Ward raise similar problems of method and meta-theory in the social sciences. I shall therefore begin with a brief account of the philosophical basis of the poststructuralist shift that they represent, both in its strengths and weaknesses. At the same time, the two chapters manifest rather different political commitments and theoretical stances behind the poststructuralist veil, which I shall take up subsequently. Lastly, the piece by Wells speaks volumes about matters of concern in our shared bailiwick of California, which I can address with some confidence about both local knowledge and the global significance of the events.

PHILOSOPHICAL FIELDS FOREVER

The authors use poststructuralist method to good effect in order to render the analysis of agrarian change more open, subtle and specific, and even dialectical. But one can justly ask whether such methodological florescence obviates the need for any theoretical commitment at all, that is to say, the need for any weighting of social forces and analytical models with systematic logic or causal process. Too many contemporary thinkers imagine that by laying out the terrain of inquiry more clearly and mapping the twists, turns, and branches of the pathways of knowledge, one creates a more complex landscape of learning more appropriate to the 'messiness' of the real world, as it were. Sara Berry (1993) among the agrarian theorists comes to mind. Yet I fear that this is not enough, and that the overgrown gardens of poststructuralist thought are just as often warrens of unrelated concepts, populated with the ruins of perfectly good theories.

Let's examine what hides behind the bushes in Lowe and Ward's approach. They cite Bruno Latour (1993) and John Law (1994) as their philosophical oracles to the effect that the social world is 'complex and messy', and attempts to impose an order on it are a modernist delusion. One must hive from an 'ostensive' view of society as the cause of people's actions and cleave to a 'performative' vision in which society is made (constructed) by human action. This is all well and good, except that in the hands of Lowe and Ward (as so many poststructuralists) it allows them to gambol merrily on without any clear notion of social structure at all. They confuse, it seems to me, the documentation of process and struggle at the micro level with the need to discover whether there has been any systematic outcome. The danger is that after reading their chapter one has no more idea of whether there is a new agrarian regime (or whether any such thing could even exist) than one had at the beginning; equally, it is difficult to discern what the macro politics of pollution is all about (or why it's so important at this time), why agriculture is caught in a vice, and so forth. Indeed, Lowe and Ward appear to have retreated from the well-tended garden of agrarian theory altogether and into the bourgeois briar-patch of methodological individualism. One can see the negation of social science at work in the percentage of methodological arm-waving versus theory-building and presentation of evidence.

Wells' chapter takes a different tack, and she has a brilliant book, *Strawberry Fields* (1996), to back it up. The latter is destined to be one of the classics of agrarian studies. Its standing makes me loath to criticise, and everything I say must be read with that initial sense of wonder at a beautiful piece of scholarship. It takes a messy and complex bit of world agriculture and makes sense out of it in a remarkably convincing way, and does so by retaining a good deal of classical Marxian and agrarian theory while opening up the theoretical system to more determinations, more agency, more politics, and a great deal of fresh air. Now she, too, cites Latour approvingly but to quite different effect, taking the view that one must walk a fine line between the performative and the ostensive characterisations of society in order to capture the *reciprocal* determinants of structure and agency. She then proceeds to unpack very carefully the mutual constitution of economics and politics, labour and capital, the local and the global, the migratory and the resident, sector and economy, law and political economy, and so forth.

That careful treatment of her subject, fresh strawberry production in California's central coast and the recrudescence of 'sharecropping' there, means that she takes agrarian scholarship forward in some very specific ways while retaining some hoary – but still germane – concepts such as the commodity sector, the physical imperatives of agriculture, class, segmented labour markets, and the like. What she adds is a strong emphasis on the *local* in the face of globalising theory, of politics over the grinding logic of economics, of moral discourse over economistic calculation, and of worker agency over totalising capitalist control. I can hardly disagree with these at one level, since they are eminently valuable counterpoints to much one-sided thinking, from jejune journalists to armchair academics. Nonetheless, it must be said that a mere statement of general philosophical purposes, however couched in dialectics, network theory, poststructuralism, and the like, never solves any real analytic problem; it only prepares the way. There is still the hard work of weighing the evidence and weighting the categories, of making theoretical simplifications (abstractions) that illuminate, and of keeping the logic straight (however messy and complex and dialectical it may be). Therefore, Miriam Wells' solutions to her problem – and to others to which we might apply her reasoning – are by no means beyond reproach. I will address these in the concluding section.

TECHNICS AND AGRARIAN CIVILISATION

Technology and the labour of transforming nature according to principles independent of all social relations are an essential part of the arguments of both Wells and Lowe and Ward. In Wells' case, she hives to the now long-standing view within agrarian studies that nature matters, from which follows the ideas of sectoral specificity (agriculture is distinctive by virtue of the way it must wrestle with a fractious nature) and of resistance of agriculture (and industry) to the pure logic of capital (agriculture is not transformed in a linear and uniform way into a field of play for big corporations, big machinery, and big operations) (Goodman *et al.* 1987, Mann 1990). I learned this lesson early in my career from the rural sociologists and geographers such as Bill Friedland (McLennan and Walker 1980) and transferred it back to industrial geography, where it was still news to many (Storper and Walker 1983, 1989). I notice with some amusement that sectoral studies and commodity chains are now hot ideas in development studies (Gereffi and Korzeniewicz 1994, Evans 1995).

Behind this conviction in the importance of sectoral specificity, however, Wells maintains a radical vision of the significance of politics and social relations of production in the unfolding of industrial history, whether urban or agrarian. Her starting point is as a 1960s New Leftist in California, whose allegiance to the United Farm Workers went hand in hand with struggles against imperialism, racism, and capitalism. I wonder what Lowe and Ward's personal history or allegiances might be, since I see in their essay a quite different set of political commitments.

The extreme methodological position of Lowe and Ward is commensurate with a more liberal political stance. It is liberal by virtue of its refusal to commit to any positionality (which is true of all relativist postmodern stands). It reminds me of Simon Schama's (1995) towering monument to environmental disbelief, *Landscape and Memory*, the purpose of which is to convince us that environmentalist commitments of the present are just so much flotsam riding on the to-and-fro of history's tides. In Lowe and Ward's cast, this is achieved through the relativising effect of the move to moral discourse as the primary object of inquiry. This is not to say that the shift in social views about farming and nature the authors point to, from an agrarian ideology of the farmer as Husbandman of Mother Nature to an environmentalist ideology of farmer as Nature-Wrecker, is unimportant. But the reasons for the shift and the social struggles behind it are not elucidated, so one is left feeling that all such morals are relative, rather than that the environmentalists have a damned good point. I suspect that the authors are probably environmentalists themselves but they are so caught up in the epistemological stratagems of Law, Latour and Co. that they forget their own commitments. Elsewhere they invoke Lewis Mumford, perhaps the greatest spokesman for the baneful effects of modern technology (e.g. Mumford 1934), and write at length on 'the ethos of productivism'. The key line is this: 'In effect, modern technology has driven a wedge between the farmer's interest in production and traditional agrarian stewardship . . . '. This is a very common environmentalist view, and one I share, in part, because I was an environmentalist before I ever read Marx. I am quite convinced that technology *is* a relatively independent force for social change (and there's good reason to think Marx thought so, too) (Walker 1985, 1988). Similarly, nuclear waste will kill you whether it's produced by capitalists or socialists, as Chernobyl proved to any remaining doubters.

Nonetheless, it is quite another thing to move from this recognition of technical determination to a historical vision of 'industrial society (and its associated modernist project)'. This is the sort of postmodern stance that is actually a reversion to modernisation theory of the 1950s (Kumar 1978). It is one thing to speak of the ethos (and practice) of productivism, which infected communist societies as well as capitalist, but another to fail to mention the influence of capitalism on the English farmers who have become so wholeheartedly producers instead of stewards of the land. To blame this on postwar technology alone, as the authors do, without any prior history or contemporary influence of capitalist relations of competition, exploitation, accumulation and dualist thinking, is an oversight that Mumford himself would denounce in thunderous terms.

Presenting environmentalism in the guise of the 'field-level bureaucrat' is another depoliticising move than takes us beyond Weber and into the arms of contemporary American political science. Again, this is not to deny the insight that field officers play a significant role in regulation as moral imperative and practical restraint (I argued this myself for years when teaching courses on pollution regulation; see Walker *et al.* 1979). But Lowe and Ward leave it at that, as if the modern British state were the sum total of its low-level officials. No neo-Weberian of any standing would countenance such a view if the topic were economic development or social welfare policy (e.g. Skocpol 1979, Johnson 1982); nor would any contemporary Marxist (e.g. Anderson 1987, Hobsbawm 1990); why should environmental policy be an exception?

HOEING THE MARXIAN ROW

Wells is remarkably orthodox in her labourist, even Marxist, views of agrarian political economy. But she is also quite unorthodox in her appreciation of geography, morality, and the law, and the agency of the weak against overwhelming top-down forces. I am very much in sympathy with Wells' insistence on local, temporal, and sectoral diversions from (and reversals of) the global logic of capitalism, and their roots in politics, culture, history, and agency. Nonetheless, I am still going to suggest some ways in which she has given insufficient attention to the larger story at several geographic and historical levels, ways that confirm more than deny the cogency of a sufficiently geographical yet orthodox political economy.

California agriculture and the logic of capital

Wells' strawberry story points up the importance of paying attention to local circumstance, but she is so focused on the narrative of labour struggles and on the micro geography of the Central Coast region that the chance to drive home a greater lesson of the local and the global is foregone. That is, the significance of California agriculture as an accumulation of wildly successful segments in strawberries, grapes, raisins, cotton, lettuce, and so forth, and extraordinarily productive regions such as the Imperial Valley, Fresno County, and the Salinas Valley, is not merely that it is different, but that its difference carries a punch felt well beyond the state's boundaries, and far around the world. Wells provides a good thumbnail sketch of California agriculture in her book (Wells 1996: 19–27) as a backdrop to the amazing postwar expansion of fresh strawberry production, but still leaves the impression that strawberries are more the exception than the rule in the most dynamic agricultural region on earth. Indeed, her stress on local climate and soil in the Central Coast, while valid in part, makes California's blessing seem to be more natural than social. Yet the most important distinction the region has is not its relatively unusual Mediterranean climate, but its exceptional degree of capitalist domination of field agriculture.

California agriculture has been thoroughly capitalist for a century and a half, as Carey McWilliams pointed out long ago in *Factories in the Fields* (1939). No need here for capitalism to gradually squeeze out the family farm or edge into fields from its strongholds in industry and finance (though it did that, too). The land was bought up in seven-league parcels by San Francisco capitalists (made rich by the gold rush and nineteenth-century expansion) and immediately converted to wheat and cattle ranches. Small farms made some inroads in the early irrigation period, but were shouldered aside by larger operators once again after 1900 (Liebman 1983). Long before migration began from Guanajuato and Oaxaca, Chinese gang labourers were contracted to build levees, plant fruit trees, and cultivate speciality crops (Chan 1986). Long before the era of fresh strawberries, California agribusiness was aggressively creating a national (and even international) market for fresh citrus, canned fruit, and white asparagus, then shunting aside southern cotton growers (many of them true sharecroppers), midwestern canners, eastern peach farms and hen-houses, and even Turkish raisin shippers (Walker 1996a).

In California, the logic of industrial capital has had full sway in agriculture, with the kinds of results Marx or Lenin would have predicted: singularly large landholdings, massive investment in water systems, rapid mechanisation wherever possible, harnessing of science, mass proletarianisation, and a choke-hold on the government in its domain. This by no means denies the refractive nature of agriculture either as a peculiar industry or an arena rich with alternative (and fiercely defended) social relations of ownership. Indeed, California's 'exceptionalism' on this score, until very recently, provides an essential counterpoint to the mainstream of agrarian history and theory. Nevertheless, it also shows quite boldly how the logic of capital, once embedded in a social production system, drives it relentlessly forward, revolutionising all methods it can lay hands on and melting everything solid in its path. And that capitalist, industrial model is spreading wildly around the globe today, partly through the competitive force of places like California and partly through mimesis, by force of example, as more and more places replicate Californian methods.

All this makes the local more than a curiosity worthy of specialised attention by researchers. It is more even than a huge anomaly for regulation theory and its universalising about Fordism and post-Fordism, whether industrial or agrarian (California was never Fordist in any of its leading sectors – see Walker 1996a). Even more, the local can be a potential hothouse of innovation, competition and transformation that can burst upon an unsuspecting world at any time, as New England's machinists did at the Crystal Palace, Japanese car-makers did in the 1970s, and Taiwanese integrated circuit companies are doing in the 1990s (Hsu 1997). That is the 'truly revolutionary road' (i.e. bottom-up) to globalisation for any locality.

The macro politics of class struggle

I cannot agree more with Wells' distrust of the conservative rhetoric of mechanical globalisation, in which the global economy is said to force nations, classes, and individuals into predetermined reactions. She restores the political to pride of place among causal factors in a sophisticated model of agrarian political economy, and makes a convincing case for the impact of labour organising and legal contests in making and breaking the 'sharecropper' regime in strawberries. The argument is straightforward enough: the United Farm Workers (UFW) made great strides in organising California field and harvest workers after the Bracero immigration programme was ended via political protest. They raised wages and worker militancy, threatening owner control at the place of production. Growers responded by redefining employment from direct wage work to indirect labour contracting, utilising the same labour force and the same methods of production, but leaving the workers freer to recruit and manage their collective labour.

Wells goes into considerable detail about the micro politics of production, differentiating among three subsets of growers and workers in separate sections of the Central Coast region. While I enthusiastically support such attention to the revealing micro geographies of labour markets and production systems (Walker 1996b), in this instance Wells might have attended more to the larger scales of class struggle and less to the small ones. In part, I do not find the fine divisions she draws carry that much explanatory weight, despite her best efforts. While she distinguishes three production subregions, she can only find two sharecropping systems, with one covering 90 per cent of output. A simplified unitary model would still get us a long way with less trouble (while, admittedly, shortchanging the North Monterey Hills zone).

More important, farm labour struggles in strawberries, while unique in some respects, have been not far out of synch with California, the US in general, and even world class politics over the last thirty years (indeed, they were most out of touch during the New Deal era of the previous thirty years, thanks to the exemption of farm labour from the National Labor Relations Act of 1935). The UFW's ascent was part of a larger political upheaval of the 1960s, and in fact played a leading role in that historical moment. Its strength peaked with the last gasp of American liberalism in the 1970s, particularly Jerry Brown's smashing victory in the governor's race of 1974, in repudiation of Ronald Reagan's preceding regime, and passage of the state Agricultural Labor Relations Act in 1976. Conversely, the farmworkers, rapidly diminishing power in the 1980s corresponds very well to state and national trends in the weakening of organised labour and erosion of real wages. Brown's liberalism came under heavy assault (as did President Carter's), by 1978, and both caved in quickly to the pressures from the New Right (Walker 1995).

For all the local particularity and even leading role of California politics in national affairs, we have to acknowledge that the tidal wash of class politics and the massive defeat of the working class by the bourgeoisie in the 1980s was in some sense a global phenomenon. Moreover, it correlates rather well with global economic trends after 1973, including falling rates of profit, inflation, increased international competition, financial globalisation, and greater cyclical fluctuations (which, while uneven, were surprisingly general) (Webber and Rigby 1996). While one did not unilaterally cause the other, the economic troubles almost certainly rendered the political attack on labour both compelling and hard to resist, and the joint result has been a very clear trajectory for real wages and job security, both heading south rapidly.

Forms of labor, divisions of labor and capitalist development

At the core of Wells' story is the regional shift from wage labour to 'share-cropping' in the strawberry fields. This reversion from wage labour to a putatively pre-capitalist form of labour is a curiosity which she seeks to explain in terms of class struggle, law, and ideology. In the book, she makes clear that this is not a peasant–landlord relation, but a sophisticated type of capital–labour relation which ought properly to be called 'share labour' (Wells 1996: 302). Growers converted their workers into labour contractors in response to UFW organising. While this peculiar form of labour contract is, legally and ideologically, distinctive and not reducible to standard employment practices in the United States, Wells may be making too much of it. After all, even the judge in the *Real* case thought it was a cover for wage labour. It certainly does not justify her retreat from Marx and Lenin, as if capitalist time had been wholly reversed, rather than hiring practices shifted between possible alternatives. Nor is this sort of internal labour subcontracting system unprecedented under capitalism, as illustrated by steel and other craft worker contracts before Taylorism (Brody 1960). Indeed, California agriculture has regularly utilised labour contractors to organise gang labour in the fields and harvests, a form that has come back with a vengeance throughout the state (Villarejo and Runsten 1993). Worse yet, Wells retreats altogether at one point from Marx's notion of the centrality of wage labour in the capitalist transformation of the modern world, to join Wallerstein's World System theorists in seeing money capital as the central actor in the global economy, making use of all kinds of labour, willy-nilly, as suits its needs. Without getting lost in that larger debate, which has merit on both sides, California agriculture, in my view, has been revolutionary precisely because it has been capitalist and based on wage labour, not family or slave labour.

Another surprise for Wells is that California agriculture, particularly the fruit and vegetable segments, is so labour intensive (currently employing something like half a million workers, more than electronics, entertainment, or aerospace in the state and over half of all hired labour in US agriculture). This is surprising for all leftists who have grasped one horn of the Marxian theory of accumulation, the drive towards mechanisation and automation, but failed to grasp the other horn, the accumulation of the proletariat (cf. Marx [1863] 1967, Chapters 20–22). There is no mystery to the reappearance of human labour in the industrial machinery once one is attentive to the way capitalism repeatedly throws up new divisions of labour, including new sectors, new work places, and new jobs. This is sometimes understood in the context of internationalisation of manufacturing, where the global working class is bigger than ever, despite cries of the 'end of work' (Watts and Boal 1995), but is quite amazing to (white) intellectuals when it shows up inside the belly of the American beast. Wells misunderstands this as simply a local or sectoral deviation from the norm rather than as a theoretical necessity of the capitalist system; that is, she scores the wrong point against orthodoxy.

Finally, the labour recruitment system of California agriculture is largely an immigrant one, utilising familial and village networks to siphon north workers from Mexico. Wells does a good job of comprehending the political dimensions of such a labour market (in terms of immigration policy and labour law), but does little with the micro dynamics of household and village structures, despite a considerable literature on Mexican migration (e.g. Massey *et al.* 1987) and the insistence of feminist scholars that the patriarchal deployment of family labour and internal struggles of women over divisions of labour, surpluses, and property are every bit as political as US labour law (e.g. Carney 1986, Scheper-Hughes 1992).[1]

Moral economy and capitalist ideologies

Wells (like Lowe and Ward) calls upon the concept of 'moral discourse' for some important analytical work. In the strawberry struggles, a key role is played by the ideology of 'free contract' under the sharecropping system. Many workers buy into the view that they are thereby rendered more independent, able to manage their own labour, and likely to gain access to farm property themselves. This is not entirely a fiction, but they are also more exposed to risk and more overworked (self-exploitation), and few, if any, escape their class station to buy their own farms. In any event, when a group of sharecroppers sues the grower corporation over its heavy-handed control of production, they are unable to make legal headway until their attorney persuades them to sue under the laws protecting wage workers. Surprisingly, they win the case and lead growers to abandon the strategy of sharecropping.

Without doubt, ideas of justice and the legal system were critical to the way the class struggle played out in the Central Coast. Nonetheless, one cannot stretch the argument too far. This is a perfect example of the power of bourgeois ideology of property and contract, which, as Marx said, 'is a very Eden of the innate rights of man' ([1863] 1967 p. 176). Workers accede to that ideology quite readily, as one might expect in a capitalist society (or, perhaps, because of characteristically peasant backgrounds). But when they try to extend their interests as petit bourgeois operators, the company and the courts quickly slap them down. Only then must they face their still overwhelmingly proletarian status – persuaded by a lawyer who is a labour advocate within the US labour law system and very likely a product of the radical sixties herself.

Adding the factor of legal discourse and ideals of justice makes the study richer but does not decisively shift the terms of causality in a complex and messy world from political economy (materialism) to consciousness (idealism). Ideology, another term redolent with classical Marxism, remains quite serviceable. And while 'discourse' can easily serve in its place, it comes loaded with freight of its own, as the leading term in the anti-Marxist doctrines of the poststructuralists and postmodernists, one which fully intends to invert the terms of analysis from political economy to cultural-literary studies. The chapter by Lowe and Ward flirts with this very danger.

I am not opposed to cultural studies and the recognition of the power of symbolism, representation, ideation and the whole edifice of human thought and imagination. Indeed, in my view political economy always had to work against the grain of imagined worlds of the mind in order to have any purchase at all. The wonder of materialist explanation for me has always been that it explained so much *in spite of* the presumption that such rude facts as prices, wages, and economistic calculation seem so vulgar and trifling to most well-off Americans and almost any creative intellectual. So I am well aware that there is no single 'ruling ideology' that structures all thought. Here, too, geography, history, agency, and the openness of human affairs render reductionist formulae useless.

It is then somewhat surprising that Wells never ventures much outside the context of US labour law in exploring the effects of ideology and culture on the strawberry workers. She observes that the ones in the marginal area of the Monterey Hills are more 'Mexican' in their sense of deference and lower wage expectations, but she takes that insight no farther. One would like to see, first of all, a greater understanding of what Mexican national culture is and how it substantially intrudes on US society through mass immigration (cf. Eagleton 1991, Sanchez 1993, Johns 1997). Likewise, one would hope for a geographer's sense of the variation within Mexico, by state, city, and country; north and south; mestizo and indio; and how that, too, infiltrates the United States. Can we assume that the Mexicans employed in the Monterey Hills come from the same background as those in the Pajaro Valley? Finally, one would like Wells to signal some appreciation that Mexican culture can be as much a source of strength as a weakness in labour organising. After all, the first farm labour union in the history of California, formed in Oxnard in 1903, was a coalition of Mexican and Japanese farm-workers, and the fiercest field labour struggles of the 1930s were led by Mexicans and Filipinos (Camarillo 1984). Mexican born and mostly illegal immigrants have also been the backbone of several key industrial organising victories in California in the last five years, most famously Justice for Janitors. And in the period since the publication of *Strawberry Fields*, the Mexican workers of the Monterey Hills have risen up and joined the UFW. This is quite unexpected in Wells' terms.[2]

In short, Miriam Wells has given us a great deal to chew on, Philip Lowe and Neil Ward have posed questions for which they seem to provide few answers. In any case the analytic task before us in agrarian studies is rendered no easier by excessive philosophising or references to the latest intellectual wave from Paris breaking upon our shores. Agrarian studies as a field has a noble history and a lot of solid insights to build on. The same can still be said of Marxism, despite the end of communist history. We have nothing to lose but our commodity chains.

NOTES

1 Thanks to Sarvar Kothavala for making this point to me.
2 Thanks once more to Sarvar Kothavala for bringing this to my attention.

BIBLIOGRAPHY

Anderson, P. (1987) 'The figures of descent', *New Left Review* 161: 20–77.

Berry, S. (1993) *No Condition is Permanent: The Social Dynamics of Agrarian Change in Sub-Saharan Africa*, Madison: University of Wisconsin Press.

Brody, D. (1960) *Steelworkers in America*, Cambridge: Harvard University Press.

Camarillo, A. (1984) *Chicanos in California: A History of Mexican Americans in California*, San Francisco: Boyd and Fraser.

Carney, J. (1986) 'Struggles over land and crop rights in the Gambia: conflict and accumulation in the household', in J. Davison (ed.) *Land, Women and Agriculture in Africa*, Boulder: Westview.

Chan, S. (1986) *This Bittersweet Soil*, Berkeley: University of California Press.

Cohen, G. (1979) *Karl Marx's Theory of History: A Defence*, Princeton: Princeton University Press.

Eagleton, T. (1991) *Ideology: An Introduction*, London: Verso.

Evans, P. (1995) *Embedded Autonomy: States and Industrial Transformation*, New York: Cambridge University Press.

Gereffi, G. and Korzeniewicz, M. (eds) (1994) *Commodity Chains and Global Capitalism*, Westport: Praeger.

Goodman, D., Sorj, B., and Wilkinson, J. (1987) *From Farming to Biotechnology: A Theory of Agro-Industrial Development*, New York: Basil Blackwell.

Hobsbawm, E. (1990) *Nations and Nationalism Since 1780*, Cambridge: Cambridge University Press.

Hsu, Jinn-Yuh (1997) 'A Late Industrial District? Learning Networks in the Hsinchu Science-based Industrial Park, Taiwan', unpublished doctoral dissertation, Berkeley: University of California, Department of Geography.

Johns, M. (1997) *The City of Mexico in the Age of Diaz*, Austin: University of Texas Press.

Johnson, C. (1982) *MITI and the Japanese Miracle*, Stanford: Stanford University Press.

Kumar, K. (1978) *Prophecy and Progress: The Sociology of Industrial and Post-Industrial Society*, Harmondsworth: Penguin (second edn 1986).

Latour, B. (1993) *We Have Never Been Modern*, Hemel Hempstead: Harvester Wheatsheaf.

Law, J. (1994) *Organizing Modernity*, Oxford: Basil Blackwell.

Liebman, E. (1983) *California Farmland: A History of Large Agricultural Landholdings*, Totowa, NJ: Rowman and Allanheld.

MacLennan, C. and Walker, R. (1980) 'Crisis and change in US agriculture: an overview', in R. Burbach and P. Flynn (eds) *Agribusiness in the Americas*, New York: Monthly Review Press (pp. 21–40).

McWilliams, C. (1939) *Factories in the Fields*, Boston: Little Brown (also Santa Barbara: Peregrine Smith, 1976 edn).

Mann, S. (1990) *Agrarian Capital in Theory and Practice*, Chapel Hill: University of North Carolina Press.

Marx, K. ([1863] 1967) *Capital: Volume I*, New York: International Publishers.

Massey, D. and Meegan, R. (1978) 'Industrial restructuring versus the cities', *Urban Studies* 15: 273–88.

Massey, D., Alarcón, R., González, H., and Durand, J. (1987) *Return to Aztland: The Social Process of International Migration from Western Mexico*, Berkeley and Los Angeles: University of California Press.

Mumford, L. (1934) *Technics and Civilisation*, New York: Harcourt, Brace & Co.

Sanchez, G. (1993) *Becoming Mexican American: Ethnicity, Culture and Identity in Chicano Los Angeles, 1900–1945*, New York: Oxford University Press.

Schama, S. (1995) *Landscape and Memory*, New York: Alfred Knopf.

Scheper-Hughes, N. (1992) *Death Without Weeping: The Violence of Everyday Life in Brazil*, Berkeley: University of California Press.

Skocpol, T. (1979) *States and Social Revolution*, London: Cambridge University Press.

Storper, M. and Walker, R. (1983) 'The theory of labor and the theory of location', *International Journal of Urban and Regional Research* 7, 1: 1–41.

—— (1989) *The Capitalist Imperative: Territory, Technology and Industrial Growth*, New York: Basil Blackwell.

Villarejo, D. and Runsten, D. (1993) *California's Agricultural Dilemma: Higher Production and Lower Wages*, Davis: California Institute for Rural Studies.

Walker, R. (1985) 'Technological determination and determinism: industrial growth and location', in M. Castells (ed.) *High Technology, Space and Society*, Beverly Hills: Sage.

—— (1988) 'The dynamics of value, price and profit', *Capital and Class* 35: 147–81.

—— (1995) 'California rages against the dying of the light', *New Left Review* 209: 42–74.

—— (1996a) 'Another round of globalisation in San Francisco', *Urban Geography* 17, 1: 60–94.

—— (1996b) 'For better or Worcester: comments on Hanson and Pratt's *Gender, Work and Place*', *Antipode* 28, 4.

——, Storper, M., and Widess, E. (1979) 'The limits of environmental control: the saga of Dow in the Delta', *Antipode* 11, 2: 1–16.

Watts, M. and Boal, I. (1995) 'Working class heroes', *Transition* 5, 4: 90–115.

Webber, M. and Rigby, D. (1996) *The Golden Age Illusion*, New York: Guilford.

Wells, M. (1996) *Strawberry Fields: Politics, Class and Work in California Agriculture*, Ithaca: Cornell University Press.

PART V

TRANSNATIONAL CAPITAL AND LOCAL RESPONSES

11

NOURISHING NETWORKS

Alternative geographies of food

Sarah Whatmore and Lorraine Thorne

> [T]he capitalism of Karl Marx or Fernand Braudel is not the total capitalism of the Marxists. It is a skein of somewhat longer networks that rather inadequately embrace a world on the basis of points that become centres of profit and calculation. In following it step by step, one never crosses the mysterious lines that divide the local from the global.
>
> (Latour 1993: 121)

INTRODUCTION

The spatial imagery of a 'shrinking world' and a 'global village' are the popular hallmarks of an understanding of the limitless compass and totalising fabric of contemporary capitalism that has become something of a social science orthodoxy, known as *globalisation* (Featherstone 1990, Sklair 1991). No less heroic than the institutional complexes which it depicts, such an understanding perpetuates a peculiarly modernist geographical imagination that casts globalisation as a colonisation of surfaces which, like a spreading ink stain, progressively colours every spot on the map. This spatial imagery suffuses the political economy of agro-food through analytical devices like 'global commodity systems' (Friedland *et al.* 1991); 'agro-food regimes' (Le Heron 1994) and 'systems of provision' (Fine *et al.* 1996). In the most cogently argued versions, globalisation is animated as a political project of world economic management orchestrated by a regiment of capitalist institutions including transnational corporations (TNCs), financial institutions and regulatory infrastructures (McMichael 1996: 112). But the most potent agro-food expression of this spatial imagery must surely be George Ritzer's notion of 'McDonaldization'. He coins the term to describe a process of social rationalisation modelled on the fast-food restaurant which he argues has 'revolutionised not only the restaurant business, but also American society and, ultimately, the world' (Ritzer 1996: xvii). This is social science at its most triumphant – a rhetorically seductive best-seller which serves up the world on a plate.

That some markets indeed have global reach is not in dispute. What we want to emphasise is that this reach makes the corporations and bureaucracies that fashion such markets both powerful and vulnerable, being woven of the same substances as the more humble everyday forms of social life so often consigned to the 'local' and rendered puny in comparison. One of the most serious consequences of orthodox accounts of globalisation, whether of the more rigorous or the more populist varieties identified above, has been the eradication of social agency and struggle from the compass of analysis by presenting global reach as a systemic and logical, rather than a partial and contested, process (Amin and Thrift 1994). TNCs and associated regulatory bureaucracies become magnified into institutional dinosaurs whose scale and mass overwhelms the paltry significance of their social fabric, at the same time as the life practices and milieux of lesser social agents are dwarfed and overshadowed in this colossal landscape. But size, as the dinosaurs discovered, isn't everything.

Our point then, is that there is nothing 'global' about such corporations and bureaucracies *in themselves*, either in terms of their being disembedded from particular contexts and places or of their being in some sense comprehensive in scale and scope. Rather, their reach depends upon intricate interweavings of *situated* people, artefacts, codes, and living things and the maintenance of particular tapestries of connection across the world. Such processes and patterns of connection are not reducible to a single logic or determinant interest lying somewhere *outside* or *above* the social fray. This distinction is the difference between systems and networks; a shift in analytical metaphor which takes up critiques of the globalisation orthodoxy, notably within geography and anthropology, as a failure of both social and spatial imagination (Strathern 1995, Thrift 1996).

Two complementary influences on the elaboration of these critiques are particularly important for our purposes here, the one concerned with rethinking *political economy* and the other with recognising *space-time*. In the first case, economic sociology and institutional economics have emphasised the embodied and routinised social practices which constitute markets, corporations and regulatory bureaucracies against accounts (Marxist and neoclassical) which tend to treat these institutional complexes as abstracted presences, or the product of some historically teleological process (Underhill 1994, Thrift and Olds 1996). Economic institutions and practices are conceived of not as some separate, and still less determinant, 'sphere' of activity which articulates with other 'spheres' of civic society or governance but as socially embedded and contingent at every turn (Smelser and Swedberg 1994, Murdoch 1995). In the second case, poststructuralist ideas have informed theoretical efforts to deconstruct the geometric landscapes – what Barnes (1996) has called the 'Enlightenment view' – of political economy. By fashioning the modern world as a single grid-like surface, such landscapes make possible the encoding of general theoretical claims as omnipresent, uni-versal rationalities. In contrast, critics point to the *simultaneity* of multiple, partial space-time configurations of social life that are at once 'global' and 'local', and to the *situatedness* of social institutions, processes and knowledges as always contextual, tentative and incomplete (Thrift 1995).

Such critiques, especially that derived from institutional economics, have been taken up already by those working in agrarian political economy (see Goodman and Watts 1994, Whatmore 1994). While it remains 'against the grain', such work marks the beginnings of an understanding of globalisation as partial, uneven, and unstable; a socially contested rather than logical process in which many spaces of resistance, alterity, and possibility become analytically discernible and politically meaningful. In this paper we want to extend these lines of critique, particularly that concerned with spatial recognition, as a basis for exploring alternative geographies of food that have been eclipsed by mainstream political economy accounts. Little work in this vein has made its way into the agro-food literature as yet (but see Arce and Marsden 1993, Cook and Crang 1996 for related forays).

The title phrase 'alternative geographies of food' signals an effort on our part to see the world differently in (at least) two senses. We begin by taking up the geographical implications of *actor-network theory* (ANT) which both of us have been exploring in work elsewhere (Thorne 1997, Whatmore 1997) and which resonates with other contributions to this volume (notably, the chapter by Ward and Lowe). As the opening quotation suggests, this involves the elaboration of a *topological* spatial imagination concerned with tracing points of connection and lines of flow, as opposed to reiterating fixed surfaces and boundaries (Thrift 1996, Bingham 1996). In particular, we draw on the work of Bruno Latour (1993, 1994) and John Law (1986, 1991, 1994) to elaborate an understanding of global networks as performative orderings (always in the making), rather than as systemic entities (always already constituted). We then go on to explore some of the analytical and political spaces which such an understanding opens up, by means of a case study of *fair trade coffee networks*. This case study illustrates the fashioning of social and environmental configurations of agro-food production and consumption that coexist with those of industrial food corporations but which in some way counter, or resist, their institutional values and practices.

GLOBAL NETWORKS OR 'ACTING AT A DISTANCE'

> The two extremes, local and global, are much less interesting than the intermediary arrangements that we are calling networks.
>
> (Latour 1993: 122)

The work of Latour and Law, and their respective notions of 'hybrid networks' and 'modes of ordering', provide ways of reconceptualising power relations in space from the flat, colonised surfaces of globalisation to the frictional lengthening of networks of remote control. In so doing, the key question becomes not that of scale, encoded in a categorical distinction between the 'local' and the 'global', but of connectivity, marking lines of flow of varying length and which transgress these categories. To put this question in the terms of ANT, what are the conditions and properties of 'acting at a distance'? Formulating inquiry in this way refuses the privileged association *a priori* between particular kinds of social institutions (notably TNCs) and global reach and, by implication, the pervasive mapping of the conventional sociological binaries of 'macro–micro' and 'structure–agency' onto that of the 'global–local'. Our account builds on the early efforts of geographers to explicate the spatial dimensions of ANT and their import for understanding power as a thoroughly relational process (Murdoch 1995, Murdoch and Marsden 1995) and for recognising the active part of non-humans in the fabric of social life (Thrift 1995, 1996).

Where orthodox accounts of globalisation evoke images of an irresistible and unimpeded enclosure of the world by the relentless mass of the capitalist machine, ANT problematises global reach, conceiving of it as a laboured, uncertain, and above all, contested process of 'acting at a distance'. Law illustrates this conception with the example of Portuguese efforts to expand the reach of European trade in the fifteenth and sixteenth centuries by capturing the spice route to India (1986). This achievement required the Portuguese to refashion contemporary navigational complexes in ways which, as Law puts it, addressed not only the question of social control but also that of

> how to manage long distance control *in all its aspects*. It was how to arrange matters so that a small number of people in Lisbon might influence events half-way round the world and thereby reap a fabulous reward.
>
> (Law 1986: 235, original emphasis)

Law's evocative case study of 'acting at a distance' centres on the technological metaphor of 'remote control' which tends to conjure the dynamics of networking in the rather conventional geographical binary of core (transmitter) and periphery (receiver). Nor are the implications of this metaphor restricted in his work to this particular case study. The imprint of 'remote control' marks his elaboration of ANT more widely. Thus, for example,

> heterogeneous socio-technologies open up the possibility of ordering distant events from a centre . . . [in which] the centre is a place which monitors and represents the periphery and then calculates how to act on the periphery.
>
> (Law 1994: 104)

A rather different, and to our mind more promising, exposition of the spatial configuration of actor-networks is that derived from Latour's notion of 'network lengthening'. Reminiscent of the nomadic cartographies of Deleuze and Guattari (1983), the idea of 'lengthening' not only problematises the process of 'acting at a distance' but also disrupts the bi-polarities of 'core' and 'periphery'. These generic spaces, like those of 'local' and 'global', enshrine a geometric vocabulary concerned with the geography of surfaces. The unilinearity encoded in their relationship makes less sense in a topologic vocabulary concerned with the geography of flows. Here, a network's capacities over space-time represent the simultaneous performance of social practices and competences at different points in the network; a mass of currents rather than a single line of force. In these terms, actor-networks are best understood as 'by nature neither local nor global, but [only] more or less long and more or less connected' (Latour 1993: 122).

By implication, the size, or scale, of an actor-network is a product of network lengthening, not of some special properties peculiar to 'global' or 'core' actors – the 'dinosaurs' of our earlier analogy. Furthermore, the power associated with global reach has to be understood as a social composite of the actions and competences of many actants; an attribute not of a single person or organisation but of the number of actants involved in its composition (Callon and Latour 1981, Murdoch and Marsden 1995). How, then, is this network lengthening achieved?

The answer advanced in ANT is that network lengthening requires the mobilisation of larger numbers and more intricately interwoven constituents, or *mediators*, to sustain a web of connections over greater distances. In so doing, it focuses analytical attention on describing this process of mediation and its agents in ways which force a challenging, and sometimes disconcerting, shift in the horizons of social research. As Law notes in relation to his Portuguese case study,

> if these attempts at long-distance control are to be understood then it is not only necessary to develop a form of analysis capable of handling the social, the technological, the natural and the rest with equal facility, though this is essential. It is also necessary that the approach should be capable of making sense of the way in which these are fitted together.
>
> (Law 1986: 235)

At once it becomes essential to talk of network mediators other than people, that is other than the human actors on whom the whole compass of conventional theories of social agency (including other social network theories) is built. To be sure, people in particular guises and contexts act as important go-betweens, mobile agents weaving connections between distant points in the network; for example, the sailors in Law's Portuguese study, or the managerial elites of corporate business today. But, insists ANT, there are a wealth of other agents, technological and 'natural', mobilised in the performance of social networks whose significance increases the longer and more intricate the network becomes. Latour calls these agents 'immutable mobiles', such as money, telephones, computers, or gene banks; objects which encode and stabilise particular socio-technological capacities and sustain patterns of connection that allow us to pass with continuity not only from the local to the global, but also from the human to the non-human. The more they have proliferated in everyday life the more, it seems, these 'objects' have been effaced in social theory leaving us awed by the subsequently fantastic properties of social entities like TNCs. By taking such objects into account 'one can follow the growth of an organisation in its entirety without ever changing levels and without ever discovering "decontextualised" rationality' (Latour 1993: 122).

It should by now be apparent that a move from 'globalisation' to global networks as a basis for understanding the conditions and properties of 'acting at a distance' is no small step. Tracing the process and agents of mediation in the way suggested by ANT implies a pretty radical re-cognition of social agency. It is worth rehearsing three major, mutually reinforcing, elements of the theory as it is advanced, in different ways, by Latour and Law (and Callon) before illustrating their implications for the analysis of agro-food networks. These elements can be identified for the sake of brevity as the qualities of hybridity, collectivity, and durability.

Breaking down the global–local binary through the idea of the lengthening of networks is intricately tied up with breaking down the nature–society binary through the idea of *hybridity*. Just as the global–local distinction serves to purify processes and entities that are not of themselves confined to any particular spatial scale, so the ontological separation of society and nature purifies the messy heterogeneity of life. Overlaid, these binaries creates four distinct regions between which nothing is supposed to take place but in which most things are happening (Latour 1993: 123). Actor-networks mobilise, and are constituted by, a multiplicity of different agents, or 'actants', human and non-human; technological and textual; organic and mechanic. More radically still, networks build and enmesh differently constituted entities/actants which combine these properties in varied and dynamic ways. These hybrids represent states of being which fall somewhere between the passive objects of human will and imagination which litter the social sciences, and the autonomous external forces favoured in natural science accounts. Following Michel Serres (see Serres and Latour 1995), Latour designates these in-between states of being as *quasi-objects* which are as 'real as nature, narrated as discourse, collective as society [and] existential as being' (Latour 1993: 89).

Returning to the Portuguese study, Law shows how a composite of agents are enjoined as emissaries of network lengthening, including documents, devices, and people fashioned in particular ways. For example, a document called the 'Regimento' inscribed a distilled and simplified instruction for navigating by stars which permitted the navigator to pass beyond the established envelope of North European travel. Devices included the 'carreira', a ship designed for carrying cargo and avoiding plunder, and 'a kind of simplified black box', the astrolabe. Similarly, this effort to act at a distance mobilised people with very particular kinds of skills or embodied social practices, including navigators, sailors, and merchants.

This example picks up and illustrates a second key step in understanding the process of network lengthening, the fundamentally relational, or *collective* conception of social agency that characterises ANT. Thus, the significance of the 'documents, devices and drilled people' in Law's Portuguese study is the way in which they hold each other in position. 'The right documents, the right devices, the right people properly drilled – put together they would create a structured envelope for one another that ensured their durability and fidelity' (Law 1986: 254). Yet the full implications of this conception of social agency are relatively underdeveloped in this early case study. It is in Law's later work and, more particularly, in the notion of the *hybrid collectif* (Callon and Law 1995) that the importance of the active properties of non-human agents in the lengthening of networks is most fully explored. For Latour, these agents are a vital part of a network's collective capacity to act 'because they attach us to one another, because they circulate in our hands and define our social bond by their very circulation' (1993: 89).

But this represents the point of greatest tension between ANT and conventional theories of social agency which centre on notions of intentionality and, by implication linguistic competence, as a peculiarly human capacity. In what sense then, are the 'agents' of the hybrid collectif to be understood? The answer provided by Callon and Latour makes the break with conventional social theory explicit. Agents in ANT are

> effects generated in configurations of different materials. Which also, however, take the form of attributions. Attributions which localise agency as singularity – usually . . . in the form of human bodies. Attributions which endow one part of the configuration with the status of prime mover. Attributions efface the other entities and relations in the collectif or consign these to a supporting and infrastructural role.
>
> (Callon and Law 1995: 503)

In short, the logocentric bias of social theory which links agency (the capacity to act or to have effects) to language-based intentionality is refused. The agency of the hybrid collectif is a bold attempt to shift the (considerable) weight of this discursive privilege to recognise other, material, forms of signification by which the specific capacities and properties of entities from x-rays to viruses make their presence felt.

Thus far we have outlined hybridity and collectivity as necessary corollaries of the process of mediation by which networks are sustained over greater distances. Of equal significance is the question of how such networks are strengthened and stabilised over time or, in ANT terms, how they are made *durable*. In Law's book *Organizing Modernity*, he adapts Foucault's notion of 'discursive practices' to propose *modes of ordering* as a way of conceptualising the durability of networks. Modes of ordering are both narrative, 'ways of telling about the world . . . what used to be, or what ought to happen', and material, 'acted out and embodied in a concrete, non-verbal, manner in a network' (Law 1994: 20). He shows how organisations perform multiple 'modes of ordering', which influence the ways in which agents are enrolled in global networks. While these organisational patternings or habits are invariably plural rather than singular, Law argues that 'only a relatively small number of modes of ordering may be instantiated in the networks of the social at a given time and place' (ibid.: 109). In other words, the durability of long distance networks requires strong fabrics of social organisation at all points in the network, making the patterning of social and environmental practices in *particular* times and places integral to the business of network enrolment.

To return to the Portuguese example, a significant 'ordering' which can be identified in the reconfiguration of the spice trade is that associated with 'reaping a fabulous reward'. This example is an antecedent to one of the most significant modes of ordering observed by Law in his later work on the organisation of laboratory science in the UK today – that of 'enterprise', which celebrates 'opportunism, pragmatism and performance'. In addition to 'enterprise', Law identifies 'vocation', 'administration' and 'vision' as orderings shaping the performance of science. Parallel modes of ordering can be imputed with respect to the activities of networks of enrolment by TNCs. Their implication is that global networks, including those of TNCs, are performative rather than structural; (de)stabilised through the creative, collective practice of inter-dependent capacities, intentionalities and relationships by numerous actants. As ANT insists, this understanding of global networks seeks to avoid the reductionism of saying that networks, however long their reach, stand outside their performances. So, for Law, organisational modes of ordering

> are patterns or regularities that may be imputed to the particulars that make up the recursive and generative networks of the social. They are nowhere else. They do not drive those networks. They aren't outside them. Rather, they are a way of talking of the patterns into which the latter shape themselves.
>
> (Law 1994: 83)

FAIR TRADE COFFEE: AN ALTERNATIVE NET-WORKING

In this context, alternative geographies of food are located in the political competence and social agency of individuals, institutions, and alliances, enacting a variety of partial knowledges and strategic interests through networks which simultaneously involve a 'lengthening' of spatial and institutional reach *and* a 'strengthening' of environmental and social embeddedness. Such networks exist alongside the corporate and state networks of orthodox accounts of globalisation, sometimes overlapping them in space-time; sometimes occupying separate sites and establishing discrete lines of connection; and sometimes explicitly oriented towards challenging their associated environmental and social practices. In the case of food, the 'devices, documents and drilled people' of Law's Portuguese example translate into a broader compass of material concerns than that associated with traditional agrarian political economy. These include the encoding of particular agricultural and dietary knowledges in the form of various technologies; the legal inscription of agro-food practices, from patents to health criteria; and the disciplining of bodies, from obese and skeletal people to industrial animals and plants.

Using this collection of ideas, and working to avoid the bias of scale in structuralist accounts (which inscribe a macro–micro division and then reify the former), we propose that modes of ordering which spin documents, devices, and living creatures (including people) as other than passive agents through multiply sited networks are both possible and extant. In the case study we discuss one such patterning explicitly oriented towards enacting an alternative commodity network, which we have identified as *a mode of ordering of connectivity*. In this mode of ordering, stories are told of partnership, alliance, responsibility and fairness, but performed in very different ways to the neoliberal encoding of these terms (Barratt-Brown 1993). This is a mode of ordering concerned with the empowerment of marginalised, dismissed, and overlooked voices, human or non-human. Implicit in this empowering performance is the knowledge that 'some network configurations generate effects which, so long as everything else is equal, last longer than others' (Law 1994: 103). The mode of connectivity therefore not only tells and performs but also tries to concretely embody a recursive effect of social, and sometimes environmental, embeddedness.

With roots in nongovernmental organisations dedicated to alleviating poverty in the 'Third World', the fair trade 'movement' has grown in the UK over the past twenty-five years. The charity Oxfam established a wholly owned trading company in 1964 with other organisations gradually emerging as fully-fledged trading companies from solidarity markets (for example, Equal Exchange Trading Limited), or educational functions (for example, Twin Trading and the Third World Information Network – TWIN). The recently formed British Association of Fair Trade Shops (BAFTS) consists of shops committed to principles of fair trade. Other organisations do not trade at all but provide support in the form of campaigning and lobbying (for example the World Development Movement). The point here is that fair trade organisations in the UK are diverse and numerous, with the physical transactions of trading only one of their component activities. The institutions, transactions, and technologies of fair trade serve to illustrate some of the key concerns highlighted by an analysis of agro-food patterns as hybrid networks. In particular, it shows how the global reach of so-called alternative agro-food networks (or their capacity to 'act at a distance') enrols coincident actants and spaces to those of 'mainstream' commercial networks. It is the modes of network strengthening (or making durable) that are analytically distinctive between the orderings of fair trade and capitalist commerce and that open up economic and political possibilities for configuring alternative geographies of food.

In this case study we discuss a hybrid network of four UK fair trade organisations and a Peruvian coffee exporting cooperative, although there are many other agents in this network, as becomes clear. In the late 1980s 'Oxfam Trading', 'Twin Trading', 'Traidcraft' and 'Equal Exchange Trading Limited' came together to create a consortium called *Cafédirect* which procures, imports, and markets a brand of coffee of the same name, available in ground and freeze-dried forms. The four partners are located in different cities in the UK – Oxford, London, Newcastle-upon-Tyne, and Edinburgh respectively. In the early days, the hybrid network of Cafédirect operated with no central office. Instead, partners had designated responsibilities, for example the buyer working for one organisation, the wholesale administration handled by another and so on, with one of the partners, 'Twin Trading', acting as the operational focal point. In 1993 the consortium became registered as a private company which has recently appointed a managing director, and the partners are now 'shareholders'. Cafédirect was the second fairly traded product in the UK to receive the Fair Trademark, which legitimises the product as fairly traded according to criteria set out by the Fairtrade Foundation, an independent organisation recently given charitable status but originally established by several fair trade organisations, including Oxfam and Traidcraft. Thus the product of the Cafédirect network is given institutional legitimacy by a hybridised form of itself.

Cafédirect's southern partners are small-scale farmers whose coffee trees cling to steep mountain slopes in Costa Rica, Peru and Mexico, in the case of the Arabica component of the brand, while the Robusta component comes from similar producers in Tanzania and Uganda. In this case study we restrict our discussion to some of Cafédirect's partners in Peru, namely the exporting cooperative of CECOOAC-Nor, located in Chiclayo on the northern coast of Peru. CECOOAC-Nor is the central cooperative of nine individual coffee producing cooperatives dotted through the northern Andean mountains. They were all established in the 1970s during a period in which the then military government supported farmers cooperatives through an Agrarian Bank. But during the 1980s subsequent governments relinquished support and the bank became defunct, leaving farmers vulnerable to commercial bank interest rates and the purchasing strategies of commercial traders (*comerciantes*).

In the 1970s cooperatives were able to provide services to their members, including medical and educational services. Since then, these support services have been eroded, adding to the exposure of the cooperatives to renewed economic pressures. As access to cheap credit dried up the cooperatives have struggled to pay for their members' coffee. The situation worsened in 1989 when the International Coffee Agreement collapsed, leaving prices unregulated. Embedded in a political, economic, and social climate of considerable turmoil, the small-scale coffee producers in the northern Andes were buffeted by the vagaries of the Cocoa, Sugar, and Coffee Exchange in New York (CSCE), which regulates the Arabica futures and spot markets, and the powerful *comerciantes* for whom credit access was not a problem.

In 1990 CECOOAC-Nor made its first sale to a fair trade organisation, a key event for its survival. The buyer was a roaster belonging to Max Havelaar, a Dutch-based hybrid network of mainstream coffee roasters who, in exchange for paying a fair price, are able to carry the Max Havelaar trademark on their coffee packaging. Sales to Cafédirect followed, the contracts negotiated with both the Cafédirect buyer and the CECOOAC-Nor export manager paying close attention to the daily market prices in New York. The key difference between fair trade buyers and commercial dealers is that the former pay a guaranteed minimum price (which protects farmers should the market go into free-fall), and a standard number of points above the CSCE price when the market price exceeds the minimum (in effect, a 10 per cent premium). The CSCE is therefore one of many coincident actants and sites in the hybrid network of fairly traded coffee and the commercial coffee networks (see Figure 11.1). Other such coincidences include the export and import authorities for whom documentation must be in order for goods to be granted passage.

Cafédirect pays CECOOAC-Nor using the international financial system (although there may be delays in the release of payments from Peruvian banks for other reasons). The stock exchange, customs officials, and banking clerks are all actants of the hybrid network of fair trade, so too are their computers, telephones, and fax machines. Just as there are coincident actants and spaces between fair trade and commercial coffee networks, so too is there a coincident mode of ordering – that of *enterprise* – pragmatic, opportunistic, and canny (Law 1994: 1). The 'Third World' partners of Northern fair trade organisation are not insulated from the disciplines of the market – delivery deadlines, contracts, and quality conditions all have to be met. Northern fair trade organisations employ the same 'just-in-time' rationale as commercial companies, importing green coffee into the UK in accordance with the statutory trading regulations. The processing of ground coffee is also carried out in the UK, while the freeze-dried coffee is processed in Germany.

However, while the mode of ordering of enterprise is present throughout the fair trade hybrid network, it is mediated and re-articulated by another mode of ordering – that of *connectivity*. The raison d'être of Cafédirect, and

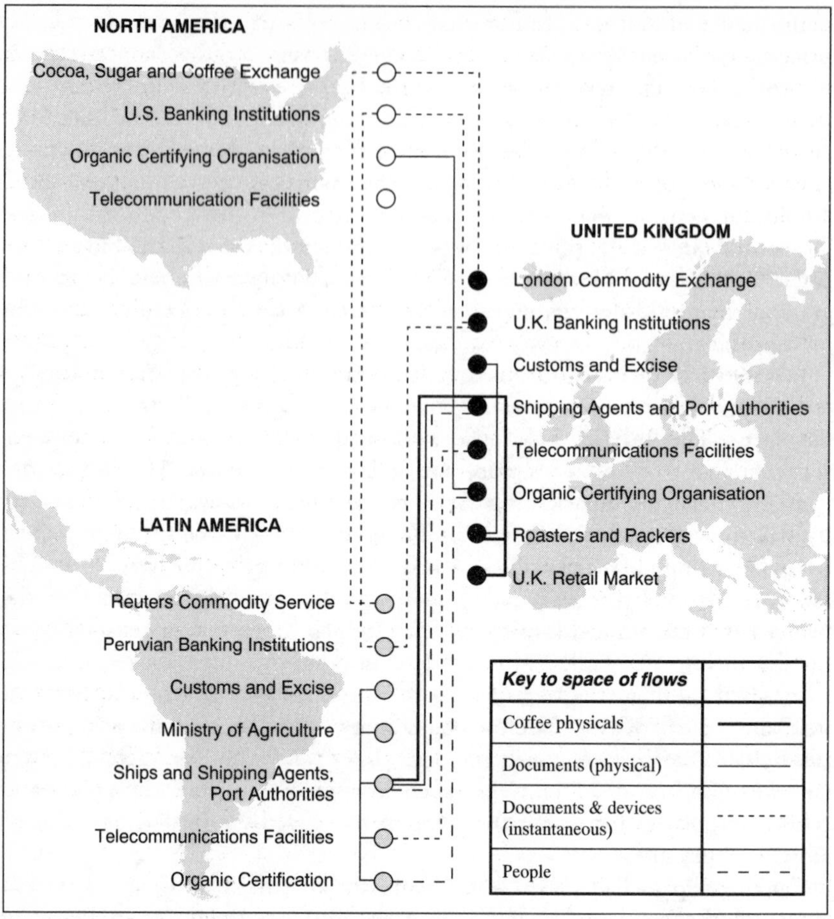

Figure 11.1 Network 'lengthening': coincident spaces of a fair trade and commercial coffee network

the social agency of the fair trade network as a whole, rests on the mobilisation of a mode of ordering of connectivity different from that of the cost-minimising, self-interested individual of neoclassical economic theory. The packaging of Cafédirect coffee products makes the discourse of connectivity explicit.

> 'This is a fair trade product. More of the money you pay for Cafédirect freeze-dried goes directly to the small-scale coffee farmers in Latin America and Africa. Fair trade means coffee growing communities can afford to invest in healthcare, education and agriculture'.

These words establish a connection between those who grow and those who buy Cafédirect coffee. *Connectivity* as a mode of ordering establishes the performance of 'fairness', rather than charity, in which the farmer gets a 'fair price' and the consumer 'gets excellent coffee'. In order to strengthen the network, fair trade organisations must make concrete the telling and performing of connectivity and fairness in the hybrid network. The farmers theoretically get a higher price for their coffee when they sell it to the cooperatives and this additional income gives them opportunities they otherwise would not have.

But the story is more complicated because another actant in this network, the coffee, is fraught with variabilities that reverberate through the network as a whole. In order to provide consumers of Cafédirect with 'excellent coffee', the cooperatives must submit only the highest quality beans. If the coffee is of low quality (reasons may include rainfall, 'pests', fermentation) it will not be suitable for the fair trade contracts negotiated by the cooperative export manager and the Cafédirect buyer, and farmers will sell to the *comerciantes*. If the price on the stock exchange is high *comerciantes* will pay well even for this low quality, and they will pay in cash. When the CSCE price is high – if, for example, the coffee harvest in Brazil is devastated by frost – farmers may see little benefit in selling to the cooperative. In such circumstances tensions between the modes of ordering of enterprise and connectivity become immediate and tangible as, for example, the warehouse goes unfilled.

At each point in the hybrid network there is instability and uncertainty, so that strengthening the embeddedness of the cooperatives is as important to the fair trade network as strengthening consumer support for fair trade coffee through marketing strategies and educational campaigns. Network strengthening is a process performed both locally and globally (see Figure 11.2). 'Strengthening' social and environmental habits of association amongst producers involves the enhancement of essential services – agricultural, medical, and educational. At the present time in the cooperatives of CECOOAC-Nor it is mainly agricultural services which are provided, facilitated by a full-time organic technician who travels between the cooperatives.

The second key aspect of the mode of ordering of *connectivity* evidenced in this case study, that of sensitivity to interactions between human and non-human actants in the network, is accomplished through organic farming practices. Making the soil fertile by cultivating earthworms (*lombrices*) and mulch, interplanting with shade trees and not burning-off makes coffee growing practices less environmentally destructive. Six of the cooperatives have organic certification granted by the Organic Crop Improvement Association (OCIA International), drawing into the hybrid network yet other actants and sites across the US, Germany and the UK. The whole process of organic certification is now regulated in law under European Union legislation (yet another coincident actant and space of fair trade and commercial networks offering organic products). The issue of environmental embeddedness, while a key part of the practices associated with the

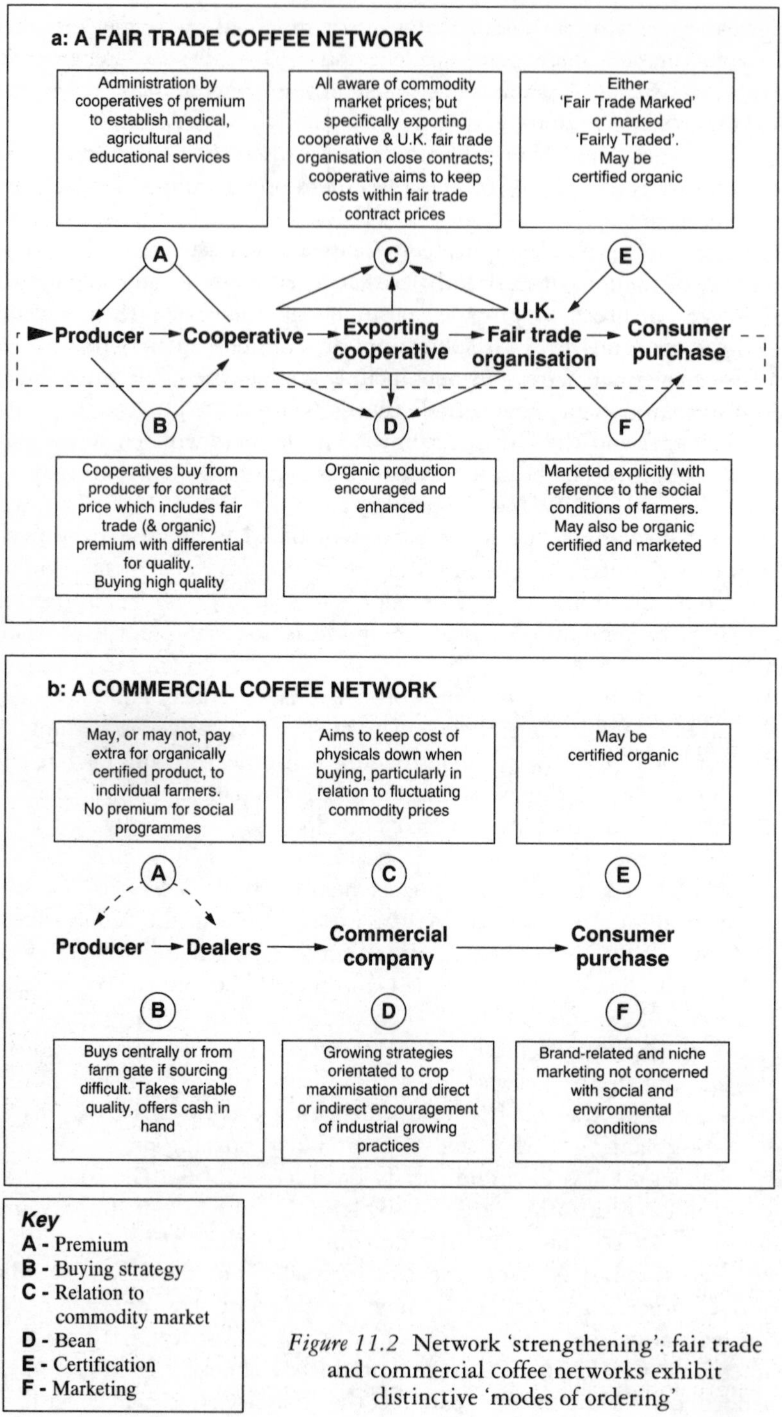

Figure 11.2 Network 'strengthening': fair trade and commercial coffee networks exhibit distinctive 'modes of ordering'

strengthening of alternative food networks, is no less dynamic than their institutional and technological aspects. For example, the land occupied by the farms of CECOOAC-Nor were pristine forest only fifty years ago, the habitat of bears and tigers.

This brief description of a fair trade network is partial not least because there are other coffee growing organisations selling to Cafédirect, creating more heterogeneity than this example is able to convey. Nonetheless, it serves to illustrate how alternative geographies of food lengthen their reach using many of the same actants and spaces as their commercial counterparts. What is analytically distinctive, however, is *how* they strengthen relationships amongst formerly 'passive' actants in commercial networks – the producers and consumers – through a mode of ordering of connectivity which works for non-hierarchical relationships framed by 'fairness'.

ALTERNATIVE GEOGRAPHIES OF FOOD

> What is to be done, then, with such sleek, filled-in surfaces, with such absolute totalities? Turn them inside out all at once, of course; subvert them, revolutionise them. The moderns have invented at one and the same time the total system, the total revolution to put an end to the system, and the equally total failure to carry out that revolution.
>
> (Latour 1993: 126)

The tendency to transform the lengthened networks of modern social life into systematic and global totalities generates heroic accounts of globalisation which do not recognise their own partiality (Thrift 1995: 24). For Latour, such accounts of the relentless logic of social rationalisation belie 'a simple category mistake, the confusion of one branch of mathematics with another' (Latour 1993: 119). Where the concepts 'local' and 'global' work well for surfaces and geometry, they mean little for networks and topology. While Latour's own style of writing is too supercilious for some – there is nothing simple about this category mistake – it should not detract from the importance of recognising the power of metaphor and language in shaping economic, as much as scientific, understandings of the world (Mirowski 1995, Barnes 1996); a power that is made flesh in numerous ways. In one sense, such vocabularies become embodied in the performance of individual and collective social identities and practices, both amongst corporate managers and oppositional political movements. In another sense, they become encoded in the authoritative texts and devices of law and science or in the engineered bodies of plants and animals (including humans).

In outlining an alternative understanding of global networks, the main points that we take from actor-network theory are that networks, unlike systems, are not self-sustaining; they rely on hundreds of thousands of people, machines, and codes to make the network. They are *collective*, that is their length and durability are woven between the capacities and practices of actants-in-relation. They are *hybrid*, combining people and devices and other living things in intricate and fallible ways in the performance of social practices. They are *situated*, inhabiting numerous nodes and sites in particular places and involving their own particular frictions (cultural and environmental) to network activity. And, finally, they are *partial* even as they are global, embracing surfaces without covering them, however long their reach.

Our treatment of 'global agro-food systems' as hybrid networks – partial and unstable orderings of numerous practices, instruments, documents, and beings – is a provocative one for those brought together under the hallmark of political economy in this book. The workshop from which this collection of papers derives provided a rare, perhaps unique, opportunity to air such ideas in an atmosphere of creative engagement. In the much less forgiving spirit of print, we are aware that we have probably overstated our case in places and have certainly left our own difficulties and qualms about ANT for another time. But if this exploratory piece has succeeded in conveying something of the possibilities for the field of agro-food research that have excited us about this theoretical project it will have served its purpose.

Rather than conceptualising the spatial orderings of economic activity in territorial terms – a globalisation of surfaces – this approach implies a conception of the spatial orderings of economic activity in mobile terms – a lengthening of flows. It shifts concern from a predictable unfolding of social structures in space to the means whereby networks of actors construct space by using certain forms of ordering which mobilise particular rationalities, technological and representational devices, living beings (including people), and physical properties. More than this, unlike the filled-in surfaces of globalisation, this approach opens up space-time to the coexistence of multiple cross-cutting networks of varied length and durability, for example in the many coincidences between the institutional spaces and geographical places inhabited by commercial and fair trade coffee networks. TNCs emerge as no longer unique in their substantive capacity for global reach. By exploring the role of the unspoken presences (or immutable mobiles) which hold such connections in place we can begin to talk about 'alternative geographies' of food in the same register, not as some pale spectre in the colossal landscape of 'capital'. It is the political competence and social agency expressed through the mode of ordering of connectivity in the fair trade network – including Cafédirect, the exporting cooperative CECOOAC-Nor, and the 3,000 small coffee farmers – which effects the difference.

ACKNOWLEDGEMENTS

We are indebted to the organisers and participants of the 'Berkeley Workshop' for intellectual (and financial) support, to colleagues in the Geography Department at the University of Bristol (particularly the 'Nature and Society reading group') who worked through many of these ideas with us, to Nick Bingham and Jon Murdoch for their comments, and to Simon Godden, the department's cartographer, for the diagrams. Finally, we should like to acknowledge the funding provided for the empirical work by a University of Bristol Scholarship, and the cooperation of people at Cafédirect and CECOOAC-Nor.

BIBLIOGRAPHY

Amin, A. and Thrift, N. (1994) *Globalisation, Institutions and Regional Development in Europe*, Oxford: Oxford University Press.

Arce, A. and Marsden, T. (1993) 'The social construction of international food: a new research agenda', *Economic Geography* 69, 3: 291–311.

Barnes, T. (1996) *Logics of Dislocation: Models, Metaphors and Meanings of Economic Space*, London: Guilford Press.

Barratt-Brown, M. (1993) *Fair Trade*, London: Zed Books.

Bingham, N. (1996) 'Objections: from technological determinism towards geographies of relations', *Society and Space* 14, 6: 635–57.

Callon, M. and Latour, B. (1981) 'Unscrewing the big leviathan', in K. Knorr-Cetina and A. Cicourel (eds) *Advances in Social Theory and Methodology*, London: Routledge and Kegan Paul.

—— and Law, J. (1995) 'Agency and the hybrid collectif', *South Atlantic Quarterly* 94, 2: 481–507.

Cook, I. and Crang, P. (1996) 'The world on a plate. Culinary culture, displacement and geographical knowledges', *Journal of Material Culture* 1, 2: 131–53.

Corbridge, S. (1993) 'Marxisms, modernities and moralities: development praxis and the claims of distant strangers', *Society and Space* 11, 4: 449–72.

Deleuze, G. and Guattari, F. (1983) (English translation) *Anti-Oedipus*, Minneapolis: University of Minnesota Press.

Featherstone, M. (ed.) (1990) *Global Culture: Nationalism, Globalisation, And Modernity*, London: Sage.

Fine B., Heasman, M., and Wright, J. (1996) *Consumption in the Age of Affluence: The World of Food*, London: Routledge.

Friedland, W., Busch, L., Buttel, F., and Rudy, A. (1991) *Towards a New Political Economy of Agriculture*, Boulder: Westview Press.

Goodman, D. and Watts, M. (1994) 'Reconfiguring the rural or fording the divide?: capitalist restructuring and the agro-food system', *Journal of Peasant Studies* 22, 1: 1–49.

Latour, B. (1993) *We Have Never Been Modern*, Brighton: Harvester Wheatsheaf.

—— (1994) 'Pragmatologies', *American Behavioral Scientist* 37, 6: 791–808.

Law, J. (1986) 'On the methods of long-distance control: vessels, navigation and the Portuguese route to India', *Sociological Review Monograph* 32: 234–63.

—— (ed.) (1991) *A Sociology of Monsters: Essays on Power, Technology and Domination*, London: Routledge.

—— (1994) *Organising Modernity*, Oxford: Basil Blackwell.

Le Heron, R. (1994) *Globalized Agriculture*, Oxford: Pergamon.

Lupton, D. (1986) *Food, the Body and the Self*: London: Sage.

McMichael, P. (1996) *Development and Social Change: A Global Perspective*, California: Pine Forge Press.

Mirowski, P. (1995) *Natural Images in Economic Thought*, Cambridge: Cambridge University Press.

Murdoch, J. (1995) 'Actor-networks and the evolution of economic forms: combining description and explanation in theories of regulation, flexible specialisation and networks', *Environment and Planning A* 27, 5: 731–57.
—— and Marsden, T. (1995) 'The spatialization of politics: local and national actor-spaces in environmental conflict', *Transactions of the Institute of British Geographers* 20, 3: 368–80.
Ritzer, G. (1996) *The McDonaldization of Society*, California: Pine Forge Press (revised edn).
Serres, M. and Latour, B. (1995) *Conversations on Science, Culture, and Time* (translated by R. Lapidus), Ann Arbor: University of Michigan Press.
Sklair, L. (1991) *Sociology of the Global System*, Baltimore: Johns Hopkins University Press.
Smelser, N. and Swedberg. R. (eds) (1994) *The Handbook of Economic Sociology*, Princeton: Princeton University Press.
Strathern, M. (ed.) (1995) *Shifting Contexts*, London: Routledge.
Thorne, L. (1997) *Towards Ethical trading space?*, unpublished PhD thesis, University of Bristol.
Thrift, N. (1995) 'A hyperactive world', in R. Johnston, P. Taylor, and M. Watts (eds) *Geographies of Global Change*, Oxford: Basil Blackwell.
—— (1996) *Spatial Formations*, London: Sage.
—— and Olds, K. (1996) 'Refiguring the economic in economic geography', *Progress in Human Geography* 20, 3: 311–37.
Underhill, G. (1994) 'Conceptualising the changing global order', in R. Stubbs and G. Underhill (eds) *Political Economy and the Changing Global Order*, Basingstoke: Macmillan.
Whatmore, S. (1994) 'Global agro-food complexes and the refashioning of rural Europe', in A. Amin and N. Thrift (eds) *Globalization, Institutions and Regional Development in Europe*, Oxford: Oxford University Press.
—— (1997) 'Dissecting the autonomous self: hybrid cartographies for a relational ethics', *Society and Space* 15, 1: 37–53.

12

REOPENING TOTALITIES

Venezuela's restructuring and the globalisation debate

Lourdes Gouveia

INTRODUCTION

Ever since globalisation became a buzzword in the popular media and an object of inquiry in academic writings, many have sought to warn us about its overstated arguments for epochal change (Gordon 1988), its empirical and regulatory limits (Hirst and Thompson 1996), or its bigger-than-life ideological discourse (Lie 1996).

These misgivings are also evident in a growing segment of social science literature concerned about recent transformations in the agro-food sector.[1] Most recently a growing body of works identified with the actor-network theoretical perspective (ANT) of writers such as Latour (1993), and confined largely to European communities of scholars responding to the hyperstructuralist writings of individuals like Althusser, has been challenging many of the assumptions associated with a political economy of globalisation theory. The first part of this chapter takes a cursory look at some of the conceptual and methodological issues that seem to divide us. In the second part of the chapter, I review various components of an empirical case study of current restructurings in Venezuelan society, and the agro-food sector in particular, in the spirit of dialoguing with other authors contributing to this volume and exploring the extent to which our analyses are truly conflictive or complementary.

The rejection of the totalising theories of modernity

The theoretical and methodological impasse between ANT and political economy of agro-food systems is most apparent in discussions about the imprecise definition of globalisation and what constitutes global and local and their connections. I will return to these later in the body of the chapter. But at this point it is important to point out that behind these immediate conceptual differences there lies a more fundamental impasse regarding the appropriate role of social theory and the particular questions which it can answer adequately.

At the risk of oversimplification, I would suggest that most globalisation theories, but especially those that follow closely in the footsteps of modernist thinkers such as Marx and Weber, continue to believe in the capacity of social theory to reveal a set of patterned interactions that eventually emerge into dominant and coherent structures and processes and become the markers of particular epochs, regions, or for that matter, 'agro-food regimes'. Within this tradition, social theory's central objective, however, is not simply to trace the contours of a particular epoch such as 'the emerging global order' or the crises that may account for its emergence, but also to reveal the social and political hierarchies that differentially benefit from these dominant structures and suggest where the potential for emancipatory practices may lie (Layder 1994). More often than not, actor-network theory, although formally breaking with much of postmodern theory, retains, among other things, the postmodern opposition to epochal or systemic modes of theorising. In this context, globalist perspectives are generally faulted by ANT for elevating a selective and disparate concatenation of allegedly 'global' actors, processes, and structures to a totalising and overly-deterministic theory of social change (see various contributions in this volume, and Amin and Thrift 1994).

There is no doubt that globalisation approaches, especially during early attempts to come to grips with the rapid changes of today's world, have often fallen into the same sort of evolutionistic arguments, static formulations of rigid historical divides, and an undifferentiated concept of a historical subject to which modern classical theory in general and vulgar Marxism in particular has periodically succumbed (Antonio and Kellner 1994). These orthodox accounts, whether in academic or popular mediums, eager to articulate and sometimes celebrate an emerging era, have in fact trampled over and minimised local histories and the diversity of social agents and their points of resistance, and have over-emphasised historical ruptures while downplaying continuities. Therein lies the strength of theoretical approaches such as actor-network theory which force us to zero in, with 'high-resolution' power, on the multiplicity of actors, often hidden from view, and their minute points of 'performative' and frictional connections along rather seamless global–local networks. As Norman Long and others have explained, emerging social patterns are not the result of an inexorable logic of capitalism, but the creation of a multitude of actors or 'active participants who process information and strategise in their dealings with various local actors as well as with outside institutions and personnel' (Long 1994: 21). The result, with which I concur, is a view of globalisation as partial, contested, and by no means inevitable.

The question still remains, however, as to how far we carry our rejection of all attempts to delineate the contours of a social epoch. Can we avoid exaggerating the solidity of our social structures, as Latour (1993) argues theories of modernity have done, and still speak reasonably of relatively resistant, durable, macro-institutional arrangements which give a certain degree of order and predictability to, as well as have a profound impact on, people's lives for generations to come? Can we be sensitive to the entire plethora of a cast of actors involved in social transformation and yet retain the contributions social scientists have made about theorising actors' positions in the social hierarchy and revealing the consequences of their power differentials within the resulting institutional arrangements? Is there a danger, as Booth (1994) warns us, of converting our eagerness to document the heterogeneity of actors 'on the ground' into an end in itself, thus suppressing the significance of theorising about social hierarchies or regularised patterns of continuity and change?

The macro–micro issue and the problem of levels of analysis

In sociology, macro and micro commonly refer to different levels of analysis and the particular focus of research, or points of entry, as I like to think of them. According to Layder (1994), micro social analysis has traditionally been associated with a focus on face-to-face interactions while macro analysis focuses on the study of more impersonal phenomena such as institutions and reproduced patterns of differential power and resources. This macro–micro distinction has long been part and parcel of a more general attack on the rigid dualisms which dot the social science literature. Those working from one position or another seldom proceed to conceptualise the linkages, and thus the tendency has been to stay at one or the other level of analysis. This has clearly been a problem within globalisation studies which have tended to concentrate on the 'macro' or global institutions and regulatory structures alleged to be emerging in today's society, while often ignoring diverse 'micro' or local responses and institutions.

Latour's view is that we abandon these dualisms which tend to privilege those institutions and actors associated with macro structural arrangements. Instead, we should think of globalisation as longer networks that can be carefully traced 'as one looks at gas lines and sewage pipes', and which are therefore neither local nor global (Latour 1993). For Latour, the main differences between networks stem from the fact that lengthened networks are more 'hybrid' – made up of a higher mixture of humans and humble non-humans such as documents, laws that must be mobilised to reach beyond one's locality.

This preoccupation with an adequate theorisation of the linkages between macro and micro, global and local, is evident also in an increasing number of works informed by political economy perspectives. McMichael (1996), for example, argues that these should be viewed in relational terms, mutually conditioning, and most importantly, continually changing. However, McMichael retains the sociological distinction between the macro, global processes and the micro as the 'local face' throught which global processes are realized and expressed. Latour (1993) adamantly opposes this emphasis on scale (of TNCs, of institutional arrangements) as relevant and criticises our 'habit of looking at events from contexts'. Instead, he proposes the 'middle' or the network, the seamless linkages, as the entry point for our analyses.

ANT's conceptualisation of global–local as single networks is an important one. However, I am again concerned as to whether this tracing of organisational networks 'without changing levels' becomes an end in itself. Starting 'in the middle' is an intriguing notion for field research. In general, ANT's research orientation is a corrective to political economy's tendency to hide the multitude of actors found at the micro level. On the other hand, ANT's mode of argumentation and research foci tend to undermine the value of macro level analysis of institutionalised social arrangements (Arce and Marsden 1993). Today, the goal of interrogating our historical cases for evidence of larger scale institutional and hierarchical arrangements, be it the state, multilateral agencies, or capitalism as a whole, and the impact they have on our lives, appears particularly relevant. While it may give us great comfort to discover that TNCs are made up of humble humans, and that capitalism has been rather inadequate at transforming entire social entities in its image, the role such forces play in producing environmental degradation and social polarisation must remain an important research focus. It is not sufficient to trace the length or reveal the hybrids that make up longer networks. It is also necessary to think about the consequences of the 'width' of particular networks and the differential capacity of different sets of actors, situated along the way, to cast a net over multiple populations.

In this sense, I am inclined to adopt a view of macro–micro and global–local linkages that is closer to the relational approach suggested by McMichael. I would insist, however, that we treat these as heuristic tools or entry points, not definitive moments, in our research. Similarly, I think we must further problematise the issue as to what constitutes 'context' and what constitutes local actions and treat these as reversible at any time, depending on the vantage point from which we are conducting our particular research.

A final approximation to the definition of globalisation

I should make clear from the start that I don't have my own definitive definition of globalisation. Instead, I suggest a series of approximations, once again relying on and beginning with McMichael and his view of globalisation as a political project. Such a notion is implicitly sensitive to the question of agency so important to ANT, but retains the political economy agenda of identifying patterned regularities emerging from this process. According to McMichael (1996: 31), globalisation is a historical project that seeks to stabilise market contradictions for corporate global expansion. An incipient ruling class of global managers represented in organisations such as regional free trade agreements, the International Monetary Fund (IMF), and the World Bank (WB) sought the opening provided by the debt crises of the 1980s to reconstruct national and global monetary relations and reformulate the role of the nation state away from the 'developmentalist project' of the previous era.

The notion of globalisation as a political project, as I will illustrate later, is useful for examining the recent processes of state and economic restructuring in the specific case of Venezuela and of Latin America and other debtor nations as a whole. However, the Venezuelan case also supports the ANT view that a multitude of actors, beyond MLA and state managers, are involved in reformulating what is essentially the relationship between the market, the state, and civil society (Peet and Watts 1993). Therefore, it is best to problematise this notion by thinking of multiple globalisation projects, without assigning any of them *a priori* significance, and then examining the manner in which they intersect, modify, or contradict one another. This way of conceptualising globalisation retains the significance of differentiated social actors negotiating and coping with current changes, while reaffirming the empirical reality of increased transnationalisation of relations, practices, and organisations (Lie 1996). The extent to which particular elements of these global relations and projects cohere into identifiable, durable patterns of social organisation can best be determined through a continuous dialogue between empirical investigation and our analytical constructs.

Finally, a word about the importance of examining globalisation as discourse. Like the modernity discourse of the developmentalist era, the neoliberal discourse of globalisation represents the 'narrative' component of a political project to reformulate social relations and material practices. It is important not to underestimate the extent to which these discourses constitute a sort of 'internal colonisation' which ultimately precludes us from envisioning and articulating alternative projects or 'modes of ordering'.

THE CONTINUING RELEVANCE OF POLITICAL ECONOMIC APPROACHES: AN ILLUSTRATION

Multilaterals, TNCs, and local actors: shaping the globalisation project(s) inside Venezuela's food and agricultural sector: the methodology

Venezuela represents an interesting case study of what has turned out to be a highly combative and contradictory process of transition towards political and economic arrangements consonant with the central tenets of a neoliberal globalisation paradigm. It is also a relevant case study for illustrating the usefulness of a political economy approach that emphasises the study of broad social-organisational changes but which, at the same time, incorporates insights from, or is consonant with, actor-oriented approaches.

The case study summarised here is a composite of three separate moments of an investigation conducted from 'different points of entry'. This concerns the restructuring of the state as well as the agro-food sector following the eruption of the Latin American debt crisis in the 1980s.[2] We began the investigation at the 'macro' level, by focusing on the restructuring of the Venezuelan state and the economy in this context of crisis and the creeping presence of multilateral institutions such as the IMF and the World Bank (Llambi and Gouveia 1994). The magnitude of the economic crisis in Venezuela and its oil-dominated economy has kept foreign investment to relatively insignificant levels and thus TNC participation far in the background of the economic restructuring process. It has instead been incumbent upon MLAs to assume the more visible and substantial role of 'transnational agent' of reform. We relied heavily on document searches and media coverage to trace the evolution of the Venezuelan state reform process and the key actors associated with its various versions.

In the next moment of our investigation, we utilised a combined 'macro–micro' approach as points of entry into the study of the interaction between global changes, state and economic restructurings, and transformations occurring at the level of rural municipalities dependent on export-oriented commodities (cacao and horticultural commodities) – the mainstay of neoliberal prescriptions for economic development. The research also included Colombia as a comparative case.[3] Similar to the methodology suggested in ANT, we made a serious attempt to trace the various actors situated along more or less connected spheres of action and their differential capacities to negotiate with their changing environment. We relied heavily on the collection of field interviews and observations combined with documents and statistical analysis of aggregate changes at the international, national, and local levels. We also relied on 'relational' constructions of macro and micro, global and local, as alternative and heuristic units of analysis. The cacao world market, regional free trade agreements, and the IMF's structural adjustment package are all forms of regulatory structures which may be conceived far from the Andean communities of Colombia and Venezuela, but their impact reverberates throughout. In this sense, restructuring constitutes a valid, and relationally constructed, 'macro' context within which to observe transformations at the level of these communities.

By the same token, however, we assumed and thus explored the presence of influences emanating and flowing upwards from the 'micro' level of the social-historical conditions and political histories of the rural communities under study, as well as from the biophysical particularities of their local production systems. The assumption here is that, historically, it is largely these types of relevant local dynamics and traits that are constantly informing, modifying, or undermining the larger regulatory structures to which they are more or less connected – albeit in more dispersed and often less tangible and visibly patterned ways. The very process of political decentralisation and erosion or modification of the space occupied by the central state in the previous era of developmentalist politics has made such influences more transparent. In this sense, 'face-to-face' observations at the 'micro' level of the rural communities constitute a rich environment for simultaneously exploring how local conditions change in the context of structural adjustment, but importantly, how these local conditions and histories are also part of the context within which the political project of globalisation takes shape.[4]

Finally, recent participation in the design of a food policy study allowed me to get a glimpse, at the 'micro' level of the local office of a multilateral agency and a Venezuelan government ministry, of how these organisations' 'social and institutional embeddedness' becomes an added set of forces determining the degree of success with which MLAs 'enrol' nation states into the neoliberal globalisation project.

The findings

In our 'macro archeology' of policy-formulation doings, we found, first of all, that the initial effort to reform the state was an 'endogenous' project, formulated in the early 1980s by a group of business representatives and young 'technocrats' within the public sector. The group eventually became a formal Commission for the Reform of the State or 'COPRE'. Its major concern was to deal with the state's 'inefficiency', clientelistic practices and the generalisation of corruption, which together were rapidly eroding the legitimacy of the entire political system. At this point, Venezuela's 'mixed economy' model was not in question, but COPRE's reform project also was not immune to the increasingly popular neoliberal notions, as was evident by its support for gradual processes of political decentralisation, a smaller 'efficient' state, and economic opening.

However, an impending debt crisis loomed large behind these narrow political reforms, auguring the end of import substitution, or ISI, the bedrock of developmentalism in much of Latin America. In Venezuela, ISI was built upon exchange and price controls, a protectionist commercial policy and a highly interventionist state subsidised by generous oil rent. This role was evident, for example, in the agricultural, food, and rural development sectors where the state had sponsored a massive land and agrarian reform in the early 1960s, a cheap food policy to subsidise industrialisation, and a complex system of cheap agricultural credit and extension services. As oil prices slumped and debt mounted, state officials found it increasingly difficult to maintain this costly interventionist role while ignoring, aiding, or abetting the looting of public and private funds and their transfer to overseas financial institutions which had become rampant as the years went by.

In this context, and that of increasing social polarisation and unrest, state reformers began to reformulate the restructuring project, moving it further away from ISI and the mixed economy model that sustained it. In 1983 the government loosened exchange controls, devaluing the currency 170 per cent, and took some modest steps towards a free market and an open economy (Llambi and Gouveia 1994).

Clearly, state restructuring threatened domestic producers and agro-industries which for years had enjoyed import subsidies for various inputs and protection from competing imports. Actors within this sector fought hard, and often successfully, to re-instate state supports and thus redirect the reform project. By the same token, however, this process opened unintended spaces for grassroots and local political organisations in rural, and also in urban communities. Depending on their own political histories and organisational capacities, local community actors appropriated and amplified the political decentralisation and diminished state intervention discourse accompanying the restructuring process. Decentralisation was and is also a central component of the MLA's restructuring project articulated later, and this fact has strengthened the calls for a political devolution among groups from all sides of the political spectrum in Venezuela today.

Local governments and newly emerging civil society organisations began to demand, often successfully, more control over government budgets and electoral processes as well as formal representation of these grassroots organisations in policy-making arenas. The form and intensity this takes was most clearly observed in the micro level studies conducted in two rural Venezuelan municipalities (and, comparatively, in Colombia as well). For example, in Yaguaraparo, the Venezuelan cacao-producing municipality, the end of the state marketing board monopoly has been translated into an increased capacity by local producers to negotiate prices among a growing number of marketing houses. Politically, former peasant organisations tied to the two major parties via an entrenched patronage system are being replaced either by opposition party structures buoyed by their new control over financial resources, or by local, albeit incipient, NGOs and Juntas de Vecinos (neighbourhood groups) attending to a more diversified set of social and economic issues. A local leader expressed it this way:

> The population is like . . . awakening. All this we are living right now is only like four or five years old . . . like the fight for human rights. There is emerging among the people the idea that we must have some say in all of those things that were like dormant.
>
> (Gouveia 1996)

Similarly, in the Andean and potato producing municipality of Pueblo Llano, Venezuela, newly formed 'irrigation committees' have taken over much of the sophisticated irrigation system previously financed and operated, and now severely neglected, by the central state. Later in 1993, when

Venezuela signed a free trade agreement with Colombia in preparation for US-supported NAFTA, Pueblo Llano producers publicly stated this was an agreement signed 'behind their back' and literally modified it at the 'local/trans-border' level through an alternative agreement of 'managed trade' signed in conjunction with their Colombian counterparts. Globalisation, as Long (1994) suggests, is 'relocalised' or reworked at the local level, thereby constantly changing what we mean by globalisation processes (Llambi, Gouveia and Arias 1995).

We labelled the conglomerate of these alternative practices the 'bottom-up' version of the endogenous state restructuring project (Llambi and Gouveia 1994). By 1989, just as the friction of these two versions was beginning to play itself out, the debt crisis accelerated and a newly elected administration signed its first letter of intent with the International Monetary Fund (IMF). The Government promised to submit the country to the standard recipe of structural adjustment measures and the export-oriented development guidelines deployed throughout Latin American and other nations as a condition for debt restructuring. We labelled this the 'exogenous restructuring project'. Power suddenly seemed to have moved upwards, towards transnational organisations and their neoliberal free market agenda.

Once again, the implementation of these measures has proven to be more elusive than multilateral agents probably ever envisioned. As we have documented elsewhere, both the social explosion and incontrovertible social costs brought about by structural adjustment measures led to substantial back-tracking on the part of the national government, such as the reinstitution of subsidies and price controls (Llambi and Gouveia 1994). The MLAS also took corrective actions: the IMF enrolled the assistance of the World Bank (WB) and the Inter-American Development Bank (IDB) to implement a series of compensatory programmes which have now accompanied most structural adjustment programmes in Latin America – and whose objectives are crafted purely in terms of moving forward those programmes or the neoliberal agenda as a whole. Most of these were designed as short term, essentially safety net programmes, to buffer the social costs of structural adjustment to the poorer sectors of the population. Others were designed to restructure the administrative and productive components of economic sectors which had been left out of the macroeconomic adjustment measures of IMF. The latter was critically important in the case of the agriculture and food sector, which severely affected rural producers. Moreover, the sector's negative growth, the poor performance of export-oriented 'non-traditional' commodities, and the continuing high cost of the sector's public administrative component, threatened to undermine financial stabilisation and debt repayment which were at the core of the IMF project.

But even the implementation of these 'kinder and gentler' compensatory programmes has proven to be nothing but messy, contradictory, and often ineffective (judged from various vantage points). As a frustrated World Bank official recently put it: 'Venezuela has the worst record of any country regarding disbursement of funds from World Bank loans' (*El Universal* 1996). Since the 1990s, bank programme staff, for example, conceived an elaborate programme for the overhaul of the agricultural and food sector according to neoliberal principles of economic opening and comparative advantage and the elimination of 'food self-sufficiency' as a worthy goal of development.[5] This was titled the Agricultural Sector Investment Loan Program, or ASIL. More recently it approved a loan for an agricultural extension project designed to help smaller producers make the transition to a free market and open economy. ASIL never got off the ground and the Venezuelan government paid annual interest on a $600 million loan whose actual disbursements were negligible and which it has now been asked, by bank officials, to return. The extension project is behind schedule. A similar fate has recently befallen loans for social safety net (food and nutrition) programmes for the poor and the corollary restructuring of government ministries charged with their implementation (Garcia 1996).

WB and IDB officials and their teams of consultants are invariably heard – not always publicly – blaming inept state managers and the lack of technical and administrative capacity of local bureaucracies for their failure to meet, with the required speed, the 'explosion of conditionality' introduced by the various multilaterals: 'a typical five year project can take up to twenty-five years in this country' (*El Universal* 1996) The reasons, however, are more complex. First, there are serious splits within and among the WB and IDB and between them and their local government counterparts regarding the form restructuring and accompanying compensatory programmes should take, and regarding the methodologies used to measure their progress. In this context, reforms are not only slow to start, but to the extent that they take place, no one can say with much certainty whether they are complying with multilateral cross-conditionality, and ultimately adhering to the general principles of a neoliberal globalisation project.

Second, and according to a member of the technical staff in the ministry charged with social programmes, 'for the members of the multilateral organisations, the experience in the social program area is also very limited, which complicates the operations' (CIES 1996: 9). Putting it simply, high-ranking multilateral bank representatives are often uninformed, if not outright incompetent, about the local administrative, legal, and even cultural norms which must be followed in order to process a particular loan disbursement or negotiate the design of a new study or programme. Other staff members contend that this annoys public servants who subsequently become less inclined to cooperate with local bank representatives. Evidence of this was that the day after local newspapers published the bank officials' statements cited above, government offices were 'instructed' to stop all negotiations with bank representatives. This cold-shoulder treatment forced the author of those statements to 'clarify' them in a letter to government ministries and cabinet members in an attempt to reaffirm the good relations that exist between the bank and local ministries.[6]

Third, beyond these micro squabbles and alleged methodological differences, disagreements about reforming the state according to minimalist criteria also result from purposeful resistance strategies by legislators, state managers as well as NGOs and their political clienteles or allies. Analysts often attribute these resistance strategies to state managers' need to preserve political constituencies and, naturally, their bureaucratic positions. But there are also profound ideological differences among factions within the state and, to a much lesser extent, within the MLAs. Those differences are evident, for example, in programmes for the agriculture and food sector; the official bank position is to reject, while many state actors support, food self-sufficiency as a desirable goal to strive for (Llambi and Gouveia 1994, World Bank 1996b). The same is evident in the more recent compensatory food and nutrition programmes for the social sector. Official bank policy views these programs as short-term assistance measures to be applied during the transition towards a free market economy. Conversely, among Venezuelan policy makers and local analysts, 'There is a wide consensus regarding the fact that to attack poverty in [these] countries . . . requires development policies which integrate the economic and the social and not simply the instrumentation of isolated projects targeting the poor' (Cartaya 1996). Some bank representatives and official policies have begun to move closer to this view as the former gain first-hand knowledge of the severity of structural poverty in Latin America (Tapia and Biton 1995, World Bank 1996b). However, this places them in the odd position of having to support a continuing, and often expanding, role of the state in the management of economic relations, which is anathema to the neoliberal project.[7]

Discussion and preliminary conclusions

The above paragraphs support many of the views held by those writing from an ANT perspective, including Whatmore (this volume). It is evident, for example, that MLAs, like TNCs, are less than formidable, socially and institutionally embedded institutions, and that their performative capacities are largely determined by the 'durability' and length of their networks. In this context the difference between MLAs such as BID, which has a long history of participation in Latin America programme lending and policy reform, and the World Bank's shorter history and dubious success record is worthy of consideration. Hence, the foregoing analyses does defy the image of globalisation as a coherent project and suggests, instead, one of a highly contingent and contested process.

Finally, the analysis is also consistent with the central hypothesis of this research which assumed globalisation to be not one, but multiple projects of global restructuring and 'orderings' initiated at various levels, and whose outcome is more heterogeneous and uncertain than orthodox global political economic views might have it. Our research in rural municipalities, for example, revealed that the impact of restructuring on local producers is far from homogenous and is mediated by a number of factors, ranging from local knowledge and resources to the historical continuity of the institutional arrangements under which traditionally globalised commodities such as cacao have been organised (Llambi, Gouveia and Arias 1995).[8]

However, from a (non-orthodox) political economic perspective the analytical task does not come to an end with the discovery of diversity, heterogeneity, or the fact that all actors have some degree of power. At the risk of offending post- or anti-modern sensibilities, it is important to complete the analytical loop and return to the macro level for a simultaneous interrogation of data and historical constructs to determine whether, despite diversity, broader socio-structural changes can still be identified. In other words, despite the rough and tumble of restructuring, has there been a major reorganisation of political and economic arrangements consonant with a neoliberal globalisation project? If so, what are the long-term consequences of this reorganisation for local populations? Is the importance of 'scale' (not only the length, but the width) of transnational networks and institutional arrangements, and of powerful actors within them, understated by ANTs? The macro level may not be as 'interesting' as the middle but I believe its investigation can yield additional analytical insights.

The broad outcomes of Venezuelan restructurings

Explicit efforts to enrol Venezuela in the globalisation project have generated a plethora of heterogeneous intended and unintended impacts. MLA-led restructuring generated 'counter-reform' movements within and outside the state apparatus. However, none of them coalesced into an alternative and coherent national or regional project designed to alleviate the crisis. On the contrary, they mostly exacerbated it, generating a series of devastating social and economic consequences. With its 'tail between its legs', the counter-reform Caldera administration, after being elected on the promise that it would never negotiate with the IMF, signed a new letter of intent with this MLA in March 1996 and promised to move the stalled, but never totally interrupted, neoliberal agenda forward. Venezuela braced for a second round of structural adjustment measures at double the cost and under still stricter cross-conditionality requirements among the various MLAs.

Changes resulting from the adjustment process are particularly salient at the macroeconomic level. The IMF, for example, has succeeded in its push towards liberalisation of exchange rates, price controls, and trade policies aimed at debt repayment and, relatedly, the creation of a foreign investment-friendly economic environment (Llambi, Gouveia and Arias 1995). Debt repayment has continued more or less on schedule, but the same cannot be said for the growth of productive foreign investment which has largely stayed away due to uncontrolled inflation and an explosive political and social climate (*Boletin Agroplan* 1996a). On the other hand, the internationalisation of Venezuela's economy is taking place in quieter ways as financial sector deregulation creates a more attractive climate for non-productive investments travelling through the hypermobile international capital markets.

Macroeconomic adjustments and their failure to restart productive sectors on their own has led to a rapid deterioration of the social sector and of non-financial sectors such as agriculture. Poverty estimates range from 50 per cent to 80 per cent; the basic food basket costs about 75 per cent of average incomes, and agriculture's rate of growth continues to decline with few exceptions (*Boletin Agroplan* 1996b). In this context, cries for state reforms, compensatory social programmes, and sectoral policies for agriculture and other economic sectors have gathered force within both MLA and state managers. Paradoxically, the strong presence of MLAs in this process, according to state reformers and a new wave of progressive sociologists hired to work in the reform, has served to uncover the high degree of government inefficiency and corruption. This, along with a deepening of the economic crisis in the late 1990s, has in turn furnished transnational reformers and their local allies with the high degree of consensus necessary not simply to reform the state, but to reformulate its mission as a promoter of market forces – a consensus that was noticeably lacking in the previous period of adjustment.[9]

As suggested earlier, the results of state restructuring appear as uneven and chaotic. However, the fingerprints of the neoliberal agenda can still be detected. For example, the state has reduced its presence in many areas. More often than not this is due to the fiscal constraints resulting in part from structural adjustment rather than to some coherent plan crafted to comply with MLA cross-conditionality (no WB or IDB loans are disbursed until the IMF's macro structural adjustments have been made). Decentralisation, by default or by design, as I illustrated earlier, is in many ways a positive outcome of state restructuring. But before we celebrate too loudly, it is important to look at the costs as well. In the potato-producing community of Pueblo Llano, for example, local farmers eager to increase their income in the context of high inflation and a domestic boom market for horticultural products have embraced the opportunities directly or indirectly opened by the neoliberal agenda and increased their production of specialised cash crops – potatoes and carrots in this case. But the end of agricultural subsidies and the retreat of agricultural extension services from their communities, together with the increasingly unregulated entry of agricultural inputs – all part of this agenda – has led them to adopt cheaper and poorly managed organic fertilisers, and the indiscriminate use of agro-chemicals now crossing freely from Colombia. Potato-processing agro-industries in Colombia and agro-chemical companies in Venezuela are becoming the de facto extension agencies and their main interest is simply, and not surprisingly, to increase productivity and sales (Llambi, Gouveia and Arias 1995).

Similarly, in the case of the cacao producing community of Yaguaraparo, the opportunities afforded to the local peasants resulting from an improvement in international prices and deregulation of the sector has resulted in increased cacao production and peasant control over production and marketing decisions. But, in the same context of a diminished state presence in areas like extension and credit, smaller producers appear vulnerable to TNCs' progressive control over productive and marketing activities. In Venezuela, Nestlé recently purchased an old chocolate factory landmark, *Savoy*. In less than four years it has come to dominate 55 per cent of the cocoa processed in the country. Also, drawing on its formidable resources and experience in global markets, it has quickly moved to the forefront of organisations taking advantage of recent regional free trade agreements. Its long-term strategy is to move into production via subcontracting with the most efficient farmers who can produce under the high quality standards expected by the company. The *Wall Street Journal* recently celebrated Nestlé's global strategy, noting it ranked number one in a UN survey of the top 100 companies with global exposure: 'There is one thing that characterises Nestlé. It dominates its markets. . . . And it can use its muscle as the world's largest food company to stand up to retailers, outspend the competition and, with persistent advertising, change consumer habits' (Steinmetz and Parker-Pope 1996). This case seems to lend some credence to the fact that, however imperfectly, neoliberal policies contribute to the opening of investment fields for TNCs.

Finally, the compensatory and agricultural sector programme loans and the delays associated with their troubled implementation have led, among other things, to budget increases and new layers of actors involved in these programmes with little to show for this effort. MLAs' own inefficiency and their need to build alliances with local actors (lengthen their networks) often contributes to the chaos and reproduces the very clientelistic practices, ad hoc policy making and, subsequently, unequal distribution of resources that it is supposed to eliminate. Again, MLA programmes may often be incapable of effecting desired changes. But the costs to the local population of these recalcitrant attempts to reformulate the state and the policy foibles that accompany them, are not negligible. And they are not simply material costs. The capacity of MLAs and their allies to return to the drawing board and insist upon a neoliberal discourse over and over again, eats away at other actors' capacities to resist and to articulate alternative views of development.[10]

In conclusion, the success of economic and political restructuring according to a neoliberal globalisation project may be uneven and at times outright ineffective. But as we contemplate the Venezuelan case, there is little doubt that a visible institutional rearrangement has taken place, and that, however ineffective, when judged by its officially stated goals, the entire population is deeply affected by the reverberations of the process – whether it is as innocent bystander or as active agent, negotiating with it, reformulating it, or resisting it. Similarly, local producers and TNCs may be part of the same 'lengthened network', but their material and non-material resource differentials, i.e. their power differences, should not be underestimated. Counter-globalisation movements and 'alternative geographies of food' are important responses. However, it seems highly unlikely that either the trickle of NGOs that make their way into these areas or the current growth of local organisations are sufficiently numerous, fast, or even dedicated to fill the gaps left by a reorganised and diminished state and catch up with the devastating effects of neoliberalism in Latin American countries.

I believe our task as researchers is to continue to problematise the dynamic that produces various assortments of winners and losers and simultaneously to highlight the ongoing erosion of rights for populations all over the world within the current context of transnationalisation of economic and political arrangements. Additionally, we must remain vigilant to the dangers posed by political devolution[11] and the nation state's shedding of responsibilities for the more vulnerable sectors of its population.

FINAL THOUGHTS ON THEORY AND GLOBALISATION

Globalisation may not represent a tidy coherent project, and multiple 'counter-globalisation' movements may undermine some its ideological underpinnings and institutional basis. But the neoliberal version is a powerful project nonetheless, and one whose institutional and material practices are far from negligible. Additionally, the agents of globalisation such as TNCs and MLAs, or the fluid and ghostly network of hyper-mobile finance capital, may by themselves not seem so formidable. But when their actions overlap into long and wide 'networks of power' and reinforce each other's neoliberal agenda, they can indeed render helpless local peasantries and undermine alternative geographies of food.

There is no question that we should be mindful of deductivist and orthodox characterisations of globalisation as an inexorable force that explains all of social reality today. But at the same time, social scientists have the responsibility to retain as an important research focus the issue of how, and at what cost, intensifying transnational relations and powerful actors are involved in transformations of the relationship between the state, the market, and civil society.

In this same line of thought, I believe the wholesale attack on modern theories and political economy is at times as overstated in ANT as it is in postmodernist and poststructuralist approaches. As Antonio and Kellner point out (1994: 131–32) most writers in this tradition understand that broad-scope theories always depend on especially precarious and imperfect balancing of the general and the particular and that they never capture the social world in all its richness, particularities, diversity, and temporality. Moreover, one should not underestimate Marx's and other modern thinkers' contributions to continuing to understand capitalism and its polarising force, which even Latour suggests is more than just another lengthened network (Latour 1993: 121, Kellner 1995). If this is so, we should be wary of abandoning our right to become indignant about its human costs, and remind ourselves that the critique of society is a rare commodity traded among a small community of social scientists.

One danger of ANT methodological orientation is that of forgetting the lesson learned from other communities of scholars about the tendency to study the less powerful, and their capacity to articulate alternative projects, while suppressing the need to keep tabs on those who wield disproportionate power in society. In the same vein, ANT may inadvertently strengthen neoliberal visions of a society made up of free-choosing and powerful actors and undermine the study of the constraints and 'frictions' within social networks while underscoring the connectivities.

NOTES

1 See Goodman and Watts (1994) for a strong critique. But even among those of us writing from a political economy of globalisation perspective, there are disagreements as to how far we can stretch a homogeneous view of globalisation as an explanation for change in various aspects of the agro-food sector (see for example Bonnano *et al.* 1994).

2 The investigation is the result of several years of collaboration with Luis Llambi and his research associates from the Instituto Venezolano de Investigaciones Cientificas (IVIC). The summary presented here owes much to their work and leadership in this process.

3 The complete outline and results of this investigation are summarised in the final report submitted to the North–South Center, from the University of Miami which funded the project (see Llambi *et al.* 1996). For partial analyses based on this study see also Llambi, Gouveia and Arias (1995), Llambi (1996)

4 Many authors have insisted on this relational approach to the study of macro–micro processes. In order to treat this seriously it is necessary to adjust our methodologies and expand our field studies in ways that allow us to complete the 'dialectic' and seriously investigate multidirectional influences between macro and micro.

5 The rejection of food self-sufficiency and endogenous food security as a programatic goal is a constant theme in bank documents and internal evaluations of project designs often disallow the infiltration of such notions (World Bank 1996b).

6 Personal observations, summer 1996. See also Mosley, Harrigan and Toye (1991), for an excellent set of in-depth case studies illustrating the same kinds

of contradictions and ultimately limited implementation of the WB's policies in loan recipient countries.

7 Several authors have referred to this apparent paradox between neoliberal notions of a minimalist state and the need MLAs have to strengthen state institutions to carry out this profound restructuring process as well as to meet unexpected challenges from the social base (Llambi and Gouveia 1994, Torre 1995, Tussie 1995)

8 Cacao is one of the oldest global commodities. The sector has historically been organised along the lines of an international division of labour whereby the riskier and more costly production tasks take place in less developed countries and value-adding processing and marketing takes place in industrialised countries. An oligopolistic structure of TNCs such as Nestlé control the latter phases (Gouveia and Llambi 1995).

9 Personal conversations with social programmes staff during August and September 1996.

10 I observed how, for example, in order to move their agenda forward, representatives from the different MLAs must ally themselves with and thus reproduce the presence of different state factions.

11 For an interesting essay on the dangers of devolution at a moment when capital power has increased dramatically, see Chomsky (1996).

BIBLIOGRAPHY

Amin, A. and Thrift, N. (1994) *Globalisation, Institutions, and Regional Development in Europe*, Oxford: Oxford University Press.

Antonio, R. and Kellner, D. (1994) 'The future of social theory and the limits of postmodern critique', in D. Dickens and A. Fontana (eds) *Postmodernism and Social Inquiry*, London: The Guilford Press.

Arce, A. and Marsden, T. (1993) 'The social construction of international food: a new research agenda', *Economic Geography* 69, 3: 293–310.

Boletin Agroplan, (1996a) 'Estamos mejor pero no vamos bien', July/August.

—— (1996b) 'Se requiere una nueva política comercial para la agricultura', July/August: 14–17

Bonnano, A., Busch, L., Friedland, B., Gouveia, L., and Mingione, E. (eds) (1994) *From Columbus to Conagra: The Globalisation of Agriculture and Food*, Lawrence: University Press of Kansas.

Booth, D. (1994) 'How far beyond the impasse? A provisional summing up', in D. Booth (ed.) *Rethinking Social Development: Theory, Research and Practice*, Essex: Longman Scientific and Technical.

Cartaya, V. (1996) 'Un ataque al presupuesto familiar', *CIES Boletin* second quarter.

Chomsky, N. (1996) 'You say you want a devolution', *The Progressive* 18 (March).

CIES Boletin (1996) 'Entrevistas: Ser osados para ser productivos', second quarter.

El Universal (1996) 'No se usa el 85 por ciento de prestamos otorgados por el Banco Mundial', 18 August.

Garcia, H. (1996) 'La compensación como la política social', *CIES Boletin* second quarter.

Goodman, D. and Watts, M. (1994) 'Reconfiguring the rural or fording the divide?: capitalist restructuring and the global agro-food system', *Journal of Peasant Studies* 22, 1: 1–49.

Gordon, D. (1988) 'The global economy: new edifice or crumbling foundations?' *New Left Review* 168: 24–65.

Gouveia, L. (1996) 'Globalisation and agrarian restructuring through the Latin American lens: lessons from Venezuela and Colombia', presentation at Cornell University, Ithaca: 26 April.

—— and Llambi, L. (1995) 'Beyond NAFTA: regional integration and the transformation of Latin America's agro-food systems', paper presented at the International Studies Convention, Chicago.

Hirst, P. and Thompson, G. (1996) *Globalisation in Question: The International Economy and the Possibilities of Governance*, London: Polity Press.

Kellner, D. (1995) 'The end of orthodox Marxism', in A. Callari, S. Cullenberg and C. Biewenner (eds) *Marxism in the Postmodern Age*, New York, London: The Guilford Press.

Latour, B. (1993) *We Have Never Been Modern*, Hemel Hempstead: Harvester Wheatsheaf.

Layder, D. (1994) *Understanding Social Theory*, London: Sage.

Lie, J., (1996) 'Globalisation and its discontents', *Contemporary Sociology* 25, 5: 585–87.

Llambi, L. (1996) 'Globalización, ajuste estructural y nueva ruralidad: una agenda para la investigación y el desarrollo rural', paper presented at the Asociación Venezolana de Sociologia Rural y Economistas Agrarios.

—— and Gouveia, L. (1994) 'The restructuring of the Venezuelan state and state theory', *International Journal of Sociology of Agriculture and Food* 4: 64–83.

Llambi, L., Gouveia, L., and Arias, E. (1995) 'Global–local links in Latin America's new ruralities: A case in the Venezuelan Andes', in *Proceedings: Agrarian Questions: The Politics of Farming* vol. 1, Wageningen: Wageningen Agricultural University.

Llambi, L., Gouveia, L., Perez, E., Rojas, M., and research associates (1996) 'Beyond NAFTA: Global–Local Links in the Restructuring of Venezuelan and Colombian Agri-food Systems', final report presented to the North–South Center, University of Miami, January.

Long, N. (1992) 'From paradigm lost to paradigm regained? The case for an actor-oriented sociology of development', in N. Long and A. Long (eds) *The Interlocking Theory and Practice in Social Research and Development*, London: Routledge.

—— (1994) 'Globalization and localization: new challenges for rural research' (manuscript).

McMichael, P. (1996) 'Globalisation: myths and realities', *Rural Sociology* 61, 1: 25–55.

Mosley, P., Harrigan, J., and Toye, J. (1991) *Aid and Power: The World Bank and Policy-Based Lending* vol. 2 (Case Studies), London: Routledge.

Peet, R. and Watts, M. (1993) 'Introduction: development theory and environment in an age of market triumphalism', *Economic Geography* 69, 3: 227–53.

Steinmetz, G. and Parker-Pope, T. (1996) 'All over the map: at a time when companies are scrambling to be global, Nestlé has long been there', *Wall Street Journal* 26 September: R4–R6.

Storper, M. (1992) 'The limits to globalisation: technology, districts and international trade', *Economic Geography* 68: 60–93.

Tapia, J. and Biton, R. (1995) 'The lessons of neoliberal adjustments and the construction of a new agenda for reforms in Latin America: the role of CEPAL and BID', paper presented at the Latin American Studies Association, Washington DC.

Thrift, N. and Olds, K. (1996) 'Refiguring the economy in economic geography', *Progress in Human Geography* 20, 3: 311–37.

Torre, J. C. (1995) 'El lanzamiento político de las reformas estructurales: Bolivia, Argentina, Brasil, Mexico', paper presented at the Latin American Studies Association, Washington DC.

Tussie, D. (1995) *The Inter-American Development Bank*, Boulder: Lynne Rienner.

World Bank (1996a) 'Proposal draft for a rural poverty study' (manuscript).

—— (1996b) 'Minutes from proposal review session' (manuscript).

COMMENTARY ON PART V

Theoretical reflections: transnational capital and its alternatives

Anthony Winson

The two chapters in this section each take a different direction from the crossroads at which the analysis of globalisation is presently situated. One chapter deals with the various shortcomings of the globalisation debate by undertaking a substantial departure from the field, and opening up a different theoretical line of inquiry via actor-network theory. The other stays within the fold, attempting to address shortcomings of past analyses and laying open the possibilities for incorporating some of the insights from the actor-network theory approach of the first chapter.

At a most fundamental level, Sarah Whatmore and Lorraine Thorne's 'Nourishing Networks . . . ' represents a critique of the highly formalised political economic discourse on the globalisation process that has become fashionable in some academic circles. It is an eloquently worded attempt to shift ground to an alternative conceptual apparatus. Actor-network theory is offered as an alternative that would force, it is claimed, 'a challenging and at times disconcerting, shift in the horizons of social research'. Whatmore and Thorne justifiably decry the over-formalised treatises on globalisation that present us with the image of an 'irresistible and unimpeded enclosure of the world by the relentless mass of the capitalist machine'. They challenge us to rethink certain basic formulations that much (but certainly not all) academic discourse on globalisation has uncritically accepted and reproduced. They problematise the 'global–local' and 'macro–micro' dualities, offering instead a 'topological spatial imagination' that focuses on 'tracing points of connection and lines of flow'. They challenge us to examine the always problematic constitution of global networks rather than limit ourselves to concepts of already constituted systemic entities. They encourage us to examine *how* social practices are made durable instead of simply assuming that they are so. Moreover, just as they problematise the global–local binary concept, so too do they question our understanding of society and nature as separate watertight conceptual compartments. They urge us to partake of a more hybrid conceptualisation of the way in which agents – human, technological, and technical, in combination and fashioned in a particular way – interact to produce agency. Finally, they bring to the discussion details of an alternative food trading network as an empirical confirmation of the utility of their approach.

Anyone concerned with genuinely furthering our understanding of the contemporary social world must sympathise with the authors' calling to task much of the language of current discourses on globalisation, a language that often imparts to conceptual entities a certain 'categorical logic' that seems to imply inevitability and to deny alternatives (see Whatmore 1994). A major failing of 'orthodox' globalisation analyses, they contend, is the 'eradication of social agency and struggle from the compass of analysis' and the failure to grasp the partial and contested nature of the process. I find myself in basic agreement with their concerns on this score. We need very much to move towards an understanding of globalisation which is, in the eloquent words of the authors, a process that is 'partial, uneven, and unstable; a socially contested rather than logical process in which many spaces of resistance, alterity, and possibility become analytically discernible and politically meaningful'. It is the tools by which they choose to overcome the perceived weaknesses with which I find myself in disagreement.

The question at hand is this: can one refashion what has often been a hyper-structuralist analysis of the globalisation of the agro-food sector *from within* the political economy paradigm, or does one need a new theoretical approach altogether? Lourdes Gouveia's chapter 'Reopening Totalities . . . ' offers a worthy defence of the political economy paradigm to the challenge offered by actor-network theory. She rightly points out, to my mind, that Whatmore and Thorne's *formal* distancing from postmodern theories still retains 'the postmodern opposition to epochal or systemic modes of theorising'. She justifiably warns us of theoretical approaches that, in privileging the discovery of the heterogeneity of actors 'on the ground', and the contingency of social processes at any given time or any given place, carries with it a real danger. It may in fact suppress the real need to theorise the means by which growing imbalances of market power are being produced, with its implications for social inequality and the integrity of the environment. As Gouveia pointedly says of actor-network theory, it is not going to be sufficient to 'trace the length or reveal the hybrids that make up longer networks', as Whatmore and Thorne enjoin us to do. Rather, we must still think about, and strive to comprehend more adequately, the different capacity and strength of networks, together with 'sets of actors' to substantially limit and determine the actions and life chances of multiple populations.

The task of conceptualising more adequately the role of social agency needs to be discussed more explicitly in analyses of globalisation of the agro-food sector (and discussion of globalisation more generally, I might add). If the postmodern riposte to Marxism was to challenge the privileged position class once had in the analysis of social agency, the status of class in Whatmore and Thorne's formulation of actor-network theory is even more ambiguous due to its virtual neglect. Even Gouveia's analysis, while defending the utility of retaining political economy's emphasis on patterned macro level institutional arrangements or 'regimes', does not treat the matter of class in more than a cursory way. Nevertheless, she leaves the door firmly open, in my view, for a nuanced class analysis to inform the debate on globalisation of the agro-food sector. Hopefully this will begin to occur, since much of the broader discussion on globalisation has been overly deterministic from an economic or technological point of view, depending upon the political orientation of the writer. The issue of social agency must be addressed, and no less so with respect to debates limited to the agro-food sector. Each of these chapters has a different approach as to *how* this will be done.

The debate around globalisation as it pertains to the agro-food sector has strong resonances with earlier debates in the literature on development and underdevelopment in the Third World. Perhaps something is to be learned from that experience.

It may be remembered that the attack on mainstream approaches to development in the Third World was not long in engendering strong criticism itself, but mainly from those within the critical tradition in the social sciences. Criticism of the 'dependency' model of development and underdevelopment as it was expressed in the work of such well known writers as Frank (1967, 1969), Cardoso and Faletto (1971) and Frank *et al.* (1972), and Thomas (1976) has some interesting parallels with criticism of contemporary conceptualisations of globalisation. The model in its various forms was found to be wanting with respect to the units of analysis it took as most significant in understanding the mechanisms by which underdevelopment actually took place. It substituted sweeping historical generalisations for careful historical scholarship that might, and indeed did, uncover previously unrecognised qualitative changes in the social organisation of these agrarian societies. Furthermore, there was considerable debate around the inadequacy of social analysis in the dependency model, and its weakness in theorising what we now have come to refer to as *social agency* and its role in determining the characteristics of broad historical transitions from one society to another. In addition to economic determinism and ahistoricism, the dependency model was criticised for lacking a truly holistic perspective, in that it typically did not adequately theorise the significance of technological, economic, and social change in the more developed colonising countries for determining the agents of underdevelopment in the periphery.

Since those early debates, the political economy of development has been refashioned, largely from *within*, and via a fresh interrogation of Marx's work and revisiting of mid-twentieth-century debates among economic historians around capitalist development in Europe (see, for example, Brenner 1976, 1977, Hilton 1978, Laclau 1971). The new, revitalised political economy of development and agrarian change that has arisen since was characterised by a move away from an overly abstract and hyper-structural model that rarely descended from a macro level of analysis, to detailed historical studies of far more geographically limited units of analysis, including studies of individual agricultural estates or plantations over a considerable time period.[1] Much closer attention was given to the process of local class formation, ideological expression and mobilisation, and inter- and intra-class conflict. As a consequence we now have a much richer understanding of the broad issue of social agency in the development process, a better idea of how what were thought to be ineluctable macro processes were in fact characterised by a notable degree of contingency, partiality, and indeed historical regression at certain times and in certain places. It was these factors which were responsible for the wide variation in social structure and economic development that characterises the contemporary Third World despite the legacy of Euro-centred mercantilism and imperialism that impacted underdeveloped countries from one part of the globe to another.

The temptation nowadays when the political economy paradigm does not 'deliver' is to jettison it entirely in favour of a new approach that typically eschews the analysis of larger structures and processes, and with it the possibility of class analysis as well. Perhaps we can take encouragement from the recent intellectual history of debates on Third World development in believing that the real weaknesses of globalisation analyses identified by Whatmore and Thorne can, in fact, be addressed by working within the paradigm. At this point I would like to indicate some tentative lines for this 'renewal' of the debate as it impacts on the agro-food sector, and specifically the more limited domain of transnational capital in the food economy, and emerging alternatives.

POSSIBLE FUTURE DIRECTIONS

Both chapters in this section incorporate empirical case studies to enrich and confirm their argument. This is to be applauded, and would seem to be an emerging trend even by some of the more notable writers on the broad process of globalisation (see McMichael 1996). My strong view is that progress will be made in the area of agro-food studies only by *deepening* this empirical research, and by enriching it with careful historical scholarship. This will be necessary to counter the tendency with much research now situated at the broadest macro levels to make sweeping generalisations on what is all too often an insufficient base of careful empirical scholarship. To give but one example from the Canadian context, it is interesting to note that despite the extremely powerful position of food retail chains within the food economy there, virtually no critical research has been undertaken that might shed light on their role in shaping the Free Trade and North American Free Trade agreements as they affect agro-food, or their strategies in exploiting the free trade zone shared with Mexico and the United States that has now been established. Synthetic studies purporting to clarify the likely future configuration of the Canadian agro-food complex, given the broader global processes at work, and given the present state of empirical research on retailers, are not proceeding on terribly firm ground, and their conclusions would likely be wanting. My suspicion is that much the same could be said for many other national contexts.

I would also argue that in a world of growing social polarisation, the terrain of the agro-food sector has already become, in certain locales, a prime site for the expression of social agency in the new global economy. The class dimensions of the struggles around food are hardly insignificant, and yet they are all too often ignored today. Nevertheless, there are dimensions to new social movements built around food issues that are not reducible to class alone, and this too must be examined. The contradictions that come to light in the food economy of today are often striking and difficult to make sense of. The nature and significance of these contradictions must be explored more thoroughly.

In addition to more serious fieldwork and historical scholarship on agro-food issues, and further emphasis on the question of social agency, the notion that globalisation is a political project, or rather, as Gouveia correctly points out, multiple political projects, must be further examined. Students of the agro-food sector will need to investigate further than they have the role of food transnationals in furthering this project in each country. While globalisation analysis has pointed out, often incessantly, the diminished stature of the nation state today, we all know, if only intuitively, that nations still do matter. Nations have been the sites where the struggle over neoliberal political projects has taken place. There is much work still to do before we understand the role of various actors within the agro-food sector – from transnationals to the various commodity groupings of agriculturalists – in the intense ideological struggle that preceded attempts to introduce various elements of a neoliberal political agenda in one country after another. The struggle over the neoliberal project was an ideological struggle first and foremost, at least in those countries where liberal democracy has a strong foothold. In countries such as Canada the ideological battles were often most heated over debates on how agricultural producers protected by the supply management apparatus would be treated by agreements to liberalise trade. The most powerful news organisation in the country waged a protracted campaign for several years to vilify supply management. While much of the neoliberal agenda is now being completed, the agriculturalists have had some significant victories, and have at least forestalled the full implementation of an agenda generally favoured by the agro-food transnationals. A fuller understanding of the secret of the partial success of this resistance awaits further research.

Finally, and perhaps most important of all, there is a pressing need for critical scholarship on the myriad of ways in which the growing, processing, and distribution of food is being refashioned outside the corporate sector. Several questions come to mind. Is the delinking of food production from agribusiness control with the expansion of organic farming likely to reshape the food economy of the future, or will such endeavours be vulnerable to absorption by corporate capital down the line? Will the once-strong cooperative movement in the agricultural sector – itself seriously challenged by globalised capitalist competition – experience a revival through the new forms of cooperative enterprise now taking shape in response to changed structural and ideological realities? Will food banks continue to serve as essential safety valves for capitalist societies that are producing ever greater levels of income disparity, or will they become one of the sites for a broader-based resistance to the policies producing such inequalities?

The contradictions of globalisation will continue to open the field for study of alternative food production and provisioning arrangements and the re-examination of struggles to refashion, if not transform, the global capitalism of tomorrow. These struggles will surely put an indelible stamp on the early years of the next millennium.

NOTE

1 This literature is now sufficiently voluminous as to be impossible to capture here. Some early manifestations of this new direction were brought together in an important volume by Duncan and Rutledge titled *Land and Labour in Latin America* (1977). See Winson (1982) for a discussion of the early examples of such scholarship.

BIBLIOGRAPHY

Brenner, R. (1976) 'Agrarian class structure and economic development in pre-industrial Europe', *Past and Present* 20, (February): 30–75.

—— (1977) 'The origins of capitalist development: a critique of Neo-Smithian Marxism', *New Left Review* 104 (July–August): 25–92.

Cardoso, F. H. and Faletto, E. (1971) *Dependencia y Desarrollo en America Latina*, Mexico City: Siglo Veintiuno Editores SA.

Duncan, K. and Rutledge, I. (1977) *Land and Labour in Latin America*, Cambridge: Cambridge University Press.

Frank, A. G. (1967) *Capitalism and Underdevelopment in Latin America*, New York: Monthly Review Press.

—— (1969) 'Sociology of development and underdevelopment of sociology', in A. G. Frank (ed.) *Latin America: Underdevelopment or Revolution*, New York: Monthly Review Press.

——, Cockcroft, J., and Johnson, D. (1972) *Dependence and Underdevelopment*, New York: Anchor Books.

Hilton, R. (1978) *The Transition from Feudalism to Capitalism*, London: Verso Books.

Laclau, E. (1971) 'Feudalism and capitalism in Latin America', *New Left Review* 67: 19–38.

McMichael, P. (1996) 'Globalisation: myths and realities', *Rural Sociology* 61, 1: 25–55.

Thomas, C. (1976) *Dependence and Transformation*, New York: Monthly Review Press
Whatmore, S. (1994) 'Global agro-food complexes and the refashioning of rural Europe', in A. Amin and N. Thrift (eds) *Globalisation, Institutions, and Regional Development in Europe*, Oxford: Oxford University Press.
Winson, A. (1982) 'The "Prussian Road" of agrarian development: a reconsideration', *Economy and Society* 11, 4 (November): 381–408.

PART VI

NATURE, SUSTAINABILITY AND THE AGRARIAN QUESTION

13

SUSTAINABILITY AND THEORY

An agenda for action

Michael Redclift

INTRODUCTION

Late industrial societies have set themselves various environmental targets for achieving 'sustainable development'. In this chapter I will not discuss whether this exercise is a genuine one or not. Nor will I examine the targets themselves. The Second Assessment of the Intergovernmental Panel on Climate Change (Working Group Three, 1995) suggests that obligations under current international agreements are extremely modest, if the objective is to stabilise greenhouse gas concentrations within the next half century. Finally, I am not proposing to undertake a review of the meaning of 'sustainable development', for which there is a voluminous literature that seems to grow in inverse proportion to its achievement.

The purpose of this chapter is to examine the way different bodies of theory approach the question of sustainability from different disciplinary standpoints, and to explore the strengths and weaknesses of these paradigms. In particular, I want to consider whether different theoretical approaches actually complement each other. Are there, indeed, any points of convergence in their trajectories, and any implications for political action for those who are both Green and on the political left? Can we identify in the discourses surrounding sustainability any of the historically informed, explanatory models of nature–society interactions which are located in the work of the young Marx?

Three principal questions are usually posed about sustainability. They are:

How do societies arrive at sustainable levels of resource exploitation? This question strikes at the balance between the production of goods and services, and the rate at which we consume resources. Is a 'sustainable' resource system compatible with given levels of consumption, and at what level (local, regional, global)? The technical discussion includes concepts like critical loads, carrying capacity, and ways of assessing biodiversity losses, and non-reversible environmental costs (critical natural capital).

The much less obvious issue, which scarcely plays any part in most of the literature debate, is not about what constitutes 'sustainable levels', but about how they might be *achieved*. In this area most suggestions point towards global environmental management (pace UNCED 1992) usually linked to local action (Agenda 21). As we shall see, arriving at 'rational' grounds for environmental action has proved difficult in practice.

How do societies arrive at ways of valuing environmental losses and gains? This question is posed by economists, most of whom answer it in a similar way: by establishing how much people would be willing to pay for the environment. The exceptions, such as Daly, Norgaard, Ekins, and Jacobs are interesting because they take issue, to varying degrees, with the assumptions of neoclassical economics.

Again, the formal question obscures a much more important primary issue: do valuation criteria reflect real social choices, and according to whose understanding? The ways we arrive at environmental valuations can take several forms, including mechanisms to try and achieve social consensus and (more commonly) the conclusions of 'expert' witnesses. It is notable that the Green agenda on decision making has been advanced almost entirely by what Beck calls 'sub-political' groups, outside conventional political parties. At the same time, mainstream political discourse has not addressed sustainability.

How do societies influence the social and economic behaviour of their citizens towards more concern with sustainability? The problems identified above, that questions are posed in ways which seek to *depoliticise the debate*, are even more evident in this case. Most of society's institutions, from the most formal (government, law, church) to the least formal (family, work, community) take no conscious account of 'sustainability'. Can we steal it in, as it were (as most commentators hope), or do we have to begin again? Where does good example begin: in the north, which has shown very little commitment so far to reduced consumption, or in the south, where most people consume to survive?

Clearly we should begin to consider the institutions that we bequeath future generations, rather than merely the environment we hand on to them. However, the fact that the question is posed in this way at all, that sustainability is increasingly viewed as an issue alongside governance, tells us something about the shortage of ready-to-wear clothes in the economic policy closet. Unlike the postwar Keynesian medicine, the remedy this time around looks more challenging, and threatens to require some fundamental shifts in our 'getting and spending', in that ubiquitous shibboleth, 'our lifestyles'. Rather than being asked to spend more (Keynes) Green economics asks us to spend less.

In the next section I want to consider the way these questions are posed, and answered, from three very different points of departure: that of *industrial metabolism*, *environmental economics*, and *recent social theory of risk*. By exploring these questions from different points of departure, it should be possible to relate sustainability to some of the issues which are emerging as part of a wider political debate. How does sustainability become incorporated in wider social goals? And what are the means of providing a radical social and political platform which is relevant to future generations, as well as to present ones?

THE COMPARATIVE DYNAMICS OF PRODUCTION AND CONSUMPTION PROCESSES

In most societies there is an ambiguous relationship between the volume of output and the extent of environmental destruction. High output provides resources for environmental protection, but the benefits of conserving resources may be outweighed by the environmental impact of production and consumption itself. On the other hand, although low output provides fewer resources for environmental protection, it may be grossly inefficient and do little to prevent environmental degradation. In the 1970s the principal fear in environmental policy circles was the supposed 'limits to growth' (Meadows *et al.* 1972). In the 1990s 'limits to sink capacities' (or output) poses an equivalent fear, particularly in the context of global problems such as the destruction of the ozone layer and the enhanced greenhouse effect.

The complex interactions between economic activities and the natural environment depend crucially on the way a society organises its relationship to the environment, and the view that different social groups take of this relationship. Many environmental changes are represented as 'demand driven', in the hands of consumers, rather than 'supply driven', in the hands of the formal economic levers dictating production. In fact it is almost impossible to separate patterns of consumption, and 'lifestyles' from economic instruments and ideologies. Most production in late industrial societies is geared to increasing volumes, rather than the life-cycle effects of goods and services.

The existence of new materials and productive processes has accelerated the changes through which environmental costs are transported in space (and frequently in time) usually to poorer developing countries. At the same time levels of air pollution, for example, in the newly developing countries of East Asia, are increasing more rapidly than their increase in Gross Domestic Product. The social processes at work are not simply in the hands of consumers. They are embodied in the global political economy of market capitalism.

Some industries in the north, aware of the costs of employing dirty technologies on their doorstep, have sought either to export their pollution and wastes, or to internalise the problems associated with unmanageable levels of waste. They have turned their attention to changing material flows and waste streams, as a focus for technological innovation itself.

Some companies are learning how to maintain or expand output, while at the same time cutting resource inputs and environmental impacts to a minimum. In similar fashion the same principles can be applied to households, rather than companies. One might then examine the use that households could make of resources to improve welfare without increasing aggregate levels of personal consumption to the point at which they are not sustainable.

Examples are the way in which reductions in product size can facilitate reductions in the time taken to transport goods and services, thereby maximising the efficiency in the use of space and time. Combining the quality of output and environmental protection in single low-impact technologies, such as combined heat and power generation for energy utilities, is another example.

Similarly, the agriculture and food sectors provide many illustrations of the huge scope for policies which make a realistic assessment of environmental impacts and benefits, replacing the 'value-added' by the industrialisation of food (packaging, expensive inputs, etc.) by benefits from local provisioning, reduced packaging, and more attention to the nutritional quality of food. Reducing the shelf-life of a food product, and the costs added by advertising and packaging, is unlikely to be possible until these values are affirmed. At the moment the fortunes of the food industry depend upon denying their relevance.

Such changes in the way products and services are valued also bring in their train significant shifts in the use of space and the allocation of time. Planning for such developments requires a radical overhaul not so much of consumer attitudes as the whole infrastructure of modern living.

Exploring the dynamics of consumption and production processes takes us well beyond environmental protection. It affords the possibility, in principle, for societies to use the smaller stock of resources available in a more efficient way. At the moment environmental management is principally concerned with modifying existing human behaviour. Policies are concerned with reducing the full environmental impact of our actions. The underlying behaviour, or social commitments, associated with lifestyles, is currently viewed as non-negotiable.

In a paper written for the World Bank, Robert Goodland has argued that achieving per capita income levels in low income countries of $1,500 to $2,000 (rather than OECD's $21,000 average) is quite possible. This would represent a significant advance in welfare for such countries. 'Moreover', he adds, 'that level of income may provide 80 percent of the basic welfare provided by a $20,000 income, as measured by life expectancy, nutrition, and education' (Goodland 1994). There might also be measurable gains in personal security and reduced social tensions in societies experiencing this type of transition. Goodland then suggests that, 'Colleagues working on Northern overconsumption should address the corollary. Can $21,000 per capita countries cut their consumption by a factor of ten and suffer "only" a 20 percent loss of basic welfare' (Goodland 1994)? This argument illustrates the way in which thinking about sustainability forces us to think also about the constituents of social welfare, and the political conditions under which gains in both sustainability and welfare might be made.

If underlying commitments are non-negotiable then there is little point in pressing on with this type of analysis. However, the problem of overconsumption might be approached from the other direction. Given that certain resource limits are binding, how can we live within them, producing and consuming goods and services in a way that carries net benefits for society in the future? Arguably, a better understanding of production and consumption flows could help societies make radical choices, at different levels of analysis and different time scales, which transferred welfare benefits between countries, as well as within them, making 'development' a much more reflexive, and redistributive, process than it is today.

At this stage it is worth pausing to consider some of the difficulties with this argument.

First, the approach from industrial metabolism argues that many environmental problems, such as the management of wastes and pollution, arise out of the inherent differences between 'natural' and 'industrial' systems. The laws of thermodynamics have precedence over those of the market. At a philosophical level, this may be true. However, efforts to quantify the life-cycle costs of the production and use of goods and services are difficult to express in alternative economic terms. It is not easy to arrive at the utilisation values which correspond with actual environmental costs and to substitute them for current market values. Without such utilisation values it is impossible to alter the economic behaviour of individuals, still less that of governments.

Second, being able to arrive at sustainable levels of exploitation in *environmental* terms – conserving resources and enhancing sinks – tells us nothing about the social and economic mechanisms that will be required to achieve these objectives. Nor does it tell us about the processes through which sustainability assumes legitimacy with the public. These factors – the social mechanisms to achieve greater sustainability, and the legitimacy of sustainability as a social goal – lie outside the provenance of industrial metabolism as understood in the literature. They are linked to wider questions of engagement in the society. As we shall see, we need to turn to other approaches to understand them better.

WAYS OF EVALUATING ENVIRONMENTAL GAINS AND LOSSES

As we have seen, the second question, concerning the way that values are attached to the environment, has been dominated by economists. The point of departure for most economists, particularly those who espouse a neoclassical position, is the trade-off between economic growth and environmental protection. Policies to correct environmental problems, it is argued, inevitably carry costs for economic growth, and with it the level of consumption. As Jacobs and Ekins (1995) point out, this concern with the cost of environmental measures serves to disguise the fact that neoclassical economics has considerable difficulty in acknowledging that distributional issues, within and between generations, lie at the heart of valuation. The 'willingness to pay' axiom, with which environmental goods are accorded market values, sets aside the central issues which beset the policy agenda: who should pay, and when?

The environmental economics literature focuses upon two related questions which assume importance in a modified neoclassical position: externalities and public goods (Bhaskar and Glyn 1995). Externalities are important because of the lack of a market in environmental goods. Two solutions are usually proposed. Neoclassical theory has developed around ways of imputing market values to environmental costs and benefits, through instruments like contingency valuation. The second solution is to 'internalise' externalities, what has come to be known, especially in Germany and the Netherlands, as ecological modernisation (Mol 1994). Quoting Joseph Huber, one of the principal exponents of ecological modernisation, Mol refers to this process as 'an ecological switch-over of the industrialization process . . . all ways *out* of the environmental crisis lead us further *into industrial society*' (Mol 1994: 11).

Both 'solutions' have problems attached to them. The modified neoclassical position, which imputes market values to nature, starts from a number of assumptions which can be challenged. First, it assumes that individuals act alone to calculate their advantage from making choices under market conditions. The 'individual rational calculator' approach, which lies at the heart of the neoclassical paradigm, provides an approach to decision making that reduces human agency to price signals. There is no place for society in this view of the economy, as several writers have observed (Jacobs 1991, Benton and Redclift 1994).

Second, this perspective turns on the relationship between prices and values. We are always in danger, as Oscar Wilde put it, of knowing the price of everything and the value of nothing. We know, for example, that the North Sea has different values for different groups of people, and for some of these groups it is not adequate to translate these as 'intrinsic values' or 'existence values', for they lie in the abandonment of the market, not its extension.

There are similar problems with ecological modernisation. Externalities are not merely environmental costs which can be refashioned into an environment-friendly good or service. They frequently have distributive consequences, and causes, which carry political implications for global markets. It is state power, and that of transnational corporations, which frequently lies behind externalities, determining who bears the costs of pollution, toxic wastes, or the effects of pesticide use. Indeed, as Murray (1994) has suggested in a highly original treatise on pesticide use in Latin America, the behaviour of companies which modify agro-chemicals serves to confirm, rather than alter, the essential relationships of power which drive the diffusion of modern, industrialised agribusiness. There are limits to the extent to which 'modernisation' incorporates ecological considerations.

Finally, by way of critique, it can be observed that some goals of neoliberal economic policy, such as the liberalisation of trade, actually serve to increase externalities. There are basic contradictions between economic growth in the late twentieth century, and the protection and conservation of the natural environment. Ecological modernisation does not alter these facts.

The paradox at the heart of the economic valuation of the environment is that of distribution. The environment can only be properly 'valued' by successive generations of people, for whom any one generation acts as stewards. Inter-generational equity, moreover, assumes more importance as problems assume 'global' importance, and are increasingly governed by uncertainty. Issues like global warming, the destruction of the ozone layer, and the loss of biodiversity make little sense within the neoclassical model (which is why so much effort is invested in demonstrating that they do not represent a problem for the model). Since unborn generations have no 'rights' it is difficult to rely on efficiency in allocating resources between future and present generations.

The problem of inter-generational equity is matched by that of intra-generational equity. Does every member of the human race count equally in their responsibility for carbon emissions? Do emissions of methane from paddy fields or livestock in poor, rural societies, count equally with carbon emissions from vehicles being driven between home and the shopping mall in California? The most revealing aspects of carbon budgets for nations is that they reveal such vast inequalities in consumption, and such different trajectories for future development, that only a complete overhaul of the global economic order would enable any meaningful and lasting international agreements to be made. Intra-generational inequities are such that we might as well abandon any pretence that development can be sustainable.

The problem with the neoclassical valuation of the environment is that it makes questionable assumptions about human behaviour. As a consequence it emphasises 'efficiency gains' for which the confidence limits are low, over vitally important distributional consequences, for which the confidence limits are high. It succeeds in turning the world on its head. Environmental economics, at least in its mainstream neoclassical version, requires that we ignore the institutional context for decision making, which itself determines whether economic models are used at all.

SOCIAL BEHAVIOUR AND SUSTAINABILITY

To meet even modest environmental goals requires significant changes in human behaviour. But how would such changes in behaviour be brought about? If differences in behaviour are based on different perceptions of risk or uncertainty, then it is necessary to begin by explaining the gap between perceptions and behaviour. We all possess enough anecdotal evidence that perceptions influence behaviour in ways that are frequently perverse. The British Meteorological Office has established, for example, that people turn on their central heating in response to cloudy, rather than cold, weather. Can people be relied upon to do what 'rational' action requires to avert environmental disasters?

This depends on what is meant by 'rational' action, the question at the heart of the debate about so-called risk society and modernity (Beck 1992). According to Ulrich Beck, late industrial societies have removed technology from the political arena. In the process such societies make it more difficult for individuals to deal with risk in everyday life. In Beck's view social problems are viewed, in late industrial societies, as problems of individual failure, rather than social competence. When people challenge their society's competence, in particular that of its 'experts' and political leaders, it becomes a challenge to public science. Human agency becomes embroiled in challenges to the authority of science.

The contrasts with mainstream neoclassical economics could not be greater. While economics has been pushing environmental assessment *towards* market-based, individualised models of human behaviour, sociological theory has been emphasising that environmental risks increasingly reflect individualisation, *socially constructed differences*. Economics has sought to bring the environment into the ambit of human decision making by allocating environmental values like any other market value. Sociology is arguing that environmental decision making actually reflects social concerns and commitments, that the environment is nothing more, nor less, than the battleground on which political interests are contested. Social theory is increasingly concerned with the problems that the environment poses *for individuals, rather than by them*.

The differences do not end there. While economics has difficulties, as we have seen, in incorporating distributional issues (time, space and class) into its methodological canon, sociology experiences difficulties in distinguishing between 'what is happening in actually-existing societies' (realist concerns) and the part played by human societies and consciousness in *inter-*

preting what is happening (constructivism). Some sociological writing gives the impression that ecological issues can be confined to the margins of society, separated from the social institutions of modern life (the family, work, religion and politics). It is as if our getting and spending were somehow divorced from its social consequences.

These charges, whatever their force, cannot be levelled at all recent social theory: witness the work of Rustin (1994). Beck argues that modernism was founded on ways of exploiting nature, and countering tradition. By contrast, postmodernism is concerned with the problems of modernism itself, it is 'becoming its own theme' (Rustin 1994). Beck grounds his approach, as Rustin avers, in the modernist vision, rather than its sequel. He believes that rational consensus is attainable once the conditions of democratic decision making have been achieved in civil society. Contrary to many, who see human society as constituting a problem for ecological goals (deep ecologists, many ecofeminists) Beck envisages a society which takes responsibility for its own development, as a real, historical possibility.

For Beck contemporary society, in the industrialised world at least, is an incomplete rationality. What we have described as industrial ecology, the search for utilisation values which adequately reflect environmental costs, is hampered by the hegemony of the market. The rationality being imposed is the 'wrong rationality', since, if sustainability is the goal, it forces us to obey the rules of industrialism, rather than those of nature.

Similarly, neoclassical economics finds rationality in its own (implicit) model of human behaviour, which facilitates the efficient allocation of resources. For postindustrial social theory it is rationality itself which is the problem. In Beck's view the *categories* of industrial society prevent the achievement of rational modernity through the three domains of gender, class, and ecology. He identifies institutional spheres in which our commitment to the *principles* of industrialism has prevented us from responding adequately to its problems and contradictions. In Marxist terms, in modern society much of the superstructure is actively out of engagement with the trajectory of the material base.

Having opened up important questions, Beck's *Risk Society* does not necessarily answer them. The 'individualized social inequality' which he sees as having replaced class, has implications for the relations between labour and the environment, which he fails to develop. He discusses the way individual rights have been invoked in the 1980s *against* forms of social provision. However, modern consumer society raises other, equally important questions. To what extent does the achievement of higher levels of social provision and collective consumption also endanger environmental goals? And to what extent are environmental goals increasingly a means of expressing political and social goals?

These issues, in a curiously coded form, are part of the current policy discourse around employment and sustainability in the European Union, encapsulated in the recent White Paper of the European Commission, *Growth, Competitiveness and Employment (1993)* (Fleming 1994).

Finally, moral imperatives appear to lie behind the broad brush strokes. Beck sees 'solidarity motivated by anxiety . . . (replacing) solidarity motivated by need', but it is not always clear whether society *should* be more anxious about ecological risks, or that it is more anxious. Similarly, the concept of 'need' is itself an interesting candidate for reflexive analysis, since it is unlikely that 'needs' can ever be fully separated from 'wants' (Redclift 1993).

Nevertheless, Beck's argument, and that of others (Benton, Dobson, Dickens, Eckersley, Merchant, and Redclift) undermines the separation between 'nature' and 'society'. As the conquest of nature becomes absolute, it is a powerful corrective to conventional class politics. Increasingly, social movements can be expected to contest the dominance of nature rather than to further it, although the political expression of these tendencies will vary over time and space. As science begets more science, and through environmental regulation seeks to 'manage' the consequences of pollution and environmental risk, so civil society comes to represent social and political barriers in environmental terms.

The problem with Beck's analysis is that his society of 'incomplete rationality' requires (and gets) an active citizenry committed to the goals of liberation from technological tyranny. Do we possess such a citizenry, or are the ecological politics of the risk society still those of a small, self-conscious minority? There is a bigger political issue to discuss here. Ironically, the social commitments that underlie our everyday consumption are generally more powerful than the democratic ideals which lead us to question the social authority of science, or the necessity of being yoked to a technological treadmill. Pervasive underlying social commitments – our everyday consumption practices – may be the other side of the same coin, facets of the emergence of uncertainty and risk as social phenomena. If they are, then they explain ambivalence towards late industrial society, rather than wholehearted rejection of it, which characterises the first 'postmodern' generation.

Superficially, the three bodies of theory on which I have drawn refer to different social and economic domains – production, processes, market values, and human agency. A politics of sustainability which reflected all of them would be heterogeneous and incoherent. But, if we regarded each domain as a circle in a Venn diagram, and concentrated on the areas where they overlap rather than their different points of departure, the conclusion might be more surprising. The search for an end to alienation, from material production, market values, and disempowerment, lies at the heart of sustainability today, and answers to the politics of the 1990s. And we have seen these issues discussed before . . . in the early writings of Marx, a century and a half ago.

BIBLIOGRAPHY

Beck, U. (1992) *Risk Society: Towards a New Modernity*, London: Sage.

Bhaskar, V. and Glyn, A. (eds) (1995) *The North, the South and the Environment*, London: United Nations University Press/Earthscan.

Daly, H. (1992) *Steady-State Economics* (second edn), London: Earthscan.

Dobson, A. (1990) *Green Political Thought*, London: Unwin Hyman.

Ekins, P. (1992) *Wealth Beyond Measure*, London: Gaia Books.

European Commission (1993) *Growth, Competitiveness and Employment*, Brussels: European Commission.

Fleming, D. (1994) 'Towards the low-output economy: the future that the Delors White Paper tries not to face', *European Environment* 12: 36–48.

Goodland, R. (1994) *Environmental Sustainability*, Working Paper, Washington DC: World Bank.

Intergovernmental Panel on Climate Change (IPCC) Working Group Three (1995) *Second Assessment*.

Jacobs, M. (1991) *The Green Economy*, London: Pluto Press.

—— and Ekins, P. (1995) 'Environmental sustainability and the growth of GDP', in V. Bhaskar and A. Glyn (eds) *The North, the South and the Environment*, London: United Nations University Press/Earthscan.

Meadows, D. C., Randers, J., and Behvene, W. (1972) *The Limits To Growth*, London: Pan.

Merchant, C. (1992) *Radical Ecology*, New York: Routledge.

Mol, A. (1994) 'Ecological Modernisation and Institutional Reflexivity', paper presented at the XIII ISA Congress, Bielefeld.

Murray, D. L. (1994) *Cultivating Crisis: the Human Costs of Pesticide Use in Latin America*, Galveston: University of Texas.

Norgaard, R. (1994) *Development Betrayed*, London: Routledge.

Redclift, M. R. (1993) 'Values, needs, rights: reassessing sustainable development', *Environmental Values* 2, 1: 3–20.

—— and Benton (eds) (1994) 'Introduction' in *Social Theory and the Global Environment*, London: Routledge.

Rustin, M. (1994) 'Incomplete modernity: Ulrich Beck's *Risk Society*', *Radical Philosophy* 67: 3–12.

UNCED (United Nations Conference on Environment and Development) (1992) Rio de Janeiro.

von Weizsacher, E. U. (1994) *Earth Politics*, London: Zed Press.

Weale, A. (1992) *The New Politics of Pollution*, Manchester: Manchester University Press.

14

SOME OBSERVATIONS ON AGRO-FOOD CHANGE AND THE FUTURE OF AGRICULTURAL SUSTAINABILITY MOVEMENTS

Frederick H. Buttel

INTRODUCTION

From the end of the Second World War through roughly the 1970s there was an overall trend to convergence of agrarian structures across the advanced industrial societies. This convergent pattern, however, was substantially at variance with that anticipated among the major classical figures in agrarian political economy. In particular, instead of peasant, family farmer, gentry-dominated, and precapitalist-estate agricultures converging slowly but surely toward very large-scale capitalist agriculture, the shift was more in the direction of multiple routes towards a predominance of moderately large, highly capitalised, petty-capitalist *family proprietor farms*.[1] Thus, this shift in some sense turned the Leninist notion of 'roads' of development of capitalist agriculture (see Goodman and Redclift 1982, Lehman 1986, for a summary) on its head.

During this mid-century heyday of family farming in the industrial world[2] there was, to be sure, widespread rural exodus and social differentiation of agriculture. But there were simultaneously a variety of state-reformist agricultural policies and other mechanisms supporting recruitment into family labour agriculture that tended to moderate somewhat the tendency to depeasantisation, or else discouraged large-scale capitalist agriculture. Commodity programmes, food import restrictions, soil conservation subsidies, rural-regional programmes, 'lender-of-last-resort' state credit policies for subcommercial and moderate-sized farms, the exclusion of agricultural commodities from the provisions of GATT, and food security programmes were some of the more common mid-century policies that contributed, at least in part,[3] to bolstering family labour agriculture. These policies typically served to restrain the loss of farms, to increase the viability of moderate-scale farms, or to encourage young persons to enter farming.[4] These policies thus attenuated rural exodus relative to the pace that might have otherwise occurred under conditions of declining real agricultural commodity prices, rapid growth of metropolitan labour markets, and rising real wages in non-farm industry. Similarly, the traditional structures of agricultural communities served to encourage entry into farming and to reinforce the restraints on farmer exodus and agrarian differentiation. The traditional demographic structure of farm households (large families, spouse and children as a labour force, and socialisation of young people to consider farming as a career) served to retain labour in agriculture and to provide large flows of actual and potential recruits into farming, even as rural-to-urban migration continued unabated.

Thus, family proprietor farms – albeit increasingly larger, more heavily capitalised, and specialised ones – were becoming clearly predominant in aggregate-sectoral terms. To be sure, farm numbers in the advanced countries became dominated by small-scale and/or part-time farm households, which in the USA still account for roughly two-thirds of all census-enumerated farms. Also, some segments of agriculture (especially broilers, fresh fruits, and vegetables) became (or continued to be) dominated by large-scale capitalist enterprises during mid-century (see, e.g. Friedland *et al.* 1981). But during roughly the 'late Fordist era' (late 1960s through the 1970s) there came to be sufficient stability of capitalised, moderately large-scale family farming so that rates of decrease in farm numbers, and of increase in average farm sizes, were substantially attenuated relative to their postwar peaks during the 1950s. From 1974 to 1982, for example, US farm numbers were virtually stable – for the first time since the Great Depression – while average farm size (in acres) was relatively constant.

Since roughly the early 1980s, however, the social bases and processes of agriculture and agro-food systems have begun to shift in significant respects. Spurred by the breakdown of the Bretton Woods international monetary system, and by subsequent stages of liberalisation of global movements of capital, the conditions for dismantling the national food economy and for the globalisation of agriculture and food have advanced in tandem (Friedmann and McMichael 1989). The public policies that helped reinforce capitalised family labour farming were rural and agricultural expressions of the larger pattern of nationally regulated economies, Keynesianism, the welfare state, and expansion of state capacity. These agrarian-reformist institutions of the mid-century, however, have been progressively terminated, scaled back, or subjected to protracted attack in a manner similar to their non-agricultural counterparts (e.g. welfare and other entitlement benefits). The international farm crisis (Goodman and Redclift 1989), which was in substantial measure precipitated by the Reagan administration's monetarist strategy for adapting to international economic integration and monetary disorder (and by the inadvertent Keynesianism associated with funding military expansion in an era of stagnation of tax revenues), destabilised industrial country agricultures. This farm crisis, which when superimposed on overproduction, environmental problems, fiscal crisis, declining or tenuous state support of agriculture, and other forces, ensured that agriculture would not return to any semblance of the relative stability of the 'late Fordist' era. The structure of rural communities has also shifted considerably. Farm families have become smaller, the availability of non-waged farm-household labour supplies has declined, agriculture has become a decreasing component of rural economies, and farming is no longer regarded unambiguously as an attractive occupation for rural youth to pursue.

As a result of these national and global shifts, there has been a growing exposure of farm and agribusiness enterprises to naked (global) market forces, a return to a more rapid decline in farm numbers, 'industrialisation' of agriculture (Welsh 1996), and the associated restructuring of commodity chains across national borders (Bonnano *et al.* 1994, McMichael 1994). Concentration of production is rapidly increasing. For example, of the 1.925 million farms enumerated in the US 1987 Census of Agriculture, 3.6 per cent, the largest of these farms, averaging roughly 2,800 acres, accounted for 50 per cent of US farm output. Only eighteen years earlier (1969 Census of Agriculture), 8.1 per cent of farms, averaging about 1,610 acres, accounted for 50 per cent of national farm output (NRC 1995: 25).

The future seems likely to hold more of the same, at least in the US and the other OECD countries in which deregulation and exposure of national agro-food systems have proceeded the farthest. There will in all likelihood be a continued trend to differentiation of the family farm to the point that the classic full-time family farm of early mid-century (a farm that occupies the labour of a household on a more or less full-time basis while yielding a household disposable income) will become a distinct minority. There will continue to be a substantial number of family farms, but increasingly these will primarily be subcommercial, part-time farms. With some exceptions (e.g. ruminant livestock in the northern temperate areas), the scale economies of livestock and fresh fruit and vegetable production will lead to near-universal 'industrialisation' (in the qualified sense of the term, à la Welsh 1996) of and contractual integration within these subsectors. Entry into commercial-scale farming will become increasingly constricted. While some sectors of agriculture, particularly the basic food crops, the food and feedgrains, and oilseeds, will not be privately profitable for large-scale capitalist enterprises, the sphere of household production in agriculture will become progressively smaller and of less and less consequence to the overall character of the agro-food system. Increased competition will place a premium on technological innovation that can reduce average costs, reduce short-term risks, or rationalise ever-larger enterprises. The structures, technoscientific practices, and ideologies of global economic integration and competition will slowly but surely tend to obviate or override the geo-social-agro-ecological specificities of agriculture – the neo-Chayanovian, 'Mann-Dickinson' (Mann 1990) or 'natural-production-process' (Goodman *et al.* 1987) factors – that have historically contributed to agricultural production tending to be relegated to household producers.

TOWARDS A MATERIALIST UNDERSTANDING OF THE DEMATERIALISATION OF SOCIAL THEORY

The analysis just presented is blunt, critical, and at variance with a good share of the scholarly and trade literature that trumpets the dynamism of modern agriculture in a global-competitive context. At another level, however, this analysis is quite unremarkable. Like most mainstream or political-economic analysts of agriculture, I have portrayed the historical tendencies of agro-food systems almost entirely in anthropocentric and social terms – mainly with respect to large-scale social and economic forces. Indeed, one of the contradictory achievements of modern social science applied to agro-food systems is that it has enabled the analyst to deploy powerful abstractions which permit one to ignore the infinite details of biota, soils, agro-ecosystem processes, physical labour, and so on. The more a conceptual scheme abstracts beyond these factors, and thus holds across time and space, the more persuasive it is regarded. Modern social science has accordingly tended to conjure up a highly dematerialised view of agro-food realities – a view that tends to regard the natural environment of agriculture as being essentially epiphenomenal.

Interestingly, this dematerialised view of agro-food systems will very likely become even more persuasive – at the same time that it becomes increasingly misleading – if the trends just discussed extend themselves into the future. The ultimate ('second')[5] contradiction of contemporary industrial country agriculture is that its structural trend is towards agro-ecological specificities being diminished or overridden at the same time that agro-food systems' vulnerabilities to ecological problems increase.

There is perhaps no better example of this contradictory reality than the experience of the 1996 grain harvest in the US Midwest. The 1996 growing season was plagued by a very wet spring and early summer, and by severe late summer drought, which led to fears that world corn and soybean supplies would be calamitously short. Corn and soybean futures prices increased to dizzying levels during the middle and late summer of 1996. But it proved to be the case that the 1996 corn and soybean harvests were actually quite large. This was due in large part to the fact that the core off-farm inputs for these crops – especially nitrogen fertiliser, pesticides, and modern varieties – enabled satisfactory levels of output to be achieved despite adverse weather conditions. The large harvest contributed to a devastating cycle of downward pressure on crop prices for the autumn 1996 harvest. The farm press began to declare that Corn Belt corn and soybean production is 'bulletproof' – that there is no natural force that can disrupt the productivity and output from the Midwestern prairie land resource. But while Midwestern agriculture's 'bulletproof' stature was being touted, in eastern Illinois and western Indiana there was emerging a mutant strain of corn rootworm that is now resistant to major insecticides, and also able to survive over soybean rotations (which in conventional corn-soybean agriculture are used to control build-up of insects). If this resistant rootworm spreads widely, it may provide the beginnings of a crisis for the basic corn-soybean cropping system of the American heartland. Topsoil is also being lost, and in the western Corn Belt aquifers are being depleted. Thus, the apparently 'bulletproof' corn-soybean row-crop grain desert of the American Corn Belt may well be occurring simultaneously with, but concealing and overriding, the appearance of a profound ecological contradiction of the spatial and biological homogeneity of modern agriculture.

Put somewhat differently, the 'ecological' forces within and limits on contemporary agriculture are, more properly speaking, 'socioecological' ones. That is, such socio-ecological contradictions and limits have biophysical foundations (e.g. resistance to chemicals, water quality degradation, land degradation), at least in part. The manifestations of these contradictions may be essentially biophysical (declining yield response in soils experiencing salinisation), but these limits may in reality or perception be mostly social or ideational (e.g. the organic food movement, resistance to industrialisation of hog production, public wariness about rapid pace of corporate concentration of control over the food system, and the declining capacity of the world food system to manage overproduction and food scarcity). Or, as is more commonly the case, the forces may be socioecological per se – that is, a composite or fusion of the socio-economic and ecological realms.[6] For example, there is a growing accumulation of evidence and data, and rising public concern and movement mobilisation, relating to the fact that infants are much more susceptible to adverse health impacts from pesticides than is taken into account in conventional toxicological protocols (Colborn *et al.* 1996); in this case, the phenomenon is biophysical, but the consequences are (tragically) social, as are the movements that are beginning to seize on this issue. Further development of theoretical understandings of historical and contemporary agro-food systems and regimes must be built on awareness of the socio-material bases of agro-food systems, and most other social institutions for that matter (see Buttel 1996).

LATE TWENTIETH-CENTURY AGRICULTURE: AGRO-ECOLOGICAL AND SUSTAINABILITY CONTRADICTIONS

This general – essentially metatheoretical – conception of the socioecological character of agro-food systems is by no means novel. It is, for example, congruent with the basic thrust of much ecological economics and environmental sociology (e.g. Norgaard 1994, Buttel 1996). But because it is metatheoretical it is also indeterminate (Dickins 1992). How, concretely, should ecological and social factors be conjoined or disaggregated in analyses? Should achievement of sustainability be regarded as an inevitable – or utopian or unachievable – outcome? Is sustainability a moral imperative that social-scientific analysts of (as well as participants and stakeholders in) agro-food systems, have an obligation to achieve?

This chapter cannot answer a set of such penetrating questions (but see Thompson 1995, who has gone as far as anyone – certainly farther than I – in developing some perspectives and answers). My approach will be to set forth some general principles, and ultimately to look at agro-food sustainability movements through these lenses.

For present purposes, the following strike me as some crucial observations about the general forms in which the ecological contradictions of agriculture are becoming manifest. First, it is important to begin with the recognition that there is essentially no such thing as a global agro-ecosystem (in the sense of a globally functioning system analogous to the atmosphere or oceans). Thus, agriculture can continue to exhibit environmentally destructive patterns and tendencies without these being widely apparent, and without their tending to be cumulative in a directly perceptible manner. Thus, agro-ecosystem destruction, much like (traditional peasant) agro-ecosystem development (Buttel 1995), is a congeries of *local* processes (which are, however, affected by political, economic, and other factors at the national to global scales). Waterlogging of irrigated lands in South Asia, for example, does not directly affect the 'health' of soils in the US Corn Belt (in contrast to the way in which greenhouse gases from across the globe will tend to be cumulative in terms of atmospheric disruption). Many agro-ecological degradation processes that are held by many analysts to be 'global' are, in fact, relatively local in terms of both antecedents and socioecological implications; desertification is a classic example of an ostensibly global agro-ecosystem change which, in reality, reflects primarily the local and regional character of agro-ecosystemic processes, and especially the regional character of climatic processes.

Second, while the ecological implications of modern agriculture are relatively localised ones, there is sufficient similarity of, or homology across, the processes that cause these problems (as well as in their consequences) that environmental challenges can become an enormous stimulus to technological innovations and technological changes that rationalise, mask, or 'patch up' these problems.[7] Technological innovation that *compensates for the contradictions of modern agriculture* is not only a short- to medium-term stabilisation system; it may also open up new vectors of capital accumulation that reinforce the extant trajectory (Krimsky and Wrubel 1996). Good examples in this latter regard are Bt-engineered crops, which address insect resistance and pesticide-residue problems caused by conventional pesticides; precision-agriculture technology, which tailors doses of agricultural chemicals to the site-specific characteristics of small grids on fields; and industrial-scale manure disposal and handling systems which reduce the odours and water pollution risks of industrial hog operations.

Third, the technological processes that are induced to rationalise industrial agriculture may do so imperfectly, and lead to other serious environmental problems, but they are typically able to *enhance output even as the foundational biophysical systems are undergoing degradation*. The earlier example of 'bulletproof' corn-soybean production systems in the Corn Belt is one such example. The fact that Bt-engineered crop varieties will tend to provide some modest yield premium is another. Augmentation of aggregate output (at the national, and especially the global, level) thus serves to mask the extent to which long-term agro-ecological sustainability (especially as expressed in terms of the quantity and quality of the land resource base and of the ecosystem services that these agro-ecosystems can provide) is being compromised. The output augmentation aspect of modern agro-food system development is of particularly great socio-political significance. Output expansion, which leads to a long-term tendency to declining commodity prices in real terms, will strike citizens, scientists and policy makers as being evidence that sustainability concerns are unwarranted or exaggerated. The Achilles' heel of agro-environmental and sustainability movements is the fact that progressive augmentation of output undermines the face validity of claims that agricultural sustainability is a critical human concern.[8]

Compensatory technological change and output augmentation thus tend to be the aggregate reflections of producers' and R&D systems' responses to the problems and opportunities caused by the long-term compromising of sustainability. This tendency suggests another crucial set of principles for socioecological analyses of agro-food systems. The highly variegated nature of agro-ecosystems, and especially the fact that they do not constitute a definite global system, means that global agro-ecological crisis per se is unlikely to be manifest. In addition, it is unlikely that straightforward reactions to ecological challenges at the consumer-economic or farm enterprise levels, or induced-innovative institutional responses, will provide meaningful solutions to these problems. Thus, somewhat ironically, there is a certain *sustainability of unsustainability* in world agricultures as we move to the twenty-first century. That is, the social relations and technoscientific practices which may be regarded as unsustainable in a long-term sense exhibit a powerful dynamic towards being preserved, or sustained, in the short-term.

Agro-ecosystem degradation processes thus tend not to create endogenous processes – either cataclysmic or evolutionary – of reform or redirection. State subsidies that help to socialise the ecological externalities of modern agriculture and output increases from the compensatory technologies that mask foundational sustainability problems have thus far served to minimise the role of endogenous processes of change. The principal ways through which these dynamics can be influenced for the better are *social and political*. This is to say that social movements are likely to be the primary mechanism for successfully effecting changes in the various political-economic feedback loops of socialisation of external costs and development of compensatory technologies that make sustainability threats so sustainable over the short- to medium-term.

UNDERSTANDING THE DYNAMICS OF AGRICULTURAL SUSTAINABILITY MOVEMENTS

The late twentieth century shows every sign of being the beginning of an era in which there is a rapid, albeit uneven and partial, industrialisation of agriculture in most of the north (Welsh 1996). This nascent shift is a cause, and in some senses an effect, of the patterns by which foundational agro-ecosystem sustainability is being compromised. In turn, as a result of global competition pressures becoming increasingly generalised in the agro-food systems of the north (especially outside the EU), there is a renewed trend towards a 'disappearing middle' in the agrarian structures of many of the advanced industrial societies (NRC 1995). This 'disappearing middle' process appears to be becoming sufficiently broad that it now often involves the marginalisation of what were formerly regarded as relatively large family proprietor farms (e.g. many producers with $200,000 to $300,000 gross annual sales of hogs and milk) (Durrenberger and Thu 1996).

I would argue that the most significant 'weak underbelly' of incipient industrial-capitalist agriculture is its political-ecological implications. But in the preceding section it was claimed that the foundational sustainability of agro-ecosystems tends not to become implicated in politics and systemic social change. How, then, can it be that mainstream agriculture exhibits political-ecological vulnerabilities? I would suggest that while long-term foundational sustainability is not at issue, many of the symptoms of sustainability problems are. These include pollution from agricultural sources, farm-structure concerns (e.g. future of family farming), equity issues concerning the fairness of subsidising the pollution abatement of industrial farms that cause the lion's share of these problems, concerns over public agricultural research priorities, impacts of liberalised trade on farm structure and the agricultural environment, and concerns regarding food quality (in the broad sense not only of wholesomeness, but also ethical concerns about the conditions under which food is produced, e.g. Kloppenburg *et al.* 1996).

How likely is it that a movement that focuses on such a range of issues closely or not so closely related to sustainability can be effective over the long term? This question can only be answered in partial ways, and again in terms of understanding the context and some general principles about movements such as environmentalism. First, under advanced capitalism, the general achievement of sustainability must inherently be a direct or indirect state regulatory practice.[9] Ongoing trends, however, suggest a general erosion of the regulatory state with regard to agriculture, food, and environment (McMichael 1994, Bonnano and Constance 1996, IISD 1996). In addition to the secular decline in national state regulatory capacity due to world economic integration (particularly through WTO and regional trade agreements such as NAFTA), the decline of agricultural commodity programmes will further erode state leverage over farmer practices (by undermining programmes such as the current 'cross-compliance' provisions of the 1985 and 1990 US Farm Bills).

The forces and trends just portrayed should not, of course, be thought of as immutable or linear. Late twentieth- and early twenty-first-century capitalism yet retains a great deal of structural flexibility that could well result in some unpredictable shifts. For example, its current dynamic seems to be a linkage between accumulation and stagnant or declining real wages – a linkage that would seem to have definite political-economic limits due to the tendency to underconsumption or realisation crises. Also, it is not implausible that many powerful major nation states (as well as dozens of geopolitically less consequential ones) could experience losses or vulnerabilities due to neoliberalism and trade liberalisation, so that in a decade or so the WTO and regional trade accords could be selectively rolled back, and food security and environmental protection be given more emphasis. Stagnation of global food output could cause the real price of agricultural commodities to increase significantly, and thus introduce a new global (and local) politics of food security. Dramatic shifts in real energy prices (or possibly very substantial energy or BTU tax increases) could also alter these dynamics.

Beyond these unknowable possibilities, the most important social forces that could provide a countervailing tide to global integration of the agro-food system, to the decline of household forms of agricultural commodity production, and to structural blockages to achievement of sustainability, will be social movements, as noted earlier. There are large numbers of people who have actual or potential grievances and interests that could be united into some kind of an omnibus coalitional agro-food system movement that contests deregulation, globalisation, and agro-ecosystem degradation. In the advanced countries there are many general types of social movements that could become directly involved in the pursuit of an alternative or countervailing agro-food political movement: farmer-oppositional organisations and movements (e.g. National Farmers' Union); farmer-support movements (predominantly non-farmer organisations such as the National Family Farm Coalition, Center for Rural Affairs, and North American Farm Alliance, which advocate for family farming and related causes such as farmland preservation); sustainable agriculture technology movements (e.g. Sustainable Agriculture Working Group); 'agro-localist' movements (especially CSA networks); food/nutrition/consumer/safety movements (e.g. Public Voice for Food and Health Policy); the organic and healthy-food movements; and the (mainstream) environmental movement. Note that these movements overlap to some degree, e.g. the Center for Rural Affairs in the US is simultaneously a farmer-support and sustainable agriculture technology movement organisation. Each of these movement types, however, has very different goals and constituencies. Coalitions among these movements are possible, though, and the nature of the coalitions that occur could make a major difference to the shape of global agro-food systems in the future. But this potential breadth of appeal must be considered alongside a number of constraints on effective mobilisation for a movement such as this. Chief among these constraints is the internal diversity of agricultural sustainability movements and the consequent barriers to movement cohesion.

Diversity of agricultural sustainability movements: a typological approach

It is useful to begin with the observation that, in an ultimate sense, there is actually no such thing as a sustainable agriculture movement. By this I mean that despite the range of possible appeals for achieving a greater degree of agricultural sustainability, there is no one underlying notion or strategy that can serve as a singular unifying focus for the movement. This observation has two important implications. First, we need to understand the diversity among agricultural sustainability movements. Many treatments of environmentalism and related movements such as agricultural sustainability tend to exaggerate these movements' coherence and thereby miss the fact that the roles they can play are highly situational. The diversity of environmental motivations and ideologies is nowhere better demonstrated than with respect to agriculture. Second, it is essential to appreciate that it will only be through coalitions – of sustainable agriculture movements, and of these movements with others with which they have shared concerns – that this social movement force can achieve the extent of meaningful impacts that are required to address fundamental agro-ecological sustainability problems.

Table 14.1 presents a typology of environmental claims, ideologies, and discourses and provides examples of how they have been brought to bear on agriculture.[10] The four-fold typology is constituted by two generic dilemmas of modern environmentalism. One dimension is that of *re-rationalism versus anti-rationalism* (Murphy 1994). As Scott (1990) has noted, the underlying motivation, or source of identity, of many environmental groups is a notion that the rationalisation/modernisation dynamic of industrial capitalist societies, undergirded by the rationality of modern science while beset by the irrationalities wrought by technology, brings the risk of large-scale ecological disruption or disaster. There is a strong feeling that science must be brought under social control and that the instrumental, dehumanising rationality of industrialisation, rampant consumerism, and competitive social relations must be tempered if not terminated. This orientation may thus be termed 'anti-rationalist'. But as Yearley (1991) has stressed, as much as this anti-rationalist *Weltanschauung* is deeply held in many environmental circles, it has not tended to be persuasive in the political arena because this discourse is readily delegitimated by the dominant rationalist/modernisation discourse. Environmental groups' public positions have therefore tended to reflect what might be called a 're-rationalist' position – one that seeks to modify industrial-technological rationalism at the margin by use of scientific data and accounting mechanisms, and by appeals to environmental or natural-resource efficiency (see Murphy 1994). The emphasis is thus on balancing the instrumental rationality of industrialisation with another instrumental rationality based on prudent management of the environment and natural capital.

The second dilemma of modern environmentalism is that of *centralisation versus localism*. Decentralisation or localism – and the ideal of harmonious, decentralised communities – is typically an integral component of the utopian ideal of much of the environmental community, particularly the more 'expressive' ecology movements discussed by Scott (1990). In other words, 'think globally, act locally'. But as much as this is the ideal for many groups and individuals, decentralist environmentalism also has political liabilities. By operating mainly at a local level, environmentalists would spread their resources too thinly across what is an almost infinite number of environmental policy fora, and thereby have little influence. In actual practice, the dominant organisations of modern environmentalism, such as the 'big ten' in the US, have tended to 'think globally, and act globally'. Global constructions of environmental problems, such as the current notion of global environmental change, can be politically persuasive because they can be based on scientific reasoning and data (e.g. climate science, conservation biology), can create an imagery of imminent world-scale environmental

Table 14.1 A typology of environmental ideologies, motivations and discourses, with agricultural examples

	Orientation to the Instrumental Rationality of Modernisation	
Politico-Spatial Orientation	Re-Rationalist	Anti-Rationalist
Centralizing	*Regulationism/ Managerialism*	*Preservationism*
	Regulation of chemical use in agriculture	Buffalo Commons
	Opposition to subsidies to industrial agriculture	Exclusion of agriculture from sensitive habitats
Localizing	*Alternative technologism*	*Indigenism*
		CSA movement
	Sustainable agriculture technology movements	Movements to preserve indigenous cultures and knowledges

Source: Buttel 1994

disaster if action is not taken to ameliorate the underlying causes of problems, and create a strong moral justification to override politics-as-usual to do so. Centralised strategies – Earth Summits, international environmental agreements, and influencing the priorities and policies of the World Bank and national development agencies – have come to be of particular importance to the political agenda of the most powerful and influential environmental organisations.

The four types of environmentalism identified in the typology are *regulationism/managerialism* (centralising, re-rationalist), *preservationism/'deep ecology'* (centralising, anti-rationalist), *alternative-technologism* (localising, re-rationalist), and *indigenism* (localising, anti-rationalist).[11] These four categories correspond to very different orientations to agro-food and agro-ecological change, and each would push agriculture in very different directions.

The two most influential versions of environmentalism with respect to agriculture today are regulationism/managerialism and alternative-technologism. Regulationism/managerialism is the dominant form of environmentalism in the 1990s on account of the fact that it benefits from the political persuasiveness of using a science-based alternative rationality, and from the strategic emphasis on the high-level national and international arenas of policy and power.[12] Regulationism/managerialism is reflected in several ways in agriculture. Environmental mobilisation aimed at regulating chemical use in agriculture and at increasing the efficacy of regulation of the food supply, opposition to industrialisation of livestock, opposition to state socialisation of the environmental costs of large-scale agriculture, and the promotion of sustainable agricultural development programmes in the developing world through the international development agency complex are some major examples.

Alternative-technologism is also an increasingly influential expression of environmentalism in agriculture. In virtually all of the advanced countries there are social movements that are actively engaged in influencing public research institutions to emphasise 'sustainable', 'low-input', or alternative agriculture. The principal rationale behind, and ideology of, this form of mobilisation is that a sustainable agriculture is a rational agriculture in that it is best able to balance environmental and socio-economic goals for agriculture. This is a technologically led vision in which the future is seen as hinging on whether sustainable or unsustainable technologies are given priority in research institutions.[13]

The kind of environmentalism I refer to as indigenism is a form that is mainly aimed at achieving ecologically stable, socially harmonious, decentralised communities, and making possible de-industrialisation of the relations between food production and consumption. This form of environmentalism is typified by efforts in the developing countries to promote extractive reserve systems in rainforest zones that are exclusive to indigenous peoples, and in general to preserve indigenous cultures and indigenous knowledge systems. But indigenism also has many industrial country and agricultural expressions. In the US, for example, there is a rapidly growing community-supported agriculture (CSA) movement and a related community food systems or foodshed movement (Kloppenburg *et al.* 1996). Both are aimed at 'building community' through the production and circulation of de-industrialised, decommodified foods (Friedmann 1994). These groups have particular concerns with food quality – understood not only, or even primarily, in the sense of being free of health-threatening micro-organisms, but rather in terms of being produced under 'natural' conditions and under non-exploitive social relations. Indigenist sustainable agriculture movements also include farm and non-farm groups that promote indigenous/local farmer knowledge (see Hassanein and Kloppenburg 1995).

The final category of environmentalism – one that, as of yet, is of minor import, at least in the West – could nonetheless play a major role in the future. Preservationism is a style of environmentalism that has one overriding goal: that of preserving natural habitats from encroachment by any human activities, including agriculture, that would disrupt the 'natural' biotic community. Preservationism is thus by and large hostile to agriculture (as Wood (1993) points out in appropriately veiled language). Thus far, preservationism has been most influential with respect to the developing countries and their tropical rainforest zones. Some international environmental groups whose agendas are focused on tropical rainforest conservation, biodiversity, and related areas are deeply distrustful that any form of agriculture, even 'sustainable' versions, represents suitable activities for sensitive rainforest zones (compare Fearnside 1993, Healey 1993, and Wood 1993). These groups therefore often attempt to promote rainforest and biodiversity conservation programmes and practices in which agents of resource extraction, including peasants as well as cattle ranchers and logging, timber, and mining interests, are excluded by statute and force of law. But note as well that a few environmentalists and conservation biology activists in the US have made proposals that essentially would eliminate agriculture from large expanses of sensitive North American eco-zones to advance the goal of preservation. The so-called Buffalo Commons proposal, aimed at removing agriculture from much of the Great Plains (because of insufficient water and environmental destruction) and at simultaneously helping to solve the wheat overproduction problem, is one such example. A recent proposal by conservation biology activists to create human-free natural corridors across most of the country to achieve conservation of biological diversity has also generated considerable controversy. While these proposals are unlikely to be of high policy relevance, the fact that they have been made and have achieved considerable attention illustrates that preservationism can be an influential environmental discourse in the industrial as well as in the developing world.

Contemporary sustainable agriculture movements thus tend to exhibit important divisions between re- and de-rationalist-oriented groups, and somewhat less so between localist- vs globalist-oriented groups. While these differences seldom cause overt schisms, they do reflect socially significant differences in interest and world view. Even so, there are occasionally some significant coalitions among these groups, especially when they tend to pull together around US farm bill deliberations. The development of a greater ability to appreciate differences and work out a division of labour will thus make a major difference in the effectiveness of these movements in the future.

Sustainable agriculture movement coalitions

The diversity and vitality of sustainable agriculture movements, both sustainable agriculture organisations per se and related groups, are the encouraging part of the story for the future of this social movement force within agro-food change. The downside of this movement structure, however, is that the issues being pursued within various agricultural sustainability movements are insufficiently immediate and persuasive to generate the widespread mobilisation necessary to achieve major policy and institutional changes. The future of agricultural sustainability movements will thus lie in the nature and extent of coalitions they are able to form with other movements and groups.

An excellent example of the possibility of some kind of omnibus coalition – one that very nearly came to pass and which could have had a major effect on the global political economy – was the labour-environmentalist-farmer-sustainable agriculture coalition against NAFTA in the US. In the several months prior to the vote on NAFTA by the US Senate, every sign was that passage of the extension of NAFTA to Mexico would fail. It was only in the last month or so prior to the vote, when most major environmental groups cut deals with the Clinton administration and large corporations that caused them to leave the coalition, that there were enough votes for passage. And if NAFTA would have failed in the US Congress, it is likely that ratification of the Uruguay Round GATT agreement would have failed in the Congress as well.

Just as the anti-NAFTA movement was an omnibus coalition, none of the types of social movements mentioned above can be the sole 'lead horse' of an agro-food sustainability movement. It is unlikely, for example, that farmer movements can do so, even though farmer participation in a countervailing agro-food system movement will be critical. While there exist small groups of 'against-the-grain' farmers and considerable discontent among farm people today (mainly about farm prices and environmental and commodity programme-related regulations), I do not anticipate that farmers will take the lead in a countervailing agro-food system movement. Despite their economic problems, the farmers who remain today are increasingly persons of property, they see their interests as being tied to enhancement of the prerogatives of petty property, and must necessarily be preoccupied with more immediate matters of competition and survival (Mooney and Majka 1995). Farmer-support movements (e.g. the Center for Rural Affairs and Institute for Agriculture and Trade Policy in the US) have been important to the very modest progress that has been achieved thus far – particularly because, in the US context at least, they have been able to blend advocacy of farmers' interests with sustainability goals. But farmer-support movements are largely 'broker' movements, and can only go as far as the groups they try to unite are willing to go.

The component of the sustainable agriculture movement that focuses on research and sustainable technology has had significant accomplishments in raising awareness of sustainability issues in public research institutions. The efficacy of these organisations is limited, however, because alternative technology, in and of itself, is not a potent lever for achieving sustainability.[14] As a result, these groups are increasingly recognising the limits of a research- and technology-driven programme and are making tentative moves to incorporate extra-scientific goals and strategies. Agro-localist movements (e.g. CSA groups) are generating excitement and attention, but their local and inward focus, their tendency to avoid the policy realm, and their preoccupation with survival, argue against them being a singular force behind agricultural sustainability.

I would argue that the future character of the 'mainstream' environmental movement will be critical in creating the conditions for whether countervailing agro-food movements can succeed. The environmental movement will not lead the countervailing agro-food movement, but environmental movement support will be important in expanding the base of agro-food activism and building farmer–citizen linkages. There are many obvious reasons why the mainstream environmental movement should care about agricultural sustainability. Agriculture is the single most important use of land in the world. Food is the first and foremost resource that bears on human survival in the biosphere.

At the same time, agriculture has not been very central to mainstream environmentalism (despite the fact that Rachel Carson's (1962) *Silent Spring* was focused on pesticides and played such a formative role in modern US environmentalism). Agriculture is such a well-defined and bounded institutional ensemble that it is difficult for 'outsider' groups to penetrate its structure authoritatively. Agricultural issues are not very glamorous ones from an environmental/ecological point of view. Environmental groups have long preferred a global framework and ideology ('population bomb' neo-Malthusianism, limits to growth, global environmental change), and agriculture is the prototypical 'local' – and, hence, within such a framework, a relatively unimportant – environmental issue. Mainstream environmentalists are generally most interested in tropical rainforests, biodiversity, and so on, and when agriculture comes to mind within a mainstream environmentalist framework it is most often on account of the fact that agriculturalists are seen as threats, who will encroach on and destroy sensitive tropical habitats. In the advanced countries the mainstream farm organisations are among the most anti-environmental pressure groups; for example, in recent years the breadth of farmer support for state 'right to farm' laws that impede environmental regulation has been impressive (DeLind 1995).

Despite the manifold reasons why mainstream environmental groups do not currently provide much support for a countervailing agro-food movement, there are some good reasons why they can and should. First, I would argue that the post-UNCED period demonstrates that the current formula of global environmentalism is failing and is heading towards crisis. Global environmentalism (in the form of movement fascination with international power) indirectly gave us NAFTA, and ultimately GATT and the WTO (because the Clinton administration was able to entice mainstream environmental groups to abandon the anti-NAFTA coalition in order to get a seat at the table at NAFTA environmental dispute tribunals). Buttressed by global trade liberalisation, there is now a general assault on environmental regulations in most countries in the Western world. Modern environmentalism has generated complacency, if not boredom, among most citizens because its lead issues (global warming, in particular) lie outside of citizens' immediate experiential realities and are a phenomenon that will mainly affect only one's grandchildren or great grandchildren. Membership in the major environmental organisations is plummetting. Mainstream global environmental groups have strained relations with local environmental organisations, which are now generating the most enthusiasm and excitement in the movement (Gottlieb 1994).

There are many critics of environmental movements (and many social scientists) who exaggerate the flexibility these movements have to develop ideologies and strategies as they see fit.[15] More concretely, we need to recognise that it is something other than mere short-sightedness that causes dominant environmental ideology to stress global environmental issues and to de-emphasise issues such as toxics and agriculture and food. We need to recognise that mainstream environmentalism faces some very formidable mobilisation constraints. The continual resort to global ideologies is due, at least in part, to the exigencies of how to appeal to large numbers of members, supporters, the media, and public officials, with modest resources. Global formulations based on science that forecast calamity and catastrophe, and that justify international agreements to override politics and business as usual, are an attractive formula in many respects for enhancing mobilisation. This overall formula will likely not be superseded entirely. But for the long-term good of the movement, as well as for the sake of agro-food sustainability, the movement needs to give more emphasis to concerns that are well within the direct experiences and realities of significant numbers of people.

While not neglecting the important global dimension of environmental issues, modern environmentalism needs to find some new formulas which have some foundation in the immediate problems and concerns of the majority of citizens. Agro-food issues alone cannot provide this reorientation, but food and land issues could be of some considerable importance. For example, toxics (an issue increasingly relegated to local, grassroots groups because it is not fashionably 'global') could be a focal point of mobilisation for which agriculture, food, and land could have important linkages (through the connection with pesticides, preservatives, groundwater contamination, etc.). Most importantly, toxics or related issues are ones that help to justify the enhancement of state environmental regulatory capacity which is necessary (though, of course, not sufficient) for addressing agro-food/sustainability concerns. Extending food concerns in ways that are relevant to the interests of the growing ranks of the poor and hungry will also be pivotal.

Counter-movements

Almost as integral as sustainable agriculture movements are to the future of agro-food systems and agro-ecosystems, is the role played by movements that actively resist agricultural sustainability. Three factors have increased the role of anti-sustainability movements. One is that, now that environmental movements have gone through several cycles of rise and decline, there is a certain legitimacy created for seeing environmental world views as being transitory, and environmental reforms as being reversible. Second, the costs to agro-food capitals of sustainability reforms have become clearer, and have created an incentive to organise in new and more effective ways. Third, in the US at least, the shift of dominant political discourse to the right and the delegitimation of states and state-regulatory power have created new spaces for anti-sustainability movements. The anti-sustainability movement has a considerable span of support, ranging from mainstream farm organisations and commodity groups to agribusiness firms and conservative think-tanks (see, e.g. DeLind 1995).

In recent years anti-sustainability movements and associated agribusiness trade organisations have arguably been more effective than sustainability movements have been. Among the major victories of anti-sustainability movements have been the passage of 'right-to-farm' legislation in many states (DeLind 1995); passage of 'food disparagement' legislation in approximately twelve states, and the ability of agribusiness firms to largely dominate the International Standards Organisation and Codex Alimentarius bodies that regulate food trade (McMichael 1994). The US 'property rights' movement has also become very influential in achieving deregulation of agriculture and land markets. Perhaps most importantly, agribusiness capitals and anti-sustainability groups have been able to triumph by pushing successfully 'product over process' policy discourse that increasingly prevails in national regulatory practice (Krimsky and Wrubel 1996). Much like food disparagement legislation, 'product over process' regulatory practice dictates that public and public-interest group objections to the conditions under which products are produced do not have standing in the absence of conclusive evidence that the product itself is harmful to health or environment. The ultimate impact of these laws is the elevation of a significant barrier to agricultural sustainability mobilisation. Product over process regulatory practice, for example, would preclude most regulation of biotechnology products, and would preclude labelling of milk from cows treated with bovine growth hormone. Product over process practice in conjunction with 'food disparagement' laws also creates future threats to sustainability movements; labelling foods as 'organic' might become illegal unless organic foods could achieve certain specific product quality standards (e.g. complete lack of pesticide residues).

CONCLUDING REMARKS

The contemporary evolution of global and national agro-food systems is occurring in a milieu in which political-economic processes are increasingly being played out in ways that cannot be captured adequately or comprehensively by theories of endogenous social change (Buttel 1995). That is, many agro-food restructuring processes are increasingly occurring outside of the institutional arenas – the formal economy, trades unions, peak associations, the state, political parties, the Bretton Woods institutions of global economic relations, and so on – that have been the traditional foci of political economy. Of growing importance to agro-food affairs are processes within civil society, particularly social movement mobilisation, that are related but not reducible to the conventional axes of political economy (see also Marsden 1992). Accordingly, this chapter has set forth a perspective on the political ecology of agro-food systems and agricultural sustainability in which the role of social movements is spotlighted. It is my argument that this emphasis on movements and the processes of civil society has applicability in areas of agro-food research other than sustainability. But particularly as far as agricultural sustainability is concerned, social movement and civil-societal mobilisation analysis must be a critical component of the expanded political-economic framework required to understand the relations between agriculture and the biosphere.

NOTES

1 While comparable data from the most recent (1992) Census of Agriculture were not reported in NRC (1995), it is very likely that these data would have indicated a continued – or more rapid – trend to concentration.

2 Araghi's (1995) impressive analysis of global depeasantisation since the Second World War shows that there are some significant parallels between trends in

north and south. This is particularly the case with respect to the resumption of more rapid rates of depeasantisation from the late 1970s or so to the present.

3 Most of these policy instruments, however, had mixed or contradictory effects. Commodity programmes, for example, have tended to enable smaller or marginal producers to remain in business, while simultaneously providing price stability which encouraged larger investments and subsequently increased barriers to entry.

4 It is thus crucial to recognise, as many theories of agrarian structure do not, that the three master processes of agrarian change are the structure and rate of entry, the structure and rate of differentiation of farm household/firms over the career and life cycle, and the structure and rate of exit. There is, in particular, growing evidence that the structure and rate of entry has made a larger contribution to structural change in farming than have the other two component processes (Jackson-Smith 1995).

5 I here refer to James O'Connor's (1994) discussion of the second contradiction of capital: the tendency for capitalists facing competition to externalise environmental costs, and over time to increase the collective costs of production because of ecosystem degradation.

6 Here the absence of a hyphen in socioecological is intentional, suggesting their sui generis joint status as social and biophysical phenomena.

7 I have noted elsewhere (Buttel 1995) that one of the most significant aspects of modern agricultural technologies is their generic character and the fact that they are able to override substantially the variability among the globe's extremely diverse agro-ecosystems. Compensatory technologies also tend to be generic technologies premised on overriding local agro-ecological specificities.

8 It should thus be noted in related fashion that my own usage notwithstanding, it is important to avoid the presumption that 'nature' is only an obstacle or constraint. Indeed, the often enormous capacity for resilience, productivity, and bounty of many of the world's agro-ecosystems – even when experiencing perturbations of human and natural origin – exemplifies the 'gift of nature' aspect of our natural world, as well as some of the reasons why unsustainability can be so sustainable.

9 Note that some persons, groups, and perhaps entire communities or regions may cultivate crops and reproduce themselves on a sustainable basis, but in the advanced countries at present these practices are usually constituted as a direct or indirect expression of opposition or resistance to the larger agrarian political economy or agro-food system. Or, in other words, sustainable production in the north is typically one among several expressions of sustainable agriculture movements. Sustainable social relations and production practices in the south, by contrast, tend to be found primarily among indigenous groups and subsocieties. But to some degree in recent years there have been instances in which continuity of these practices is pursued, in part, as an element of resistance to dominant societies or global society.

10 It should be stressed that regardless of the typology of environmentalism or sustainable agriculture movements one prefers, the crucial fact is that these movements exhibit very considerable diversity that militates against movement consolidation. This typology was originally reported in Buttel (1994).

11 Note that this is a typology of environmental claims, discourses, and ideologies, rather than a typology of environmental groups. Thus, while particular environmental groups can clearly be located within this typology, many such groups will employ more than one type of claim or discourse, depending on the forum and nature of the issues that are being contested. It should also be

stressed that conceptualisation of environmental groups and movements employed above is an intentionally broad one. This definition encompasses not only conventional environmental groups, but also other new social movement-type groups (see Scott 1990) that actively employ environmental and technological risk symbols and claims, e.g. much of the consumer lobby (Beck 1992).

12 This category is essentially identical to the 'managerialist' category of Redclift (1987).

13 It should be noted that this category of environmental claims and discourse is based on the US context (in which public agricultural research policy is largely shaped at the state level) and is not intrinsically one that privileges science and technological discourse. The fact that such discourse is privileged is because the fora that are most penetrable by these groups are ones in which research and empirical data are crucial. This category has similarities to Leff's (1995) notion of 'ecotechnological rationality'.

14 Sustainable technologies, for example, may not be widely embraced if the institutional conditions (e.g. relaxed regulation of chemical use, low fossil fuel prices) are unfavourable. Some sustainable technologies under these conditions will actually be relatively attractive to very large industrial-agriculture firms.

15 One interesting reflection of this is the recent publication of numerous books which analyse environmental movement strategies – and which, in most cases, blame the lack of progress achieved by the movement on its having chosen the wrong ideologies and strategies (see, e.g. Gottlieb 1994).

BIBLIOGRAPHY

Araghi, F. A. (1995) 'Global depeasantisation, 1945–1990', *Sociological Quarterly* 36: 337–68.

Beck, U. (1992) *Risk Society*, Beverly Hills: Sage.

Bonnano, A. and Constance, D. (1996) *Caught in the Net*, Lawrence: University Press of Kansas.

——, Busch, L., Friedland, B., Gouveia, L., and Mingione, E. (eds) (1994) *From Columbus to Conagra: The Globalisation of Agriculture and Food*, Lawrence: University Press of Kansas.

Buttel, F. H. (1994) 'Agricultural change, rural society, and the state in the late twentieth century: some theoretical observations', in D. Symes and A. J. Jansen (eds) *Agricultural Restructuring and Rural Change in Europe*, Wageningen: Wageningen Studies in Sociology, Wageningen Agricultural University.

—— (1995) 'Twentieth-century agricultural-environmental transitions: a preliminary analysis', in *Research in Rural Sociology and Development* 6: 1–21.

—— (1996) 'Environmental and natural resource sociology: theoretical issues and opportunities for synthesis', *Rural Sociology* 61: 56–76.

Carson, R. (1962) *Silent Spring*, New York: Houghton Mifflin.

Colborn, T., Dumanoski, D., and Myers, J. P. (1996) *Our Stolen Future*, New York: E. P. Dutton.

DeLind, L. (1995) 'The state, hog hotels, and the "right to farm": a curious relationship', *Agriculture and Human Values* 12: 34–44.

Dickins, P. (1992) *Society and Nature*, Philadelphia: Temple University Press.

Durrenberger, E. P. and Thu, K. M. (1996) 'The industrialisation of swine production in the United States: an overview', *Culture and Agriculture* 18: 19–22.

Fearnside, P. M. (1993) 'Forests or fields: a response to the theory that tropical forest conservation poses a threat to the poor', *Land Use Policy* 10 (April): 108–21.

Friedland, W. H., Barton, A. E., and Thomas, R. J. (1981) *Manufacturing Green Gold: Capital, Labor, and Technology in the Lettuce Industry*, New York: Cambridge University Press.

Friedmann, H. (1994) 'Distance and durability: shaky foundations of the world food economy', in P. McMichael (ed.) *The Global Restructuring of Agro-Food Systems*, Ithaca: Cornell University Press.

—— and McMichael, P. (1989) 'Agriculture and the state system: the rise and decline of national agricultures, 1870 to the present', *Sociologia Ruralis* 29: 93–117.

Goodman, D. and Redclift, M. (1982) *From Peasant to Proletarian*, New York: St Martin's Press.

—— (eds) (1989) *The International Farm Crisis*, London: Macmillan.

—— (1991) *Refashioning Nature*, London: Routledge.

Goodman, D., Sorj, B., and Wilkinson, J. (1987) *From Farming to Biotechnology*, Oxford: Basil Blackwell.

Gottlieb, R. (1994) *Forcing the Spring*, Washington, DC: Island Press.

Hassanein, N. and Kloppenburg, J., Jr (1995) 'Where the grass grows again: knowledge exchange in the sustainable agriculture movement', *Rural Sociology* 60, 721–40.

Healey, R. G. (1993) 'Forests or fields: a land allocation perspective', *Land Use Policy* 10: 122–26.

International Institute for Sustainable Development (IISD) (1996) *The World Trade Organisation and Sustainable Development*, Winnipeg: IISD.

Jackson-Smith, D. B. (1995) 'Understanding the microdynamics of farm structural change: entry, exit, and restructuring among Wisconsin family farmers in the 1980s', unpublished PhD dissertation, Madison: Department of Sociology, University of Wisconsin.

Kloppenburg, J., Jr, Hendrickson, J., and Stevenson, G. W. (1996) 'Coming into the foodshed', *Agriculture and Human Values*, 13: 33–42.

Krimsky, S. and Wrubel, R. (1996) *Agricultural Biotechnology and the Environment*, Urbana: University of Illinois Press.

Leff, E. (1995) *Green Production*, New York: Guilford.

Lehman, D. (1986) 'Two paths of agrarian capitalism, or a critique of Chayanovian Marxism', *Comparative Studies in Society and History* 28: 601–27.

McMichael, P. (ed.). (1994) *The Global Restructuring of Agro-Food Systems*, Ithaca: Cornell University Press.

Mann, S. A. (1990) *Agrarian Capitalism in Theory and Practice*, Chapel Hill: University of North Carolina Press.

Marsden, T. K. (1992) 'Exploring a rural sociology for the Fordist transition: incorporating social relations into economic restructuring', *Sociologia Ruralis* 32: 209–30.

Mooney, P. H. and Majka, T. J. (1995) *Farmers' and Farm Workers' Movements*, New York: Twane Publishers.

Murphy, R. (1994) *Rationality and Nature*, Boulder: Westview Press.

Norgaard, R. B. (1994) *Development Betrayed*, London: Routledge.

National Research Council (NRC) (1995) *Colleges of Agriculture at the Land Grant Universities*, Washington, DC: National Academy Press.

O'Connor, J. (1994) 'Is sustainable capitalism possible?' in M. O'Connor (ed.) *Is Capitalism Sustainable?*, New York: Guilford.

Redclift, M. (1987) *Sustainable Development*, London: Methuen.

Scott, A. (1990) *Ideology and the New Social Movements*, London: Unwin Hyman.

Thompson, P. B. (1995) *The Spirit of the Soil*, London: Routledge.

Wood, D. (1993) 'Forests to fields: restoring tropical lands to agriculture', *Land Use Policy* 18 (April): 91–107.

Yearley, S. (1991) *The Green Case*, London: HarperCollins.

Welsh, R. (1996) *The Industrial Reorganization of US Agriculture*, Greenbelt, MD: Henry A. Wallace Institute for Alternative Agriculture.

COMMENTARY ON PART VI

Sustainability and institution building: issues and prospects as seen from New Zealand

Richard Le Heron and Michael Roche

Michael Redclift poses three questions which go to the heart of debates about sustainability. How do societies arrive at sustainable levels of resource exploitation? How do societies arrive at ways of valuing environmental gains and losses? How do societies influence the social and economic behaviour of citizens towards more concern with sustainability? His argument is powerfully simple. We should be spending more time sorting out the institutional fabric that might keep open sustainability options, so bequeathing institutions rather than environment to future generations. Fred Buttel explores similar concerns, arguing that the achievement of sustainability is intimately connected to direct and indirect state regulatory practice. Buttel's analysis of US agricultural sustainability movements, against the backdrop of new realities of the agro-food sector, reveals mobilisation of sustainability movements in civil society as a missing dimension in political economy thought. The sketch of issues in the two chapters highlights *both* regulation and governance as central to *any* sustainability problematic. Indeed, Redclift emphasises development as a potentially reflective and redistributive process which is distinguished by social/societal awareness of institutional options, shaped from a better understanding of production and consumption flows. Knowledge of the latter, he contends, would help societies make radical choices, at different levels of analysis and over different time scales.

Recent official promotion of sustainable agriculture in New Zealand in the 1990s offers a specific context to explore the Redclift-Buttel institution-building 'thesis'. The New Zealand experiment (Kelsey 1995, Le Heron and Pawson 1996) involving wholesale restructuring of the regulatory fabric and the rise of new governance practices driven by neo-liberal/new right ideologies, is a very distinctive institutional frontier in which sustainability discourses have emerged (McDermott 1997). The New Zealand situation is an especially useful national context because an historical dependence on export earnings from agro-commodity systems requires sustainability questions to be contextualised with reference to production-consumption issues (Le Heron and Roche 1996a, 1996b). New Zealand's restructuring has moved from an initial dismantling of interventionist state structures (a deregulation phase in the mid-1980s) so conferring new conditions, into a period of new links and interactions within and beyond New Zealand. Two threads of institutional transformation are apparent – the appearance of new private sector-oriented regulatory frameworks (reregulation) and the advent of an extraordinary variety of locally initiated governance arrangements in the new context (governmentality). These two relatively autonomous threads, spanning the state-economy-civil spheres, form a confusing and often contradictory setting for discussion and action relating to sustainability. Yet, it is in this context, and to a degree shaping the context, that significant and insistent moves have been made to introduce sustainability principles into New Zealand agriculture.

This short review collates information on one aspect, the extent to which ideas about institution building for sustainability have been incorporated in thinking and operational practice with respect to regulatory arrangements and governance experiments around agriculture. The review is thus an explicit attempt to highlight levels of awareness of some institutional constraints and opportunities inherent in the present New Zealand context. For the review, three interlocking dimensions are explored, awareness of structural realities impinging on institutional innovation, world-views and value positions which affect valuing processes, and the limits to existing organisational capacities. These dimensions have been selected because they are highly pertinent to the Redclift-Buttel thesis and because they are currently the subject of considerable discussion in New Zealand.

Foremost, agricultural sustainability in New Zealand is a discourse-complex (Le Heron 1996), resulting from the conjuncture of state policy and social movements in particular conditions. The discourse-complex is distinguished by a 'unifying' idea, that of sustainability, which has been appropriated and developed by coalescing, often temporary, sets of stakeholder-promoters. Deepening concern about agricultural sustainability can be traced to several triggers during the past decade. The main stimuli were concerns about impacts of climate change on agriculture (spawning studies in the late 1980s); the passage of omnibus legislation in the form of the Resource Management Act (RMA) 1991 which substituted a 'wise land use' for an 'effects' or 'outcomes' planning framework (Gleeson 1996); New Zealand's signature to international agreements in the early 1990s (Agenda 21, the Framework Convention on Climate Change, Convention on Biological Diversity); apprehension on the part of Maori (expressed through the Treaty of Waitangi framework) about the terms under which resources (land, water, sea) might be used (Stokes 1996); and growing awareness in individual ministries, trade associations, producer groups and the corporate sector of the necessity to substantiate New Zealand's 'clean green' image in the market place (Ministry for the Environment 1993, Smith 1993, 1996a). Although a sense of linearity might be implied by the listing, the concerns and the subsequent responses have been mutually interdependent and reinforcing. As a general point, at least three interwoven discourses are now discernible. These embrace sustainable resource management, sustainable land management, and sustainable land-based production. The discourses spring very directly from the new regulatory-governance milieu that distinguishes the New Zealand scene.

In terms of agriculture, the advent of the RMA was an important instrument in establishing an ethos of sustainability. The act was the first legislation in the world to define sustainable land management, albeit narrowly, as 'Managing the use, development, and protection of natural and physical resources in a way, or at a rate, which enables people and communities to provide for their social, economic and cultural well being and for their health and safety'. Reforms relating to resource management were accompanied by local government reform. Under the RMA , regional and district councils have land management responsibilities, which include preparation of policy statements and the conduct of such research as necessary to carry out functions under the act. The RMA, however, focuses on mediating environmental conflicts through recourse to the market rather than coercive instruments. The RMA quickly became contested interpretive terrain over whether planning was permissible in the marketplace, was limited to physical matters within an 'anthropomorphic envelope', and necessarily involved local participation (McDermott 1997). Newly established resource consent procedures enable different stakeholders, especially affected land owners and users, to identify and represent competing views of and agendas relating to sustainability. Variation is found in Regional Council interpretation of the RMA. Significantly, the 'effects' regime of the RMA, which has stripped away the politico-regulatory determination of 'wise' land use decisions under the old legislation, has begun to force consideration of local management frameworks to bring parties together into long-term relationships. This trend should not be overstated, however. McDermott (1997) suggests that many planners have failed to appreciate the new conditions and the different procedural requirements of the RMA. Professional practices are only slowly embracing the implications of effects-based planning procedures.

At the same time as the RMA framework was being cemented in the early 1990s, government signalled an integrated approach to sustainable land management. Several ministries organised conferences and working parties on sustainable agriculture (Ministry for the Environment and Foundation for Research, Science and Technology 1993, Ministry of Agriculture and Fisheries 1993, Strategic Consultative Group on Sustainable Land Management Research (SCSLMR) 1995, Ministry for the Environment 1995, 1996a). The collaborative process associated with the events was wide ranging and intensive, holding out the promise for a policy framework which could accommodate the range of agendas brought forward during discussions. The Strategic Consultative Group, for instance, invited views on

'An agenda for research on sustainable land management', defining key issues for comment. Amongst the issues canvassed were '*boundaries* and *categories* for sustainable land management from a research and science perspective', the 'approaches and frameworks needed for research', 'gaps and overlaps', 'priorities', 'maximising use of research' and 'coordination'. There is little doubt that insights from the consultative process have fed into sectoral policy formulation and science funding processes. As a result, four principles were identified to guide policy on sustainable land management. They are:

1 Primary responsibility for achieving sustainable land management rests with individual land owners and managers;
2 land management must proceed on an ecologically sustainable basis and the pursuit of economic and social needs should be modified accordingly;
3 land management decisions should recognise the biological and physical characteristics of the land resource and fully account for climatic risks and scientific uncertainty; and
4 government should exercise its powers to ensure that property rights are clearly specified, monitored and enforced in such a way that the market economy operates sustainably (SCSLMR 1995).

The principles reflect a simplistic conception of New Zealand society. They assume sustainability practices *can* materialise under capitalist property relations and in a market regime. Thus, private land owners and managers are seen as accepting the need for, are able to, and will actually bring about, sustainable land management practices, providing the state imposes an appropriate regulatory order of property rights. The obvious tension with Maori *iwi*-based (tribal) approaches to land management are ignored. In this respect, the possibilities of co-management with Maori, incorporating Maori principles of *matauranga Maori* (Maori understanding) and *kaitiakitanga* (stewardship), are overlooked (Urlich Cloher 1996). The priority of ecological over economic and social needs assumes land owners and managers are in a position to make changes. Again, this is indicative of land-centred thinking which isolates decisions about the land from the way in which land is incorporated into production and consumption processes. In stating the principles, however, the Strategic Consultative Group sought to break out of the straightjacket of interventionist thinking that had permeated the agricultural sector since the late 1950s. Ironically, the very commitment to individual responsibility is unsustainable without invoking state regulation of relevant practices, such as property rights, and depends on empowered and knowledgeable decision makers, 'who might be more accepting of government intervention when the need arises' (New Zealand Local Government Association, cited in Ministry for the Environment 1996b: 4)

The co-promotion of sustainability by policy and science funding organisations in the early 1990s saw the funding of several major research programmes dedicated to aspects of agricultural sustainability. The 'Definition and analysis of sustainable land-based production' programme at the University of Auckland (Blunden *et al.* 1995, Scott *et al.* 1996) is illustrative of how, by means of the Public Good Science Fund, research teams have attempted to explore the sustainability problematic in New Zealand (Le Heron 1994). The objectives of the research, which is focused on Northland, embrace an assessment of factors influencing the sustainable development of land-based production systems at the regional scale, with a focus on the conditions underlying, and the constraints on, sustainable development at the community and farm levels. The investigation has probed land-based production in one locality, featuring a mix of sheep and beef, and dairying and forestry. The operational scale of the research is thus consistent with the devolution by central government of responsibility for development to regions, and in regional contexts, to groups, on a voluntary basis. Research has sought to document through ethnography the changing social relations and materiality of land use change, at the same time as contextualising the structural realities which constrain or enable land use, farm, and community options. This conception rests on two premises, that sustainability must be considered in terms of social, economic, and environmental dimensions, and that factors underpinning sustainability intersect variably at different geographical scales.

The programme has yielded important insights, particularly about the changing nature of the institutional milieu in which the locality and the research is set. The conception of the research as a social process in its own right has meant the research team has found itself incorporated, in varying degrees, into a succession of power networks, within and beyond the locality, both as part of investigative strategies and in terms of the transfer of information to stakeholders in the locality and communities and agencies of government at the local and regional levels. These include exploration of sustainability issues at the farm scale with Northland Federated Farmers, discussions with Northland local authorities, liaising with Northland *iwi* agencies, and consultation with government departments present in Northland. Research is examining irrigation and forestry proposals, which cannot be divorced from the experiences and aspirations of local Maori communities. The irrigation work is considering questions surrounding the interpretation of sustainable management under the RMA as well as difficulties encountered by Maori in contesting resource consents (FORST Achievement Report 1996).

An invaluable source of insight on the sustainability topic have been workshops with other New Zealand research groups (AgResearch, Landcare, Forest Research Institute, Ministry of Agriculture), where a 'dawning awareness' of the institutional impediments in the New Zealand context has been collectively explored. From these meetings several clear pointers about preconditions for neoliberal-style institution building have crystallised.

First, there is a general recognition that words like 'sustainability', 'sustainable land management' and 'sustainable development' often mean little to local agents, particularly on farms and in the community, as distinct from those in government or corporate agencies. There is, however, a growing consciousness amongst farmers of the impacts of farming activities on the environment, derived in part from sensitivity to the RMA. The appearance of land and river care groups (modelled on Australian experience) are illustrative.

Second, many farmers as individuals may remain unable to address issues which impinge on sustainability at a larger scale, because economic survival of the farm or household is likely to take precedence (Taylor and McCrostie Little 1995). Blunden *et al.* (1996: 32) argue that the greater uncertainty and heightened risk inherent in farming in the present era has diverted farmers away from environmental and social objectives towards income generation. Their conclusion is salutary: 'implementing sustainability (on pastoral farms) has become much more difficult'. To this must be added the recognition that factors beyond the local underlie local economic difficulties. Yet, while the Northland research made commodity system relationships an integral element of the research from the outset, consideration of this dimension by other research groups is recent (AgResearch 1996, Kain 1996).

Third, much anecdotal evidence shows that farm-based issues are seen as individual issues, not necessarily requiring networking or the formation of groups. The ingredients of 'what gets groups together' appear to have little to do with sustainability of the environment and a great deal to do with economic survival of family capital and the farm. The crucial exception to this pattern is Maori, who already have a group ethos. The capacity of Maori to mobilise locally over particular issues and redefine local issues as issues pertaining to sustainable development for Maori is a source of frustration to many individual farmers, who are unable to conceive of negotiated partnerships and co-management options at the local level.

Fourth, participatory research is regarded by researchers as a means to bridge the gap between researcher and researched, a mechanism for building trust and group cohesion, and a framework to include those who do not normally engage in consultative processes. While participatory research may not be a total response to the perennial scepticism of locals who ask 'What's it (the research) doing for me?', it does open doors for communication and mutual understanding and can bring onto the agenda differences between individualistic and collaborative solutions to local problems. It also affords an opportunity for physical and social scientists to join together in interdisciplinary research. The commitment to participatory research is thus a research (but very social) technology of context, which could, were it widely implemented, facilitate information flows to and from policy makers. But big and unexplored questions remain. Might participatory research collapse into a vehicle for communicating information about matters such as risk, or at worst, be little more than PR? Is such research likely to be promoted as a substitute for grassroots democratic movements? Can participatory research effectively reveal the social origins of local knowledge and expertise?

Perhaps the most crucial aspect to have emerged from the cross-research team discussions is the realisation that sustainability initiatives are knowledge intensive (Arnoux 1993, Le Heron 1994, Blunden *et al.* 1996, Smith 1996a). The social model for change in New Zealand at the moment is that informed and responsible decision makers will bring about sustainable outcomes through adjustments on the land. This model of social action rests on many assumptions. Most tellingly, who might make decisions about how the land might be incorporated into production (and with this, consumption), remains very much contested. The policy principles espoused by the Strategic Consultative Group endorse a social world made up of 'farmers and their land', when the realities are that each local is simultaneously made of several (even many) communities and that each scale of analysis and each site of investigation are intersections of multiple communities. On the other hand, the role of the expert in the research process has been fundamentally altered. Whereas the previous regulatory regime delegated the power to 'prescribe' outcomes to experts, the status of expert is being challenged and redefined. With respect to sustainability, 'expert' is increasingly consonant with both information provider and facilitator.

The sketch of institutional developments relating to sustainable agriculture in the New Zealand scene exposes ongoing tensions over whose choices might prevail, who will benefit or lose from change and how and with what levels of coercion the behaviour of people might be modified 'in the interests' of sustainability. This of course politicises the debate. Sustainability is about political and social processes to achieve futures. And this leads back to the Redclift-Buttel institution-building thesis. The New Zealand experience reaffirms the basic contention that any progress on sustainability has to be grounded in a composite of regulatory and governance arrangements. The difficulty in New Zealand is that the political rhetoric of agricultural sustainability has not been widely picked up at the grassroots level, and farmer groups and agricultural movements have had minimal success in reshaping the policy framework. On an optimistic note, researchers are probing the realities of the agricultural sector more reflexively than previously, transcending traditional canons of research. The next decade will be a decisive period. In the present context the paramount goal is institutional innovation, giving new content to governance and regulation of agriculture.

ACKNOWLEDGEMENTS

Funding from the Foundation for Research, Science and Technology (FRST) under contract number UOA-509 and the Massey University Research Fund is gratefully acknowledged.

BIBLIOGRAPHY

Arnoux, L. (1993) 'Indicators of sustainable development', in Ministry for the Environment and Foundation for Research, Science and Technology, *Sustainable Development: A Social Perspective*, proceedings of a workshop, Wellington, New Zealand.

AgResearch (1996) *Annual Report 1996*, Hamilton: AgResearch.

Blunden, G., Cocklin, C., Smith, W., and Moran, W. (1996) 'Sustainability from the paddock', *New Zealand Geographer* 52, 2: 24–34.

Blunden, G., Cocklin, C., Davis, P., Moran, W., Smith, W., Laituri, M., and Le Heron, R. (1995) 'Land-Based Production in Northland', Department of Geography, University of Auckland, Occasional Paper No. 30.

FORST Achievement Report (1996) *1995/1996 Contract UOA-509*, Department of Geography, University of Auckland.

Gleeson, B. (1996) 'Political economy of the RMA 1991', in R. Le Heron and E. Pawson (eds) *Changing Places: New Zealand in the Nineties*, Auckland: Longman Paul.

Kain, B. (1996) 'Introductory Paper' presented to Joint University of Auckland-AgResearch Workshop, Hamilton, 21 November.

Kelsey, J. (1995) *The New Zealand Experiment*, Auckland: University of Auckland Press.

Le Heron, R. (1994) 'Complexity, the rural problematic and sustainability', in D. Hawke (ed.) *Proceedings of the 17th New Zealand Geography Conference*, Christchurch: New Zealand Geographical Society.

—— (1996) '"Organising industrial spaces": shifting metaphors, narratives and knowledge about industrial geographies', State-of-the-Art lecture to the Commission on the Organisation of Industrial Space, C21, International Geographical Union Congress, The Hague, 6 August.

—— and Pawson, E. (eds) (1996) *Changing Places. New Zealand in the Nineties*, Auckland: Longman Paul.

Le Heron, R. and Roche, M. (1996a) 'Eco-commodity systems: historical geographies of context, articulation and embeddedness under capitalism', in D. Burch, R. Rickson, and G. Lawrence (eds) *Globalisation and Agri-food Restructuring: Perspectives from the Australasian Region*, Aldershot: Avebury.

—— (1996b) 'Globalisation, sustainability and apple orcharding, the Hawke's Bay, New Zealand', *Economic Geography* 72, 4: 416–32.

McDermott, P. (1997) 'Positioning planning in a market economy', *Environment and Planning A*, forthcoming.

Ministry of Agriculture and Fisheries (1993) *Sustainable Agriculture*, MAF Policy Position Paper 1, Wellington: Ministry of Agriculture and Fisheries.

Ministry for the Environment (1993) *Cleaner Production at Work*, Wellington: Ministry for the Environment.

—— (1995) *Environment 2010 Strategy: A Statement of the Government's Strategy on the Environment*, Wellington: Ministry for the Environment.

—— (1996a) *Sustainable Land Management. A Strategy for New Zealand*, Wellington: Ministry for the Environment.

—— (1996b) *Learning to Care for Our Environment. Perspectives on Environmental Education: A Discussion Document*, Wellington: Ministry for the Environment.

Ministry for the Environment and Foundation for Research, Science and Technology (1993) 'Sustainable Development: A Social Perspective', proceedings of a workshop, 18 February.

Scott, K., Park, J., Cocklin, C., and Kearns, R. (1996) 'Community, Sustainability and Land-based Production', Department of Geography, University of Auckland, Working Paper No. 4.

Smith, W. (1993) 'Sustainable development: the economic and environmental case for policy reform', *New Zealand Geographer* 49, 2: 69–74

—— (1996a) 'Issues Paper: Planning a Research Agenda for the Social Sciences in Sustainable Land Management', paper presented to the Workshop on Sustainable Land-based Production, University of Auckland, 12–13 June.

—— (1996b) 'The environmental implications of socio-economic and structural change: what we need to know from environmental science', extended abstract for Royal Society Workshop, University of Auckland, 21 October.

Stokes, E. (1996) 'Maori identities', in R. Le Heron and E. Pawson (eds) *Changing Places: New Zealand in the Nineties*, Auckland: Longman Paul.

Strategic Consultative Group on Sustainable Land Management Research (1995) *Science for Sustainable Land Management: Towards a New Agenda and Partnership*, Wellington: Ministry of Research, Science and Technology.

Taylor, N. and McCrostie Little, H. (1995) *Means of Survival? A Study of Off Farm Employment*, Christchurch: Taylor Baines and Associates.

Urlich Cloher, D. (1996) 'Maori perspectives on sustainability', in R. Le Heron and E. Pawson (eds) *Changing Places: New Zealand in the Nineties*, Auckland: Longman Paul.

INDEX